Unless Recalled Earlier

SUPERVISED
AND
UNSUPERVISED
PATTERN
RECOGNITION

Feature Extraction and
Computational Intelligence

INDUSTRIAL ELECTRONICS SERIES

Series Editor
J. David Irwin, *Auburn University*

TITLES INCLUDED IN THE SERIES

**Supervised and Unsupervised Pattern Recognition:
Feature Extraction and Computational Intelligence**
Evangelia Micheli-Tzanakou, *Rutgers University*

Handbook of Applied Computational Intelligence
Mary Lou Padgett, *Auburn University*
Nicholas Karayiannis, *University of Houston*
Lofti A. Zaden, *University of California Berkeley*

Handbook of Applied Neurocontrols
Mary Lou Padgett, *Auburn University*
Charles C. Jorgensen, *NASA Ames Research Center*
Paul Werbos, *National Science Foundation*

Handbook of Power Electronics
Tim L. Skvarenina, *Purdue University*

INDUSTRIAL ELECTRONICS SERIES

SUPERVISED AND UNSUPERVISED PATTERN RECOGNITION

Feature Extraction and Computational Intelligence

EVANGELIA MICHELI-TZANAKOU

RUTGERS UNIVERSITY
PISCATAWAY, NEW JERSEY

CRC Press
Boca Raton   London   New York   Washington, D.C.

Library of Congress Cataloging-in-Publication Data

Micheli-Tzanakou, Evangelia, 1942-
 Supervised and unsupervised pattern recognition: feature extraction and computational intelligence /Evangelia Micheli-Tzanakou, editor/author
  p. cm.-- (Industrial electronics series)
 Includes bibliographical references and index.
 ISBN 0-8493-2278-2
 1. Pattern recognition systems. 2. Neural networks (Computer science) I. Title. II. Series.

TK7882.P3 M53 1999
006.4--dc21

99-043495
CIP

Dedication

---

*To my late mother for never being satisfied with my progress
and for always pushing me to better things in life.*

PREFACE

This volume describes the application of supervised and unsupervised pattern recognition schemes to the classification of various types of waveforms and images. An optimization routine, ALOPEX, is used to train the network while decreasing the likelihood of local solutions. The chapters included in this volume bring together recent research of more than ten authors in the field of neural networks and pattern recognition. All of these contributions were carried out in the Neuroelectric and Neurocomputing Laboratories in the Department of Biomedical Engineering at Rutgers University. The chapters span a large variety of problems in signal and image processing, using mainly neural networks for classification and template matching. The inputs to the neural networks are features extracted from a signal or an image by sophisticated and proven state-of-the-art techniques from the fields of digital signal processing, computer vision, and image processing. In all examples and problems examined, the biological equivalents are used as prototypes and/or simulations of those systems were performed while systems that mimic the biological functions are built.

Experimental and theoretical contributions are treated equally, and interchanges between the two are examined. Technological advances depend on a deep understanding of their biological counterparts, which is why in our laboratories, experiments on both animals and humans are performed continuously in order to test our hypotheses in developing products that have technological applications.

The reasoning of most neural networks in their decision making cannot easily be extracted upon the completion of training. However, due to the linearity of the network nodes, the cluster prototypes of an unsupervised system can be reconstructed to illustrate the reasoning of the system. In these applications, this analysis hints at the usefulness of previously unused portions of the spectrum.

The book is divided into four parts. The first part contains chapters that introduce the subjects of neural networks, classifiers, and feature extraction methods. Neural networks are of the *supervised* type of learning. The second part deals with *unsupervised* neural networks and *fuzzy* neural networks and their applications to handwritten character recognition, as well as recognition of normal and abnormal visual evoked potentials. The third part deals with advanced neural network architectures, such as modular designs and their applications to medicine and three-dimensional neural networks architectures simulating brain functions. Finally, the fourth part discusses general applications and simulations in various fields. Most importantly, the establishment of a brain-to-computer link is discussed in some detail, and the findings from these human experiments are analyzed in a new light.

All chapters have either been published in their final form or in a preliminary form in conference proceedings and presentations. All co-authors to these papers were mostly students of the editor. Extensive editing has been done so that repetitions of algorithms, unless modified, are avoided. Instead, where commonality exists, parts have been placed into a new chapter (Chapter 4), and references to this chapter are made throughout.

As is obvious from the number of names on the chapters, many students have contributed to this compendium. I thank them from this position as well. Others contributed in different ways. Mrs. Marge Melton helped with her expert typing of parts of this book and with proofreading the manuscript. Mr. Steven Orbine helped in more than one way, whenever expert help was needed. Dr. G. Kontaxakis, Dr. P. Munoz, and Mr. Wei Lin helped with the manuscripts of Chapters 1 and 3. Finally, to all the current students of my laboratories, for their patience while this work was compiled, many thanks. I will be more visible—and demanding—now.

Dr. D. Irwin was instrumental in involving me in this book series, and I thank him from this position as well. Ms. Nora Konopka I thank for her patience in waiting and for reminding me of the deadlines, a job that was continued by Ms. Felicia Shapiro and Ms. Mimi Williams. I thank them as well.

Evangelia Micheli-Tzanakou, Ph.D.
Department of Biomedical Engineering
Rutgers University
Piscataway, NJ

Contributors

Ahmet Ademoglu, Ph.D.
Assistant Professor
Institute of Biomedical Engineering
Bogazici University
Bebek, Istanbul, Turkey

Sergey Aleynikov, M.S.
IDT
Hackensack, NJ

Jeremy Bricker, Ph.D. Candidate
Environmental Fluid Mechanics
  Laboratory
Department of Civil and Environmental
  Engineering
Stanford, CA

Tae-Soo Chon, Ph.D.
Professor
Department of Biology
College of Natural Sciences
Pusan National University
Pusan, Korea

Woogon Chung, Ph.D.
Assistant Professor
Department of Control and
  Instrumentation
Sung Kyun Kwan University
Kyung Gi-Do, South Korea

Lt. Col. Timothy Cooley, Ph.D.
USAF Academy
Department of Mathematical Sciences
Colorado Springs, CO

Timothy J. Dasey, Ph.D.
MIT Lincoln Labs
Weather Sensing Group
Lexington, MA

Cynthia Enderwick, M.S.
Hewlett Packard
Palo Alto, CA

Faiq A. Fazal, M.S.
Lucent Technologies
Murray Hill, NJ

Raymond Iezzi, M.D.
Kresge Institute
Detroit, Michigan

Francis Phan, M.S.
Harmonix Music Systems, Inc.
Cambridge, MA

Seth Wolpert, Ph.D.
Associate Professor
Pennsylvania State University —
  Harrisburg
Middletown, PA

Daniel Zahner, M.S.
Data Scope Co.
Paramus, NJ

Contents

Section I — Overviews of Neural Networks, Classifiers, and Feature Extraction Methods—Supervised Neural Networks

Chapter 1   Classifiers: An Overview

1.1   Introduction 3
1.2   Criteria for Optimal Classifier Design 3
1.3   Categorizing the Classifiers 4
     1.3.1   Bayesian Optimal Classifiers 4
     1.3.2   Exemplar Classifiers 5
     1.3.3   Space Partition Methods 6
     1.3.4   Neural Networks 7
1.4   Classifiers 7
     1.4.1   Bayesian Classifiers 7
           1.4.1.1   Minimum ECM Classifers 8
           1.4.1.2   Multi-Class Optimal Classifiers 9
     1.4.2   Bayesian Classifiers with Multivariate Normal Populations 11
           1.4.2.1   Quadratic Discriminant Score 11
           1.4.2.2   Linear Discriminant Score 11
           1.4.2.3   Linear Discriminant Analysis and Classification 12
           1.4.2.4   Equivalence of LDF to Minimum *TPM* Classifier 14
     1.4.3   Learning Vector Quantizer (LVQ) 14
           1.4.3.1   Competitive Learning 14
           1.4.3.2   Self-Organizing Map 15
           1.4.3.3   Learning Vector Quantization 15
     1.4.4   Nearest Neighbor Rule 18
1.5   Neural Networks (NN) 19
     1.5.1   Introduction 19
           1.5.1.1   Artificial Neural Networks 19
           1.5.1.2   Usage of Neural Networks 19
           1.5.1.3   Other Neural Networks 20
     1.5.2   Feed-Forward Neural Networks 20
     1.5.3   Error Backpropagation 22
           1.5.3.1   Madaline Rule III for Multilayer Network with Sigmoid Function 25
           1.5.3.2   A Comment on the Terminology 'Backpropagation' 25

1.5.3.3   Optimization Machines with Feed-Forward
           Multilayer Perceptrons 25
1.5.3.4   Justification for Gradient Methods for Nonlinear
           Function Approximation 26
1.5.3.5   Training Methods for Feed-Forward Networks 27
1.5.4   Issues in Neural Networks 28
1.5.4.1   Universal Approximation 28
1.5.5   Enhancing Convergence Rate and Generalization of an
Optimization Machine 29
1.5.5.1   Suggestions for Improving the Convergence 30
1.5.5.2   Quick Prop 31
1.5.5.3   Kullback-Leibler Distance 32
1.5.5.4   Weight Decay 33
1.5.5.5   Regression Methods for Classification Purposes 34
1.5.6   Two-Group Regression and Linear Discriminant Function 34
1.5.7   Multi-Response Regression and Flexible Discriminant Analysis 36
1.5.7.1   Powerful Nonparametric Regression Methods for
           Classification Problems 37
1.5.8   Optimal Scoring (OS) 37
1.5.8.1   Partially Minimized ASR 39
1.5.9   Canonical Correlation Analysis 40
1.5.10   Linear Discriminant Analysis 41
1.5.10.1   LDA Revisited 41
1.5.11   Translation of Optimal Scoring Dimensions into
Discriminant Coordinates 42
1.5.12   Linear Discriminant Analysis via Optimal Scoring 44
1.5.12.1   LDA via OS 45
1.5.13   Flexible Discriminant Analysis by Optimal Scoring 46
1.6   Comparison of Experimental Results 48
1.7   System Performance Assessment .... : 49
1.7.1   Classifier Evaluation 50
1.7.1.1   Hold-Out Method 51
1.7.1.2   K-Fold Cross-Validation 51
1.7.2   Bootstrapping Method for Estimation 52
1.7.2.1   Jackknife Estimation 53
1.7.2.2   Bootstrap Method 54
1.8   Analysis of Prediction Rates from Bootstrapping Assessment 54
References 56

Chapter 2   Artificial Neural Networks: Definitions, Methods, Applications

2.1   Introduction 61
2.2   Definitions 62
2.3   Training Algorithms 64

2.3.1   Backpropagation Algorithm...65
2.3.2   The ALOPEX Algorithm 69
2.3.3   Multilayer Perceptron (MLP) Network Training with ALOPEX......71
2.4   Some Applications 72
2.4.1   Expert Systems and Neural Networks.............................72
2.4.2   Applications in Mammography 73
2.4.3   Chromosome and Genetic Sequences Classification 74
References 75

Chapter 3   A System for Handwritten Digit Recognition

3.1   Introduction 79
3.2   Preprocessing of Handwritten Digit Images 79
3.2.1   Optimal Size of the Mask for Dilation 85
3.2.2   Bartlett Statistic...85
3.3   Zernike Moments (ZM) for Characterization of Image Patterns....87
3.3.1   Reconstruction by Zernike Moments 90
3.3.2   Features from Zernike Moments 92
3.4   Dimensionality Reduction...96
3.4.1   Principal Component Analysis 96
3.4.2   Discriminant Analysis 98
3.5   Analysis of Prediction Error Rates from Bootstrapping Assessment........100
3.6   Summary 105
Acknowledgments 105
References 105

Chapter 4   Other Types of Feature Extraction Methods

4.1   Introduction 109
4.2   Wavelets 110
4.2.1   Discrete Wavelet Series...111
4.2.2   Discrete Wavelet Transform (DWT)..............................112
4.2.3   Spline Wavelet Transform..112
4.2.4   The Discrete B-Spline Wavelet Transform.....................114
4.2.5   Design of Quadratic Spline Wavelets............................114
4.2.6   The Fast Algorithm 117
4.3   Invariant Moments 119
4.4   Entropy 122
4.5   Cepstrum Analysis 122
4.6   Fractal Dimension...123
4.7   SGLD Texture Features 126
References 130

Section II   Unsupervised Neural Networks

Chapter 5   Fuzzy Neural Networks

5.1   Introduction 135
5.2   Pattern Recognition 135
    5.2.1   Theory and Applications 135
    5.2.2   Feature Extraction 137
    5.2.3   Clustering 138
5.3   Optimization 138
    5.3.1   Theory and Objectives 138
    5.3.2   Background 139
    5.3.3   Modified ALOPEX Algorithm 141
5.4   System Design 144
    5.4.1   Feature Extraction 144
        5.4.1.1   The Karhunen-Loève Expansion 145
        5.4.1.2   Application by a Neural Network 147
5.5   Clustering 153
    5.5.1   The Fuzzy c-Means (FCM) Clustering Algorithm 153
    References 159

Chapter 6   Application to Handwritten Digits

6.1   Introduction to Character Recognition 163
6.2   Data Collection 165
    6.2.1   Preprocessing 166
    6.2.2   Noise Thresholding 166
    6.2.3   Center of Mass Adjustment 168
    6.2.4   Line Thinning 168
    6.2.5   Fixing to Size 168
    6.2.6   Rotation 168
    6.2.7   Reducing Resolution 169
    6.2.8   Blurring 170
6.3   Results 170
6.4   Discussion 177
6.5   Summary 181
    References 182

Chapter 7   An Unsupervised Neural Network System for Visual Evoked Potentials

7.1   Introduction 185
7.2   Data Collection and Preprocessing 186
7.3   System Design 187
7.4   Results 188

7.5   Discussion 191
       References 194

Section III   *Advanced Neural Network Architectures/Modular Neural Networks*

Chapter 8   Classification of Mammograms Using a Modular Neural Network

8.1 Introduction 197
8.2 Methods and System Overview 199
       8.2.1   Data Acquisition...199
       8.2.2   Feature Extraction by Transformation...........................200
8.3   Modular Neural Networks 202
8.4   Neural Network Training 203
8.5   Classification Results 203
8.6   The Process of Obtaining Results 207
8.7   ALOPEX Parameters 209
8.8   Generalization 213
8.9   Conclusions 218
       Acknowledgments 218
       References 218

Chapter 9   Visual Ophthalmologist: An Automated System for Classification of Retinal Damage

9.1   Introduction 221
9.2   System Overview 221
       9.2.1   Image Processing 223
       9.2.2   Feature Extraction Methods...223
       9.2.3   Image Classification..223
9.3   Modular Neural Networks 223
9.4   Application to Ophthalmology 224
9.5   Results...226
9.6   Discussion 227
       References 227

Chapter 10   A Three-Dimensional Neural Network Architecture

10.1 Introduction 229
10.2 The Neural Network Architecture 229
10.3 Simulations..230
       10.3.1 Visual Receptive Fields..231
       10.3.2 Modeling of Parkinson's Disease 235
10.4 Discussion 238
       References 238

Section IV   General Applications

Chapter 11   A Feature Extraction Algorithm Using Connectivity Strengths and Moment Invariants

11.1 Introduction 241
11.2 ALOPEX Algorithms 242
    11.2.1 Original Algorithm 242
    11.2.2 Reinforcement Rules 242
    11.2.3 A Generalized ALOPEX Algorithm 243
        11.2.3.1 Process I 244
        11.2.3.2 Process II 245
11.3 Moment Invariants and ALOPEX 246
11.4 Results and Discussion 249
    Acknowledgments 262
    References 262

Chapter 12   Multilayer Perceptrons with ALOPEX: 2D-Template Matching and VLSI Implementation

12.1 Introduction 265
    12.1.1 Multilayer Perceptrons 265
12.2 Multilayer Perceptron and Template Matching 268
12.3 VLSI Implementation of ALOPEX 270
    References 275

Chapter 13   Implementing Neural Networks in Silicon

13.1 Introduction 277
13.2 The Living Neuron 278
13.3 Neuromorphic Models 280
13.4 Neurological Process Modeling 292
    References 299

Chapter 14   Speaker Identification through Wavelet Multiresolution Decomposition and ALOPEX

14.1 Introduction 301
14.2 Multiresolution Analysis through Wavelet Decomposition 303
14.3 Pattern Recognition with ALOPEX 306
14.4 Methods 306
    14.4.1 Data Acquisition 306
    14.4.2 Data Preprocessing 307
    14.4.3 Representing the Wavelet Coefficients for Template Matching 308
14.5 Results 310
14.6 Discussion 313
    Acknowledgments 314

References 314

Chapter 15   Face Recognition in Alzheimer's Disease: A Simulation
15.1 Introduction 317
15.2 Methods...317
15.3 Results...318
15.4 Discussion 321
References 321

Chapter 16   Self-Learning Layered Neural Network
16.1 Introduction 323
16.2 Neocognitron and Pattern Classification 325
    16.2.1 Training Algorithm 328
16.3 Objectives...329
16.4 Methods...329
16.5 Study A..330
    16.5.1 Network Description...330
    16.5.2 Results from Study A 331
16.6 Study B..332
    16.6.1 Results from Study B 332
16.7 Summary and Discussion 334
References 346

Chapter 17   Biological and Machine Vision
17.1 Introduction 347
17.2 Distributed Representation...347
17.3 The Model 348
17.4 A Modified ALOPEX Algorithm...348
17.5 Application to Template Matching...350
17.6 Brain to Computer Link..351
    17.6.1 Global Receptive Fields in the Human Visual System 351
    17.6.2 The Black Box Approach 353
17.7 Discussion 355
References 358

Index..359

Introduction—Why this Book?

The potential for achieving a great deal of processing power by wiring together a large number of very simple and somewhat primitive devices has captured the imagination of scientists and engineers for many years. In recent years, the possibility of implementing such systems by means of electro-optical devices and in very large scale integrations has resulted in increased research activities.

Artificial neural networks (ANNs) or simply Neural Networks (NNs) are made of interconnected devices called neurons (also called neurodes, nodes, neural units, or simply units). Loosely inspired by the makeup of the nervous system, these interconnected devices look at patterns of data and learn to classify them. NNs have been used in a wide variety of signal processing and pattern recognition applications and have been successfully applied in such diverse fields as speech processing, handwritten character recognition, time series prediction, data compression, feature extraction, and pattern recognition in general. Their attractiveness lies in the relative simplicity with which the networks can be designed for a specific problem along with their ability to perform nonlinear data processing.

As the neuron is the building block of a brain, a neural unit is the building block of a neural network. Although the two are far from being the same, or performing the same functions, they still possess similarities that are remarkably important. NNs consist of a large number of interconnected units that give them the ability to process information in a highly parallel way. An artificial neuron sums all inputs to it and creates an output that carries information to other neurons. The strength by which two neurons influence each other is called a synaptic weight. In an NN all neurons are connected to all other neurons by synaptic weights that can have seemingly arbitrary values, but in reality, these weights show the effect of a stimulus on the neural network and the ability or lack of it to recognize that stimulus. All NNs have certain architectures and all consist of several layers of neuronal arrangements. The most widely used architecture is that of the perceptron first described in 1958 by Rosenblatt.

A single node acts like an integrator of its weighted inputs. Once the result is found it is passed to other nodes via connections that are called synapses. Each node is characterized by a parameter that is called threshold or offset and by the kind of nonlinearity through which the sum of all the inputs is passed. Typical nonlinearities are the hardlimiter, the ramp (threshold logic element) and the widely used sigmoid.

NNs are specified by their processing element characteristics, the network topology and the training or learning rules they follow in order to adapt the weights, W_i. Network topology falls into two broad classes: feedforward (nonrecursive) and feedback (recursive). Nonrecursive NNs offer the advantage of simplicity of implementation and analysis. For static mappings a nonrecursive network is all one needs to specify any static condition. Adding feedback expands the network's range of behavior since now its output depends upon both the current input and network states. But one has to pay a price — longer times for teaching the NN to recognize its inputs. The most widely used training algorithm is the backpropagation algorithm. The backpropagation algorithm is a learning scheme where the error is backpropagated layer by layer and used to update the weights. The algorithm is a gradient descent method that minimizes the error between the desired outputs and the actual outputs calculated by the MLP.

The original perceptrons trained with backpropagation are examples of supervised learning. In this type of learning the NN is trained on a training set consisting of vector pairs. One of these vectors is used as input to the network, the other is used as the desired or target output. During training the weights of the NN are adjusted in such a way as to minimize the error between the target and the computed output of the network. This process might take a large number of iterations to converge, especially because some training algorithms (such as backpropagation) might converge to local minima instead of the global one. If the training process is successful, the network is capable of performing the desired mapping.

Section I

Overviews of Neural Networks,
Classifiers, and Feature
Extraction Methods—Supervised
Neural Networks

1 Classifiers: An Overview

Woogon Chung and Evangelia Micheli-Tzanakou

1.1 INTRODUCTION

One way to better understand a subject is to classify or categorize it among related subjects. Many classifiers result from different approaches to classification problems. The purpose of this article is to categorize the well-known classifiers in the literature according to how they learn to classify.

Lippmann's tutorial paper[1] described various classifiers as well as neural networks in detail after his first discussion[2] on the general application of neural networks. Another general overview on this subject is found in a paper by Hush and Horne[3] in which neural networks are reviewed in the broad dichotomy of stationary vs. dynamic networks. Weiss and Kulikowski's book[4] generally touches the classification and prediction methods from the point of view of statistics, neural networks, machine learning, and expert systems.

The purpose of this article is not to give a tutorial on the well-developed networks and other classifiers but to introduce another branch in the growing classifier tree, that of nonparametric regression approaches to classification problems. Recently Hastie, Tibshirani, and Buja[5] introduced the Flexible Discriminant Analysis (FDA) in the applied statistics literature, after the unpublished work by Breiman and Ihaka.[6]

Canonical Correlation Analysis (CCA) for two sets of variables is known to be a scalar multiple equal to the Linear Discriminant Analysis (LDA). Optimal Scaling (OS) is an alternative to CCA, where the classical Singular Value Decomposition (SVD) is used to find the solutions. OS brings the *flexibility* obtained via nonparametric regression and introduces this flexibility to discriminant analysis, hence the name *Flexible Discriminant Analysis*.

A number of recently developed multivariate regressions are used for classification, in addition to other groups of classifiers for a data set obtained from handwritten digit images. The software is contributed mainly from the authors or active researchers in this area. The sources are described in later sections after the description of each classifier.

1.2 CRITERIA FOR OPTIMAL CLASSIFIER DESIGN

We start with a general description of the classification problem and then proceed to a discussion of simpler cases in which assumptions are made. Which criterion should be used is application specific. Expected Cost for Misclassification (*ECM*) is applied to problems in which the cost of misclassification differs among the cases. For example, one may expect to assign a higher cost for misdiagnosing a patient

0-8493-2278-2/00/$0.00+$.50
© 2000 by CRC Press LLC

with a serious disease as healthy than for misdiagnosing a healthy person as unhealthy. If a meteorologist forecasts fine weather for the weekend but a heavy storm strikes the town, the cost of the misclassification will be much more than if the opposite situation occurs.

Sometimes we do not care about the resulting cost of misclassification. The cost for misclassification for a pattern recognition system to misclassify pattern 'A' as pattern 'B' may be considered the same as the cost to misclassifying pattern 'B' as pattern 'A'. In this situation we can disregard the cost information or assign the same cost to all cases. An optimal classification procedure might also consider only the probability of misclassification (from conditional distributions) and its likelihood to happen among different classes (from the *a priori* probabilities). Such an optimal classification procedure is referred to as the Total Probability of Misclassification (*TPM*). The *ECM*, however, requires three kinds of information, that is, the conditional distribution, the *a priori* probabilities, and the cost for misclassification.

In the simplest case, we also ignore the *a priori* probabilities or assume that they are all equal. In this case we only wish to reduce misclassification for all the classes without considering the class proportion of the given data. It should be noted, however, that it is relatively simple to estimate the *a priori* probabilities from the sample at hand by the frequency approximation. Thus the *TPM* is often the choice as a criterion in which the class conditional distribution and *a priori* probabilities are considered.

1.3 CATEGORIZING THE CLASSIFIERS

1.3.1 BAYESIAN OPTIMAL CLASSIFIERS

Bayesian classifiers are based on probabilistic information on the populations from which a sample of training data is to be drawn randomly. Randomness in sampling is assumed, and it is necessary for a better representation of the sample of the underlying population probability function. An optimal classifier would be one that minimizes the criterion, *ECM*, which consists of three probabilistic types of information. Those are the class conditional probabilities $p_i(\mathbf{x})$, *a priori* probabilities P_i, and cost for misclassification $C(i|j)$, $i \neq j$ *for* $i \in$ G. Another criterion of an optimal Bayesian classifier is ignoring the cost for different misclassifications or using the same cost for all the different misclassifications. Then the probabilistic information used is $p_i(\mathbf{x})$ and P_i for $i \in$ G. This minimum *TPM* classifier is the *Maximum A Posterior* classifier which may be familiar. This will be shown in section 1.4.1. For the minimum *ECM* and *TPM* optimal classifiers, we need to estimate the class conditional densities for different classes which is usually difficult for $q \gtrsim 2$. This difficulty in density estimation is related to the *curse of dimensionality* caused by the fact that a high-dimensional space is mostly empty.

A simplified Bayesian classifier can be obtained by assuming a normal distribution for the class conditional density functions. With the normal distribution assumption, the conditional density functions are parameterized by the mean vector μ_i and the covariance matrices Σ_i for $i \in \mathcal{G}$ where $\mathcal{G}$ is the set of class labels. Depending on the assumption of the covariance matrices we have a *quadratic discriminant classifier* or a *linear discriminant classifier*.

1.3.2 EXEMPLAR CLASSIFIERS

The most simple-minded nonparametric classifier is to use the label information of the training data to allocate the unknown input **x**. The idea is to find the distribution of the labels in a neighborhood of a new observation **x** in the training sample and pick the label whose occurrence is maximum. The well-known classifier in this group is the K-nearest neighbors (KNN) classifier. This classifier is justified either via nearest-neighbor density estimation, or using the nearest-neighbor nonparametric regression.[7]

Practical issues in the KNN includes the choice of a metric to measure the distance between the K nearest points and the unknown pattern point, and fast searches for neighbors. Advanced data structures such as K-D trees[8] are suggested for faster searches at the expense of complications in training and adaptation.

Other examples are the feature-map classifier,[9] Learning Vector Quantization (LVQ),[10] Adaptive Resonance Theory (ART) classifier,[11] and others that are found in the survey paper by Lippmann.[1]

Vector Quantization (VQ)[12,13] is another classical representative exemplar finding algorithm that has been used in communications engineering for the purpose of data reduction for storage and transmission. The exemplar classifiers (except for the KNN classifier) cluster the training patterns via unsupervised learning then followed by supervised learning or label assignment. A Radial Basis Function (RBF) network[14] is also a combination of unsupervised and supervised learning. The basis function is radial and symmetric around the mean vector, which is the centroid of the clusters formed in the unsupervised learning stage, hence the name radial basis function. The RBF networks are two-layer networks in which the first layer nodes represent radial functions (usually Gaussian). The second layer weights are used to combine linearly the individual radial functions, and the weights are adapted via a linear least squares algorithm during the training by supervised learning. Figure 1.1 depicts the structure of the RBF networks.

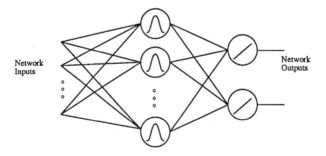

FIGURE 1.1   RBF network. Two-layer network with first layer node being any radial functions imposed on different locations and second layer node being linear.

The LMS algorithm,[15] a simple modification for the linear least squares, is usually used during training for the output layer weights. Any unsupervised clustering algorithm, such as K-means algorithm (i.e., LBG algorithm[13]) or Self-Organizing Map[10] may be used in the first clustering stage.

The most common basis is a Gaussian kernel function of the form:

$$\theta_i = \exp\left[-\frac{(x-m_j)'(x-m_j)}{2\sigma_j^2}\right] j = 1, 2, \ldots, n \tag{1.1}$$

where m_j is the mean vector of the jth cluster found from a clustering algorithm, and x is the input pattern vector. The σ_j^2 is the normalization factor which is a spread measure of the points in a cluster. The average squared distance of the points from the centroid is the common choice for the normalization factor:

$$\sigma_j^2 = \frac{1}{M_j}\sum_{x\in w_j}(x-m_j)'(x-m_j) \tag{1.2}$$

where w_j is the set of the points in the jth cluster and M_j is the number of the points in the jth cluster.

A generalization of the radial function utilizes the variance of an individual variable and covariance among the variables in the training sample. The *Mahalanobis distance* in the Gaussian kernel has the form:

$$\theta_j = \exp\left[-(x-m_j)'\Sigma_j^{-1}(x-m_j)\right] j = 1, .2, \ldots, n \tag{1.3}$$

where Σ_j is the covariance matrix in the jth cluster. The localized distribution function is now ellipsoidal rather than a radial function. A more extensive study on the RBF networks can be found in Hush and Horne.[3]

1.3.3   SPACE PARTITION METHODS

The input space X is recursively partitioned into children subspaces such that the class distributions of the subspaces become as *impure* as possible: impurity of class distribution in a subspace measures the partitioning of the input space by classes.

There are a number of different schemes for estimating trees. Quinlan's *ID3*[16] is well known in the machine learning literature. The citations for some of its variants can be found in a review paper by Ripley.[17] The most well-known partitioning method is the Classification and Regression Tree (CART),[18] which is used to build a binary tree partitioning the input space. At each split of the subspace, each variable is considered with a separating value, and the separating variable with the best separating value is chosen to split the subspace into two children subspaces.

The main issue in this CART algorithm is how to 'grow' it to fit the given training data well and 'prune' it to avoid over-fitting, i.e., to improve the regularization.

1.3.4 NEURAL NETWORKS

Neural networks are popular, and there are numerous textbooks and journals devoted to the topic. Lippmann (1987)[2] is recommended for a general overview of neural networks for classification and (auto)associative memory applications. A statistician's view on using neural networks for multivariate regression and classification purposes is found in extensive review papers by Ripley.[19,17] Different learning algorithms with historical aspects in learning can be obtained from a reference by Hinton.[20]

In this chapter we are mainly interested in multivariate regression and classification properties of neural networks, usually in the form of feed-forward multilayer perceptrons. Chapter 2 deals mainly with neural network architectures and algorithms.

1.4 CLASSIFIERS

1.4.1 BAYESIAN CLASSIFIERS

For simplicity we would like to start with a two-class classification problem and develop it for multi-class cases in a straightforward way. Three kinds of information for an optimal classification design procedure in Bayesian sense are denoted as

$C(2|1), C(1|2)$ cost of misclassification

P_1, P_2 *a priori* probabilities

$p_1(\mathbf{x}), p_2(\mathbf{x})$ class conditional probability density functions

where $C(i|j)$ is the cost for misclassification of j as i. With the notations introduced, the probability that an observation is misclassified as w_2 is represented by the product of the probability that an observation comes from w_1 but falls in w_2 and the probability that the observation comes from w_1:

$$P(\text{misclassified as } w_2)$$
$$= P(\mathbf{X} \in R_2)P(w_1) = P(2|1)P_1 \tag{1.4}$$

where the regions R_2 and $P(2|1)$ (i.e., the integration of $p_i(\mathbf{x})$ in the region R_2) are depicted in Figure 1.2.

$R_i, i \in \{1,2\}$ is an optimum decision region in the input space such that minimum error results are obtained. $P(i|j), i \neq j \in \{1,2\}$ is the integration of the conditional probability function in the region of the other class, thus measuring the possibility of error due to the regions and the conditional probability functions.

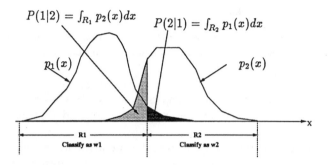

FIGURE 1.2   Misclassification probabilities and decision regions R_1 and R_2.

1.4.1.1   Minimum *ECM* Classifiers

When the criterion is to minimize the *ECM* (Expected Cost for Misclassification), the optimal resulting classifier is called a *Minimum ECM classifier*. The cost for correct classification is usually set to zero, and positive numbers are used for misclassification costs. The whole supporting region is the input space X and is divided into two exclusive and exhaustive subregions: $X = R_1 \cup R_2$.

By the definition, the Minimum *ECM* classifier for class 1 is formed as follows:

$$
\begin{aligned}
ECM &= C(2|1)P(2|1)P_1 + C(1|2)P(1|2)P_2 \\
&= C(2|1)P_1\int_{R_2} p_1(\mathbf{x})dx + C(1|2)P_2\int_{R_1} p_2(\mathbf{x})dx
\end{aligned}
\tag{1.5}
$$

$$
\begin{aligned}
&= C(2|1)P_1\left(1-\int_{R_1} p_1(\mathbf{x})dx\right) + C(1|2)P_2\int_{R_1} p_2(\mathbf{x})dx \\
&= \int_{R_1}\left(C(1|2)P_2 p_2(\mathbf{x}) - C(2|1)P_1 p_1(\mathbf{x})\right)dx + C(2|1)P_1
\end{aligned}
\tag{1.6}
$$

with all the individual quantities being positive. The minimization is achieved as close to zero as possible by having the integration in Equation 1.6 to be equal to a negative quantity. Thus the *ECM* is minimized if the region R_1 includes those values **x** for which the integrand becomes as negative as possible with which the absolute value is equal to the last quantity $C(2|1)P_1$:

$$
\left\{ C(1|2)P_2 p_2(\mathbf{x}) - C(2|1)P_1 p_1(\mathbf{x}) \right\} \le 0
\tag{1.7}
$$

and excludes those **x** for which this quantity is positive. That is, R_1, the decision region for class 1, must be the set of points **x** such that

$$
C(1|2)P_2 p_2(\mathbf{x}) \le C(2|1)P_1 p_1(\mathbf{x}) \quad \text{or}
\tag{1.8}
$$

$$\frac{p_1(\mathbf{x})}{p_2(\mathbf{x})} \geq \frac{C(1|2)P_2}{C(2|1)P_1} \tag{1.9}$$

Here we have chosen to express the region as the set of solution $\mathbf{x}$ of the inequality. The fractional form of Equation 1.9 for the region R_1 is the preferred format, since it reduces to a simple form (which will be shown) when the conditional distribution function $p_i(\mathbf{x})$, $i = 1,2$ is assumed to be normal (and thus assuming the same covariance matrix for the two conditional distributions) for simple Bayesian classifiers.

Assuming the same cost for each misclassification reduces the criterion ECM to *Total Probability of Misclassification (TPM)*:

$$P_2 p_2(\mathbf{x}) \leq P_1 p_1(\mathbf{x}) \quad \text{or} \tag{1.10}$$

$$\frac{p_1(\mathbf{x})}{p_2(\mathbf{x})} \geq \frac{P_2}{P_1} \tag{1.11}$$

from Equation 1.9. Due to the Bayes theorem:

$$P(w_k|\mathbf{x}) = \frac{P_k p_k(\mathbf{x})}{\sum_{i=1}^{2} P_i p_i(\mathbf{x})} \quad \text{for all } k \in \{1,2\} \tag{1.12}$$

the corresponding decision rule (Equation 1.10) becomes the *Maximum A Posteriori (MAP)* criterion, that is to allocate $\mathbf{x}$ into w_1 if

$$P(w_2|\mathbf{x}) \leq P(w_1|\mathbf{x}). \tag{1.13}$$

1.4.1.2　Multi-Class Optimal Classifiers

The boundary regions of the minimum ECM optimal classifier for a multi-class classifier are obtained in a straightforward manner from Equation 1.6 by minimizing

$$ECM = \sum_{i=1}^{J} P_i \left[\sum_{\substack{k=1 \\ k \neq i}}^{J} P(k|i)C(k|i) \right] \tag{1.14}$$

The probability of misclassification of $\mathbf{x} \in w_i$ into w_k is represented as

$$P(k|i) = \int_{R_k} p_i(\mathbf{x})d\mathbf{x} \tag{1.15}$$

The optimal regions $\{R_i\}$ that minimize the ECM are the set of the points $\mathbf{x}$ for which the allocation of $\mathbf{x}$ to a group w_k, $k = 1, 2, \ldots, J$ results in the least cost. It can be shown that an equivalent form of Equation 1.14 can be represented without the integral term $P(k|i)$. The equivalent minimizing ECM' is interpreted intuitively* as

The minimizing ECM is equivalent to minimizing the *a posteriori* probabilities for the wrong classes with the corresponding costs.

That is, the equivalent ECM' has the form

$$ECM' = \sum_{\substack{k=1 \\ k\neq i}}^{J} P(k|\mathbf{x})C(i|k)$$

$$= \sum_{\substack{k=1 \\ k\neq i}}^{J} \frac{P_k p_k(\mathbf{x})}{\sum_{j=1}^{J} P_j p_j(\mathbf{x})} C(i|k) \tag{1.16}$$

and since the denominator is a constant independent of the indices j, this can be further simplified as

$$ECM' = \sum_{\substack{k=1 \\ k\neq i}}^{J} P_k p_k(\mathbf{x})C(i|k) \tag{1.17}$$

In other words, the optimal minimum ECM classifier assigns $\mathbf{x}$ to w_k such that Equation 1.17 is minimized. The minimum ECM (ECM') classifier rule determines mutually exclusive and exhaustive classification regions $R_1, R_2,..., R_J$ such that Equation 1.14 (Equation 1.17) is a minimum.

If the cost is not important (or the same for all misclassifications), the minimum ECM rule becomes minimum TPM. The resulting classifier is, again as in the two-class case, a MAP classifier:

Assign unknown $\mathbf{x}$ to w_k:

$$\mathbf{x} \in w_k = \arg\min_{i \in G} \sum_{\substack{i=1 \\ i\neq 1}}^{J} P_i p_i(\mathbf{x}) \tag{1.18}$$

$$= \arg\max_{i \in G} P_i p_k(\mathbf{x}) \tag{1.19}$$

$$= \arg\max_{i \in G} P(w_k|\mathbf{x}) \tag{1.20}$$

The Bayesian classification rule which is based on the conditional probability density functions for each class, $p_i(\mathbf{x})$, is the optimal classifier in the sense that it minimizes the cost of the probability of error.[22] However, the class conditional

---

* The fact that ECM and ECM' are equivalent is shown analytically in the text.[21]

probability density function $p_i(\mathbf{x})$ needs to be estimated. The density estimation is realizable and efficient if the dimensionality is low, such as $1 \sim 2$ or 3, at most. The parametric Bayesian classification, even if it renders the optimal result in the sense that probability of error is minimized, is difficult to realize in practice. Alternatively, we look for other simple approximations using a normality assumption on the class conditional distributions.

1.4.2 BAYESIAN CLASSIFIERS WITH MULTIVARIATE NORMAL POPULATIONS

If the conditional distribution of a given class is assumed to be p-dimensional multivariate normal,

$$p_i(\mathbf{x}) = \frac{1}{(2\pi)^{p/2}|\Sigma_i|^{1/2}} \exp\left(-\frac{1}{2}(\mathbf{x} - \mu_i)'\Sigma_i^{-1}(\mathbf{x} - \mu_i)\right), \ i = 1, 2, \ldots, J \quad (1.21)$$

with mean vectors μ_i and covariance matrices Σ_i, then, the resulting Bayesian classifiers are easily realized.

1.4.2.1 Quadratic Discriminant Score

With the assumption of having the same cost for all misclassifications added to the multivariate normality, we get a simple classification rule directly from Equation 1.19. Then the minimum *TPM* decision rule can be expressed as follows:

Allocate $\mathbf{x}$ to the class w_k:

$$\begin{aligned}
\mathbf{x} \in w_k &= \arg\max_{i \in G}\{\ln P_i p_i(\mathbf{x})\} \\
&= \arg\max_{i \in G}\{d_i^q(\mathbf{x})\}
\end{aligned} \quad (1.22)$$

where the quadratic discriminant score is defined as

$$d_i^q(\mathbf{x}) = -\frac{1}{2}\ln|\Sigma_i| - \frac{1}{2}(\mathbf{x} - \mu_i)'\Sigma_i^{-1}(\mathbf{x} - \mu_i) + \ln P_i \quad (1.23)$$

and consists of contributions from the generalized variance $|\Sigma_i|$, the *a priori* probability P_i, and the squared distance from $\mathbf{x}$ to the population class mean μ_i. Note that $d_i^q(\mathbf{x})$ is the quadratic form of the unknown $\mathbf{x}$.

1.4.2.2 Linear Discriminant Score

If we further assume that the population covariance matrices Σ_i are all the same, we can simplify the quadratic discriminant score (Equation 1.23) into the linear discriminant score:

$$d_i(\mathbf{x}) = \mu_i' \Sigma_i \mathbf{x} - \frac{1}{2} \mu_i' \Sigma^{-1} \mu_i + \ln P_i \tag{1.24}$$

Then the optimal minimum *ECM* classifier with the assumptions that

1. the multivariate normal distribution in the class conditional density function is $p_i(\mathbf{x})$,
2. we have equal misclassification cost (thus a minimum *TPM* classifier), and that
3. we have equal covariance matrices Σ_i for all classes,

reduces to the simplest form with a linear discriminant score as follows:

$$\mathbf{x} \in w_k = \arg\max_{i \in G} \left\{ d_i(\mathbf{x}) = \mu_i' \Sigma_i \mathbf{x} - \frac{1}{2} \mu_i' \Sigma^{-1} \mu_i + \ln P_i \right\} \tag{1.25}$$

where $\mathbf{x}$ was assigned to class w_k.

As the name indicates, the linear discriminant score $d_i(\mathbf{x})$ for a class i used in the special case of the minimum *TPM* classifier Equation 1.25 is a *linear* functional of the input $\mathbf{x}$. The boundary regions R_1, R_2,..., R_J are hyper-linear, e.g., lines in two-dimensional, planes in three-dimensional input space, etc. However, the minimum *TPM* classifier with different covariances for the classes is given by the *quadratic* form of $\mathbf{x}$ as in Equation 1.22.

1.4.2.3   Linear Discriminant Analysis and Classification

The Fisher's Discriminant function is basically for description purposes. With new lower dimensional discriminant variables, multidimensional data may be visualized to find some interesting structures; hence, the linear discriminant analysis is exploratory. The objective of this section is to relate the linear discriminant analysis to Bayesian optimal classifiers based on *normal theory*.

The linear transform by which the discriminant variates is obtained is defined by the $q \times q$ matrix F in the transform:

$$\mathbf{x} = F\mathbf{y} \tag{1.26}$$

where q is the dimensionality of vector $\mathbf{x}$ and the matrix F consists of $s = \min\{q, J - 1\}$ eigenvectors of $W^{-1} B$ whose corresponding eigenvalues are nonzero. This result is obtained by maximizing the quadratic form of the quadratic expression of matrix W. W and B are the sample versions of pooled within and between covariance matrices, respectively defined as

$$W = \frac{1}{N-J} \sum_{i=1}^{J} \sum_{j=1}^{n_i} \left(\mathbf{y}_{ij} - \overline{\mathbf{y}}_i \right) \left(\mathbf{y}_{ij} - \overline{\mathbf{y}}_i \right)'$$

$$B = \frac{1}{J-1} \sum_{i=1}^{J} n_i \left(\overline{\mathbf{y}}_i - \overline{\mathbf{y}} \right) \left(\overline{\mathbf{y}}_i - \overline{\mathbf{y}} \right)'$$

where $N = \sum_{i=1}^{J} n_i$ is the size of the sample and J is the number of classes.

In the transformed domain or in the discriminant coordinate space (CRIMCOORD), the class mean vectors are given by

$$\mu_{i,x} = \left[\mu_{i,x_1}, \mu_{i,x_2}, \ldots, \mu_{i,x_s} \right]^t = F\mu_{i,y}$$

for $\mathbf{x} \in w_i$, and by the definition of the LDA cov $(X) = \mathbf{I}$. Thus it is appropriate to consider a Euclidean distance in order to measure the separation of the discriminant variates. The classification rule from the discriminants is now to allocate $\mathbf{x}$ into class w_k:

$$\mathbf{x} \in w_k = \arg \min_{i \in G} \left\{ \left\| \mathbf{x} - \mu_{i,x} \right\|^2 \right\} \tag{1.27}$$

Here the dimensionality of $\mathbf{x}$ is $s \leq \min\{q, J - 1\}$. The dimensionality of the transformed variables, i.e., the discriminant variates, become s and the classification rule needs only s variables in the linear discriminant classification rule (Equation 1.27).

The reason for only s variables needed for this classification purpose follows. The sample pooled *within* covariance matrix W and the *between* covariance matrix B have full ranks, hence the $W^{-1}B$, $(q \times q)$-matrix, has full rank. The number of nonzero eigenvalues should not be greater than the full rank:

$$s \leq q \tag{1.28}$$

And the class mean vectors span a multidimensional space with dimensionality:

$$p \leq J - 1 \tag{1.29}$$

which is obvious since by definition $\sum_{i=1}^{J} (\mu_i - \overline{\mu}) = 0$. From Equation 1.28 and Equation 1.29 we can conclude that $s = \min\{q, J - 1\}$. The remaining $(q - s)$-dimensional subspace is called the *null* space of the linear transformation represented by the matrix F and consists of all the vectors $\mathbf{y}$ that are mapped into $\mathbf{0}$ by the linear transformation of Equation 1.26.

1.4.2.4   Equivalence of LDF to Minimum *TPM* Classifier

It is interesting to observe the equivalence of the linear discriminant classification rule Equation 1.27 with that of the minimum *TPM* classification rule, with the assumption that all covariances $\Sigma_i = \Sigma$ are the same for all classes $i \in G$.

The argument of the minimization quantity of Equation 1.27 becomes

$$
\begin{aligned}
\left\| F\left(\mathbf{y}-\mu_{i,y}\right) \right\|^2 &= \left\| \mathbf{x}-\mu_{i,x} \right\|^2 \\
&= \left(\mathbf{y}-\mu_{i,y}\right)' \Sigma^{-1}\left(\mathbf{y}-\mu_{i,y}\right) \\
&= 2d_i(\mathbf{y})+\mathbf{y}'\Sigma^{-1}\mathbf{y}+2\ln P_i
\end{aligned}
\tag{1.30}
$$

where the last equation is due to:

$$
d_i(\mathbf{y})-\frac{1}{2}\mathbf{y}'\Sigma^{-1}\mathbf{y} = -\frac{1}{2}\left(\mu'_{i,y}\Sigma^{-1}\mu_{i,y}-2\mu'_{i,y}\Sigma^{-1}\mathbf{y}+\mathbf{y}'\Sigma^{-1}\mathbf{y}\right)+\ln P_i
\tag{1.31}
$$

The minimization of the squared distance in the Fisher's discriminant variate domain is equivalent to the maximization of the linear discriminant score $d_i(\mathbf{y})$, which results in the equivalence of the 'linear discriminant classification rule' to the 'minimum *TPM* optimal classifier.'[23]

This is an interesting observation or justification of Fisher's LDF. Even though the derivation of the Fisher's discriminant functions do not require the 'multivariate normality' assumption, the same classification rule is obtained from the minimum *TPM* criterion Bayesian classification rule in which normality is assumed.

1.4.3   LEARNING VECTOR QUANTIZER (LVQ)

Learning Vector Quantization (LVQ) is a combination of the self-organizing map and of supervised learning.[10] The self-organizing map is a typical competitive learning method and results in a number of new vectors, called *codebook vectors*, $\mathbf{m}_l$, $i = 1, 2,...,L$. The codebook vectors represent an input vector space with a small number of representative vectors (codebook M). It is a quantization of the given data set $\{\mathbf{x}_i, g_i\}_1^N$ to get a quantized codebook $\{\mathbf{m}_l, g\}_1^L$.

1.4.3.1   Competitive Learning

Given a training vector $\{\mathbf{x}_i, g_i\}_1^N$ and a size L of a randomly chosen codebook $\{\mathbf{m}_l\}_1^L$, an input of time instance k, $\mathbf{x}^{(k)}$, is compared to all the code vectors, $\mathbf{m}_l$, in order to find the closest one, $\mathbf{m}_c$, by a distance measure such that:

$$
d\left(\mathbf{x}^{(k)}, \mathbf{m}_c\right) = \min_l \left\{ d\left(\mathbf{m}_l, \mathbf{x}^{(k)}\right)\right\}
\tag{1.32}
$$

L_2-norm is a common choice, and the competitive learning with this measure utilizes the *steepest descent gradient* step optimization.[10] Once the closest code vector $\mathbf{m}_c$ is found, the competitive learning (or the steepest descent gradient optimization) updates the closest code vector, $\mathbf{m}_c$, but it does not change the other code vectors, $\mathbf{m}_l$ $l \neq c$.

$$\mathbf{m}_c^{(k+1)} = \mathbf{m}_c^{(k)} + \alpha(k)\left(\mathbf{x}^{(k)} - \mathbf{m}_c^{(k)}\right) \tag{1.33}$$

$$\mathbf{m}_l^{(k+1)} = \mathbf{m}_l^{(k)} \ \text{ for } \ l \neq c \tag{1.34}$$

with $\alpha(k)$ being suitable constant $0 < \alpha < 1$, or monotonically decreasing sequence, $0 < \alpha(k) < 1$, for which the optimization LVQ (or OLVQ that will be discussed later) is concerned with.

1.4.3.2   Self-Organizing Map

This is an algorithm for finding a codebook $\mathcal{M}$ (or a set of feature-sensitive detectors) in the input space X. It is known that the internal representations of information in the brain are generally organized spatially, and the self-organizing map mimics the spatial organization of the cells[10] in its structure. A self-organizing map enforces the logically inspired network connections, with *"lateral inhibition"* in a general way by defining a neighborhood set N_c; a time-varying monotonically decreasing set of code vectors:

$$N_c^{(k)} = \left\{\mathbf{m}_l^{(k)}\middle|d\left(\mathbf{m}_l^{(k)},\mathbf{m}_c^{(k)}\right) \leq r(k)\right\} \tag{1.35}$$

where $r(k)$ represents the radius of the $N_c^{(k)}$. Once the winning code vector (or cell) is found from Equation 1.32, all the code vectors in the neighborhood N_c, which is centered on the winning code vector $\mathbf{m}_c$, are undated and the others remain untouched. It has been suggested[10] that the $N_c^{(k)}$ be very wide in the beginning and shrink monotonically with time as $r(k)$ is a function of time, k.

Thus the updating has a similar form to simple competitive learning as in Equation 1.33,

$$\mathbf{m}_l^{(k+1)} = \begin{cases} \mathbf{m}_l^{(k)} + \alpha(k)\left(\mathbf{x}^{(k)} - \mathbf{m}_l^{(k)}\right) & \text{if } \mathbf{m}_l \in N_c^{(k)} \\ \mathbf{m}_l^{(k)} & \text{if } \mathbf{m}_l \notin N_c^{(k)} \end{cases} \tag{1.36}$$

where $\alpha(k)$ is a scalar-value "adaptation gain" $0 \leq \alpha(k) \leq 1$.

1.4.3.3   Learning Vector Quantization

If we now have a codebook that represents the input vector space X by a set of quantized vectors, i.e., a codebook $\mathcal{M}$, then the *Nearest Neighbor* rule can be used for classification problems, provided that the codebook vectors $\mathbf{m}_l$ have their labels in the space to which each codebook vector belongs. The labeling process is similar to the K-nearest neighbor rule in which (a part of) the training data are used to find the majority labels among the K closest patterns to a codebook vector $\mathbf{m}_l$. Thus the LVQ, a form of supervised learning, follows the unsupervised learning, self-organizing map, as shown in Figure 1.3:

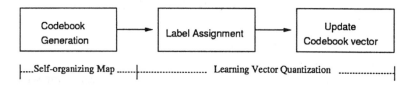

FIGURE 1.3   Block diagram for a system of Self-organizing Map and Learning Vector Quantization.

The last two stages in the figure are called LVQ, and researchers[10,24] have come up with different updating algorithms (LVQ1, LVQ2, LVQ3, OLVQ1) from different methods of updating the codebook vectors. The LVQ1 and its optimization version OLVQ1 are considered in the next sections.

1.4.3.3.1   LVQ1

This is similar to simple competitive learning (Equation 1.33), except that it includes pushing off any wrong closest codebook vector in addition to pulling operations (Equation 1.33 and Equation 1.36).

Let $\mathcal{L}(\mathbf{x}^{(k)})$ be an operation to get the label information; then the codebook updating rule LVQ1 has the form (Figure 1.4)

$$\mathbf{m}_c^{(k+1)} = \mathbf{m}_c^{(k)} + \alpha(k)\left(\mathbf{x}^{(k)} - \mathbf{m}_c^{(k)}\right) \text{ for } \mathcal{L}\left(\mathbf{x}^{(k)}\right) = \mathcal{L}\left(\mathbf{m}_c\right) \tag{1.37}$$

$$\mathbf{m}_c^{(k+1)} = \mathbf{m}_c^{(k)} - \alpha(k)\left(\mathbf{x}^{(k)} - \mathbf{m}_c^{(k)}\right) \text{ for } \mathcal{L}\left(\mathbf{x}^{(k)}\right) \neq \mathcal{L}\left(\mathbf{m}_c\right)$$
$$\mathbf{m}_i^{(k+1)} = \mathbf{m}_i^{(k)} \text{ for } i \neq c \tag{1.38}$$

Here, $0 < \alpha(k) < 1$ is a gain, which is decreasing monotonically with time, as in the competitive learning, (Equation 1.33). The authors suggest a small starting value, i.e., $\alpha(0) = 0.01$ or 0.02.

1.4.3.3.2   Optimized LVQ1 (OLVQ1)

For fast convergence of the LVQ1 algorithm in Equation 1.37 and Equation 1.38, an optimized learning rate for the LVQ1 is suggested.[24] The objective is to find an optimal learning rate $\alpha_l(k)$ for each codebook vector $\mathbf{m}_l$, so that we have individually optimized learning rates:

$$\mathbf{m}_c^{(k+1)} = \mathbf{m}_c^{(k)} + \alpha_c(k)\left(\mathbf{x}^{(k)} - \mathbf{m}_c^{(k)}\right) \quad \text{for} \quad \mathcal{L}\left(\mathbf{x}^{(k)}\right) = \mathcal{L}\left(\mathbf{m}_c\right) \tag{1.39}$$

$$\mathbf{m}_c^{(k+1)} = \mathbf{m}_c^{(k)} - \alpha_c(k)\left(\mathbf{x}^{(k)} - \mathbf{m}_c^{(k)}\right) \qquad \text{for} \qquad L\left(\mathbf{x}^{(k)}\right) \neq L\left(\mathbf{m}_c\right)$$

$$\mathbf{m}_l^{(k+1)} = \mathbf{m}_l^{(k)} \quad \text{for} \quad l \neq c \tag{1.40}$$

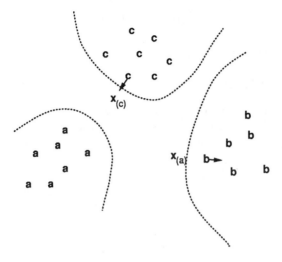

FIGURE 1.4   LVQ1 learning, or updating the initial codebook vectors a, b, c.

Equation 1.39 and Equation 1.40 can be stated with a new sign term $s(k) = 1$ or -1 for the right class and the wrong class, respectively, as follows:

$$\mathbf{m}_c^{(k+1)} = \left[\left(1 - s(k)\alpha_c(k)\right)\right]\mathbf{m}_c^{(k)} + s(k)\alpha_c(k)\mathbf{x}^{(k)} \tag{1.41}$$

It can be seen that $\mathbf{m}_c$ is directly independent but is recursively dependent on the input vector $\mathbf{x}$ from Equation 1.41.

The argument on the learning rate[10] is that:

Statistical accuracy of the learned codebook vectors $\mathbf{m}_c^{(*)}$ is optimal if the effects of the corrections made at different times are of equal weight.

The learning rate due to the current input $\mathbf{x}^{(k)}$ is $\alpha_c(k)$ from Equation 1.41, and due to the previous input $\mathbf{x}^{(k-1)}$, the current learning rate is $(1 - s(k) \alpha_c(k)) \cdot \alpha_c(k-1)$. According to the argument, the effects to the learning rates are to be the same for two consecutive inputs $\mathbf{x}^{(k)}$ and $\mathbf{x}^{(k-1)}$:

$$\alpha_c(k) = \left[1 - s(k)\alpha_c(k)\right]\alpha_c(k-1). \tag{1.42}$$

If this condition is to hold for all k, by induction, the learning rates from all the earlier $\mathbf{x}^{(k)}$, for $k = 0,1,\ldots,k$ should be the same. Therefore, due to the argument, the optimal values of learning rate $\alpha_c(k)$ are determined by the recursion from Equation 1.42 for the specific code vector $\mathbf{m}_c$ as:

$$\alpha_c(k) = \frac{\alpha_c(k-1)}{1 + s(k)\alpha_c(k-1)} \tag{1.43}$$

with which the OLVQ1 is defined as in Equation 1.39 and Equation 1.40.

1.4.4   NEAREST NEIGHBOR RULE

The Nearest Neighbor (NN) classifier, a nonparametric exemplar method, is the natural classification method one can first think of. Using the label information of the training sample, an unknown observation $\mathbf{x}$ is compared with all the cases in the training sample. N distances between a pattern vector $\mathbf{x}$ and all the training patterns are calculated, and the label information, with which the minimum distance results, is assigned to the incoming pattern $\mathbf{x}$. That is, the NN rule allocates the $\mathbf{x}$ to w_k if the closest exemplar $\mathbf{x}_c$ is with the label $k = \mathcal{L}(\mathbf{x}_c)$:

$$\begin{aligned} \mathbf{x}_c &= \arg\min_i \{d(x_0, \mathbf{x}_i)\}, \quad i = 1, 2, \ldots, N \\ \mathbf{x}_0 &\in w_k = \mathcal{L}(\mathbf{x}_k) \end{aligned} \tag{1.44}$$

The distance measure between the unknown and the training sample has a general quadratic form:

$$d(\mathbf{x}, \mathbf{x}_k) = (\mathbf{x}_0 - \mathbf{x}_k)^t M (\mathbf{x}_0 - \mathbf{x}_k) \tag{1.45}$$

With $M = \Sigma^{-1}$, the inverse of the covariance matrix in the sample, the result is the *Mahanalobis* distance. Euclidean distance is obtained when $M = I$, i.e., the identity matrix. Another choice may be the measure considering only the variance for which $M = \Lambda$, where Λ is a diagonal matrix with its elements $(\lambda_i)^{1/2} = \mathrm{var}(x_i)$ and $\mathbf{x} = (x_1, x_2, \ldots, x_p)^t$.

The K-Nearest Neighbor (KNN) rule is the same as the NN rule except that the algorithm finds K nearest points within the points in the training set from the unknown observation $\mathbf{x}$ and assigns the class of the unknown observation to the majority class in the K points.

Recent VLSI technology advances have made memory cheaper than ever; thus, the KNN rule is becoming feasible. Some modified versions of the original KNN rules are reported in what follows. These approaches interpolate between outputs of nearest neighbors stored during training to form complex nonlinear mapping functions.[25,26] Much of the work with the modified KNN rules is in designing effective distance metrics.[1] Some modified KNN are developed for parallel machine implementation, called the connectionist machine,[27] as well as for serial computing.[25]

1.5   NEURAL NETWORKS (NN)

1.5.1   INTRODUCTION

Neural networks have been a much-publicized topic of research in recent years and are now beginning to be used in a wide range of subject areas. One of the strands of interest in neural networks is to explore possible models of biological computation. Human brains contain about 1.5×10^{12} neurons of various types, with each receiving signals through 10 to 10^4 synapses. The response of a neuron is known to be happening in about $1 \sim 10$ milliseconds.[28] Yet we can recognize an old friend's face and call him in about 0.1 seconds. This is a complex pattern recognition task which must be performed in a highly parallel way, since the recognition is done in about $100 \sim 1000$ steps. This suggests that highly parallel systems can perform pattern recognition tasks more rapidly than current conventional sequential computers. As yet our VLSI technology, which is essential planar implementation with at most two- or three-layer cross-connections, is far from achieving these parallel connections that require three-dimensional interconnections.

1.5.1.1   Artificial Neural Networks

Even though originally the neural networks were intended to mimic a task-specific subsystem of a mammalian or human brain, recent research has been mostly concentrated on the *Artificial Neural Networks* which are only vaguely related to the biological system. Neural networks are specified by the (1) net topology, (2) node characteristics, and (3) training or learning rules.

Topological consideration of the artificial neural networks for different purposes can be found in review papers.[2,3] Since our interests in the neural networks are in classification, only the feed-forward multilayer perceptron topology is considered, leaving the feedback connections to the references.

The topology describes the connection with the number of layers and the units in each layer for feed-forward networks. Node functions are usually nonlinear in the middle layers but can be linear or nonlinear for output layer nodes. However, all of the units in the input layer are linear and have fan-out connections from the input to the next layer.

Each output y_j is weighted by w_{ij} and summed at the linear combiner represented by a small circle in Figure 1.5. The linear combiner thresholds its inputs before it sends them to the node function ϕ_j. The unit functions are (non-)linear, monotonically increasing and bounded functions as shown on the right of Figure 1.5.

1.5.1.2   Usage of Neural Networks

One use of a neural network is *classification*. For this purpose each input pattern is forced, adaptively, to output the pattern indicators that are part of the training data; the training set consists of the input covariate $\mathbf{x}$ and the corresponding class labels. *Feed-forward* networks, sometimes called multilayer perceptrons (MLP), are trained adaptively to transform a set of input signals, $\mathcal{X}$, into a set of output signals, $\mathcal{G}$. *Feedback* networks start with an initial activity state of a feedback system, and after

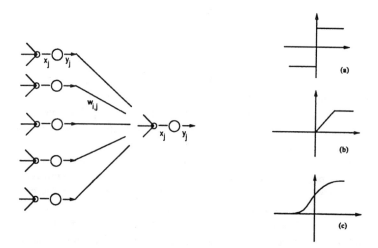

FIGURE 1.5   (I) The linear combiner output $x_j = \sum_{i=1}^{n} y_i w_{ij}$ is input to the node function ϕ_j to give the output y_j. (II) Possible node functions. Hard limiter (a), threshold (b), and sigmoid (c) nonlinear functions.

state transitions have taken place, the asymptotic final state is identified as the outcome of the computation. One use of the feedback networks is the case of *associative memories*: on being presented with pattern near a prototype X it should output pattern X', and as *autoassociative memory* or *contents-addressable memory* by which the desired output is completed to become X.

In all cases the network *learns* or *is trained by* the repeated presentation of patterns with known required outputs (or pattern indicators). Supervised neural networks find a mapping $f: \mathcal{X} \to \mathcal{Y}$ for a given set of input and output pairs.

1.5.1.3   Other Neural Networks

The other dichotomy of the neural networks family is *unsupervised learning*, that is clustering. The class information is not known or it is irrelevant; the networks find the groups of the similar input patterns.

The neighboring code vectors in a neural network compete in their activities by means of mutual lateral interactions and develop adaptively into specific detectors of different signal patterns. Examples are the Self-Organizing Map[10] and the Adaptive Resonance Theory (ART)[11] networks. ART is different from other unsupervised learning networks in that it develops new clusters by itself; the network develops a new code vector if there exist sufficiently different patterns. Thus the ART is truly adaptive, whereas others require the number of clusters to be specified in advance.

1.5.2   FEED-FORWARD NETWORKS

In feed-forward networks the signal flows only in the forward direction; no feedback exists for any node. This is perhaps best seen graphically in Figure 1.6. This

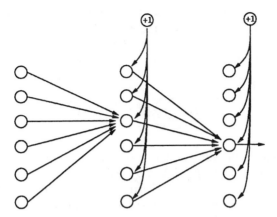

FIGURE 1.6   A generic feed-forward network with a single hidden layer. For bias terms the constant input with 1 are shown and the weights of the constant inputs are the bias values which will be learned as training proceeds.

is the simplest topology and has been shown to be good enough for most practical classification problems.[19]

The general definition allows more than one hidden layer, and also allows 'skip-layer' connections from input to output. With this skip-layer, one can write a general expression for a network output y_k with one hidden layer,

$$ y_k = \phi_k \left(b_k + \sum_{i \to k} w_{ik} x_i + \sum_{j \to k} w_{jk} \, \phi_j \left(b_j + \sum_{i \to j} w_{ij} x_i \right) \right) \qquad (1.46) $$

where the b_j and b_k represent the thresholds for each unit in the jth hidden layer and the output layer, which is the kth layer. Since the threshold values b_j, b_k are to be adaptive, it is useful to have a threshold for the weights for constant input value of 1 as in Figure 1.6. The function $\phi()$ is almost inevitably taken to be a linear, sigmoidal ($\phi(x) = e^x / (1 + e^x)$) or threshold function ($\phi(x) = I \ (x > 0)$).

Rumelhart, Hinton, and Williams[29] showed that the feed-forward multilayer perceptron networks can learn using gradient values obtained by an algorithm, called *Error Backpropagation*.* This contribution is a remarkable advance since 1969, when Minsky and Papert[30] claimed that the nonlinear boundary, required for the XOR problem, can be obtained by a multilayer perceptron. The learning method was unknown at the time.

Since Rosenblatt (1959)[31] introduced the one-layer, single perceptron learning method, called the *perceptron convergence* procedure, the research on the single

---

* A comment on the terminology 'backpropagation' is given in section 1.5.3. There, the backpropagation is interpreted as a method to find the gradient values of a feed-forward multilayer perceptron network, not as a learning method. A pseudo-steepest descent method is the learning mechanism used in the network.

perceptron had been widely active until the counter-example of the XOR problem was introduced which the single perceptron could not solve.

In multilayer network learning the usual objective or error function to be minimized has the form of a squared error:

$$E(\mathbf{w}) = \sum_{p=1}^{P} \left\| \mathbf{t}^p - f\left(\mathbf{x}^p; \mathbf{w}\right) \right\|^2 \tag{1.47}$$

that is to be minimized with respect to $\mathbf{w}$, the weights in the network. Here p represents the pattern index, $p = 1,2,...,P$, and $\mathbf{t}^p$ is the target (or desired) value when $\mathbf{x}^p$ is the input to the network. Clearly this minimization can be obtained by any number of unconstrained optimization algorithms; gradient methods or stochastic optimization are possible candidates.

The updating of weights has a form of the steepest descent method:

$$w_{ij} \leftarrow w_{ij} - \eta \frac{\partial E}{\partial w_{ij}}, \tag{1.48}$$

where the gradient value $\partial E/\partial w_{ij}$ is calculated for each pattern being present; the error term $E(\mathbf{w})$ in the on-line learning is not the summation of the squared error for all the P patterns.

Note that the gradient points are in the direction of maximum increasing error. In order to minimize the error it is necessary to multiply the gradient vector by minus one (-1) and by a learning rate η.

The updating method (Equation 1.48) has a constant learning rate η for all weights and is independent of time. The original Method of Steepest Descent has the time-dependent parameter, η_k, hence η_k needs to be calculated as iterations progress.

1.5.3 ERROR BACKPROPAGATION

The backpropagation was first discussed by Bryson and Ho (1960),[32] later by Werbos (1974),[33] and Parker[34] but was rediscovered and popularized later by Rumelhart, Hinton, and Williams (1986).[29] Each pattern is presented to the network, and the input x_j and output y_j is calculated as in Figure 1.7. The partial derivative of the error function with respect to weights is

$$\nabla E(t) = \left[\frac{\partial E(t)}{\partial w_1(t)}, ..., \frac{\partial E(t)}{\partial w_n(t)} \right]^T \tag{1.49}$$

where n is the number of weights, and t is the time index representing the instance of the input pattern presented to the network.

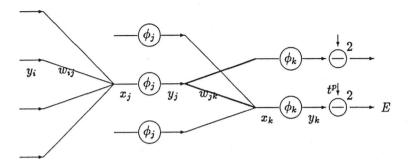

FIGURE 1.7   Error-backpropagation. The δ_j for weight w_{ij} is obtained, δ_k's are then backward propagated via thicker weight lines w_{jk}'s.

The former indexing is for the 'on-line' learning in which the gradient term of each weight does not accumulate. This is the simplified version of the gradient method that makes use of the gradient information of all training data. In other words, there are two ways to update the weights by Equation 1.49:

$$w_{ij}^{(p)} \leftarrow w_{ij}^{(p)} - \eta \left(\frac{\partial E}{\partial w_{ij}}\right)^{(p)} \quad \text{temporal learning} \tag{1.50}$$

$$w_{ij} \leftarrow w_{ij} - \eta \sum_p \left(\frac{\partial E}{\partial w_{ij}}\right)^{(p)} \quad \text{epoch learning} \tag{1.51}$$

One way is to sum all the P patterns to get the sum of the derivatives in Equation 1.51 and the other way (Equation 1.50) is to update the weights for each input and output pair temporally without summation of the derivatives. The temporal learning, also called on-line learning, (Equation 1.50), is simple to implement in a VLSI chip because it does not require the summation logic and storing each weight, while the epoch learning in Equation 1.51 does require to do so. However the temporal learning is an asymptotic approximation version of the epoch learning which is based on minimizing objective functions (Equation 1.47).

With the help of Figure 1.7 the first derivatives of E with respect to a specific weight w_{jk} can be expanded by the chain rule:

$$\frac{\partial E}{\partial w_{jk}} = \frac{\partial E}{\partial x_k}\frac{\partial x_k}{\partial w_{jk}} = \frac{\partial E}{\partial x_k}y_j = \phi_k'(x_k)\frac{\partial E}{\partial y_k}y_j \tag{1.52}$$

$$= \frac{\partial \phi_k(x_k)}{\partial x_k}\frac{\partial E}{\partial y_k}y_j = \delta_k y_j \tag{1.53}$$

For output units, $\partial E/\partial y_k$ is readily available, i.e., $2(y_k - t^p)$, where y_k and t^p are the network output and the desired target value for input pattern $\mathbf{x}^p$. The $\phi_k'(x_k)$ is straightforward for the linear and logistic nonlinear node functions; the hard limiter on the other hand is not differentiable.

For the linear node function:

$$\phi'(x) = 1 \qquad \text{with } y = \phi_x = x$$

and for the logistic unit the first order derivative becomes

$$\phi'(x) = \frac{e^x(1+e^x)-(e^x)^2}{(1+e^x)^2} \tag{1.54}$$

$$= y(1-y) \quad \text{when } \phi(x) = \frac{e^x}{1+e^x} \tag{1.55}$$

The derivative can be written in the form

$$\frac{\partial E}{\partial w_{ij}} = \sum_p y_i^p \delta_j^p \tag{1.56}$$

which has become known as the generalized delta rule.

The δ's in the generalized delta rule, Equation 1.56, for output nodes, therefore becomes

$$\delta_k = 2y_k(1-y_k)(y_k - t^p) \quad \text{for a logistic output unit}$$
$$\delta_p = 2(y_p - t^p) \qquad\qquad \text{for a linear output unit} \tag{1.57}$$

The interesting point in the backpropagation algorithm is that the δ's can be computed from output to input through hidden layers across the network. δ's for the units in earlier layers can be obtained by summing the δ's in the higher layers. As shown in Figure 1.7, the δ_j are obtained as

$$\delta_j = \phi_j'(x_j)\frac{\partial E}{\partial y_j}$$
$$= \phi_j'(x_j)\sum_{j \to k} w_{jk}\frac{\partial E}{\partial x_k} \tag{1.58}$$
$$= \phi_j'(x_j)\sum_{j \to k} w_{jk}\delta_k$$

The δ_k's are available from the output nodes. As the updating (or learning) progresses backwards, the previous (or higher) δ_k are weighted by the weights w_{jk}'s and summed to give the δ_j's. Since Equation 1.58 for δ_j only contains terms at higher layer units, it is clear that it can be calculated backwards from the output to the input of the network; hence the name backpropagation.

1.5.3.1 Madaline Rule III for Multilayer Network with Sigmoid Function

Widrow took an independent path in learning in as early as the 1960s.[35,36] After some 20 years of research in adaptive filtering, Widrow and colleagues returned to the neural network research,[36] and extended the Madaline I with the goal of developing a new technique that could adapt multiple layers of adaptive elements, using the simpler hard-limiting quantizer. The result was Madaline Rule II (or simply MRII), a multilayer linear combiner with a hard-limiting quantizer.

Andes (1988, unpublished) modified the MRII by replacing the hard-limiting quantizer resulting in MRIII by a sigmoid function in the Adaline, i.e., a single-layer linear combiner with a hard-limiting quantizer. It was proven later that MRIII is in essence equivalent to backpropagation. The important difference from the gradient based backpropagation method is that the derivative of the sigmoid function is not required in this realization; thus the analog implementation becomes feasible with this MRIII multilayer learning rule.

1.5.3.2 A Comment on the Terminology 'Backpropagation'

The terminology 'backpropagation' has been used differently from what it should mean. To get the partial derivatives of the error function (at the system output node) with respect to the weights of the units in lower than the output unit, the δ terms in the output unit are propagated backward, as in Equation 1.58. However, the network (actually the weights) learns (or weights are updated) using the Pseudo Steepest Descent method, (Equation 1.48); it is *pseudo* because a constant term is used, whereas the Steepest Descent method requires an optimal learning rate for each weight and time instance, i.e., $\eta_{ij}(k)$. The error backpropagation is indeed to find the necessary gradient values in the updating rule. Thus it is not a good idea to call the backpropagation a learning method; the learning method is a simple version of the Steepest Descent method, which is one of the classical minimizer finding algorithms. Backpropagation is an algorithm to find the gradient ∇E in a feed-forward multilayer perceptron network.

1.5.3.3 Optimization Machines with Feed-forward Multilayer Perceptrons

Optimization in multilayer perceptron structures can be easily realized by gradient-based optimization methods with the help of backpropagation. In the multilayer perceptron structure the functions can be minimized/maximized via any gradient-based unconstrained optimization algorithm, such as Newton's method or Steepest Descent method.

The optimization machine has the functional description depicted in Figure 1.8 and consists of two parts, gradient calculation and weight (or parameter) updating.

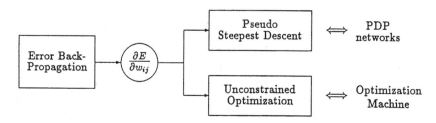

FIGURE 1.8   Functional diagram for an Optimization Machine.

The gradient ∇E of the multilayer perceptron network is obtained by error backpropagation. If this gradient is used in an on-line fashion with the constant learning rate η as in Equation 1.48, then this structure is the neural network used earlier.[29] This on-line learning structure possesses a desirable feature in VLSI implementation of the algorithm since it is temporal: no summation over all the patterns is required but the weights are updated as the individual pattern is presented to the network. It requires little memory but sometimes the convergence is too slow.

The other branch in Figure 1.8 shows unconstrained optimization of the nonlinear function. The Optimization Machine gets the gradient information as before, but various and well-developed unconstrained optimizations can be used for finding the optimizer. The unconstrained nonlinear minimization is divided basically into two categories, gradient methods and stochastic optimization. The gradient methods are deterministic and use the gradient information to find the direction for the minimizer. Stochastic optimization methods such as ALOPEX are discussed in another section of this book as well as in References 37, 38, and 39. Comparisons of ALOPEX with backpropagation are shown in References 37 and 40.

1.5.3.4   Justification for Gradient Methods for Nonlinear Function Approximation

Getting stuck in local minimizers is a well-known problem for gradient methods. However, the size of the weights (or the dimensionality of the weight space in the neural networks) is usually much larger than the dimensionality of the input space: $X \subset R^p$ that we like to search for optimization. The employed redundant degrees of freedom in the ways to find the better minimizer is a good reason or the justification for the gradient methods used in neural networks.

Another justification for the gradient method in optimization may be due to the approximation by the Taylor expansion of highly nonlinear functions[28] where the first and second order approximation, i.e., a quadratic approximation to the nonlinear function, is used. The quadratic function in a covariate $\mathbf{x}$ has a unique minimum or maximum.

1.5.3.5   Training Methods for Feed-Forward Networks

There exist two basic ways to train the feed-forward networks. They are gradient-based learning and stochastic learning. Training or learning is essentially an unconstrained optimization problem. Abundant algorithms in optimization can be applied to the function approximated by the network in a structured way defined by the network topology.

In the gradient-based methods, the most popular learning is the steepest descent/ascent method with Error Backpropagation algorithm to get the required gradient of the minimizing/maximizing error function with respect to the weights in the network.[29,41] Another method using the gradient information is Newton's method, which is basically used for zero finding of a nonlinear function. The function optimization problem is the same as the zero finding of the first derivative of the function; hence, the Newton's method is valid.

All the deterministic (as opposed to stochastic) minimization techniques are based on either or both the steepest descent and Newton's method. The objective function to be optimized is usually limited to a certain class in the network optimization. The square of the error $\|t - \hat{y}\|^2$ and the information theoretic measure, the Kullback-Leibler distance, are objective functions used in the feed-forward networks. This is due to the limitation in calculating the gradient values of the network utilized by the Error Backpropagation algorithm.

The recommended 'method of optimization' due to Broyden, Fletcher, Goldfarb, and Shannon (BFGS) is the well-known Hessian matrix update in the Newton's method of unconstrained optimization.[42] It requires gradient values. For the optimization machine of Figure 1.8 the feed-forward network with backpropagation provides the gradients, and the Hessian approximation is obtained by the BFGS method.

The other dichotomy of the minimization of an unconstrained nonlinear multivariate function is grouped into the so called 'stochastic optimization.' The representative algorithms are Simulated Annealing,[43] Boltzman Machine Learning,[44] and ALgorithm Of Pattern EXtraction (ALOPEX).[45,46] Simulated Annealing[43] has been used successfully in combinatoric optimization problems, such as the traveling salesman problem, VLSI wiring, and VLSI placement problems. An application of feed-forward network learning has been reported[47] with the weights being constrained to be integers or discrete values rather than continuum of the weight space.

Boltzman Machine learning by Hinton and Sejnowski[44] is similar to Simulated Annealing except that the acceptance of randomly chosen weights is possible even when the energy state has decreased. In Simulated Annealing the weights yielding the decreased energy state are always accepted, but in the Boltzman Machine, probability is used in accepting the increased energy states.

The Simulated Annealing and the Boltzman Machine Learning (a general form of Hopfield Network[48] for the associative memory application) are mainly for combinatoric optimization problems with binary states of the units and the weights. Extension from binary to M-ary in the states of the weights has been reported for classification problems[47] in Simulated Annealing training of the feed-forward perceptrons.

ALOPEX was originally used for construction of the visual receptive field but with some modifications was later applied to the learning of any type of network, not restricted to multilayer perceptrons. It is a random walk process in each parameter in which the direction of the constant jump is decided by the correlation between the weight changes and the energy changes.[46] Since the stream of this chapter consists of the gradient-based optimization methods and the scope of the stochastic optimization is examined elsewhere in this book,[37] we do not include the other important optimization stream of stochastic methods in this chapter.

1.5.4   ISSUES IN NEURAL NETWORKS

1.5.4.1   Universal Approximation

In the introduction section of the article by Hornik, Stinchcombe, and While (1989)[49] previous work about the approximation capability of multilayer perceptrons is summarized and is referenced here. More than 20 years ago, Minsky and Papert (1969)[30] showed that simple two-layer (no hidden layers) networks cannot approximate the nonlinearly separating functions (e.g., XOR problems) but a multilayer neural network could do the job. Many results on the capability of the multilayer perceptron have been reported. Some theoretical analyses for the network capability of the multilayer perceptron as a universal approximator are listed below and are extensively discussed in Reference.[49]

Kolmogorov (1957)[50] tried to answer the question of Hilbert's 13th problem, i.e., the multivariate function approximation by a superposition of the functions of one variable. The superposition theory sets the upper limit of the number of hidden units to $2n + 1$ units, where n is the dimensionality of the multivariate function to be approximated. However, the functional units in the network are different for the different functions to be approximated, while one would like to find an adaptive method to approximate the function from the given training data at hand. Thus Kolmogorov's superposition theory says nothing about the capability of a multilayer network nor which method to be used.

More general views were reported. Le Cun (1987)[51] and Lapedes and Farber (1988)[52] showed that monotone squashing functions can be used in the two hidden layers to approximate the functions. Fourier series expansion of a function is realized by a single layer network by Gallant and White (1988)[53] with cosine functions in the units. Further related results using the sigmoidal (or logistic) units are shown by Hecht-Nielsen (1989).[54] Hornik, Stinchcombe, and White (1989)[49] presented a general approximation theory of one hidden layer network using arbitrary squashing functions such as cosine, logistic, hyperbolic tangent, etc., provided that sufficiently many hidden units are available. However the number of hidden units is not considered to attain any given degree of approximation in Hornik, Stinchcombe, and White.[49]

The number of hidden units obviously depends on the characteristics of the training data set, i.e., the underlying function to be estimated. It is intuitive to say that the more complicated the functions to be trained, the more hidden units are required.

For the number of hidden units, Baum and Haussler[55] limit the size of general networks (not necessarily the feed-forward multilayer perceptrons) by relating it to the size of the training sample. The authors analytically showed that if the size of the sample is N and we want to correctly classify future observations with at least a fraction $1 - \frac{\epsilon}{2}$ correctly, then the size of the sample has a lower bound given by

$$N \geq O\left(\frac{W}{\epsilon} \log \frac{N}{\epsilon}\right)$$

where W is the number of the weights and N the number of the nodes in a network. This, however, does not apply to the interesting feed-forward neural networks, and the given bound is not useful for most applications.

There seems to be no rule of thumb for the number of hidden units.[19] The size of the hidden units can usually be found by cross-validation or any other resampling methods. Usual starting value for the size is suggested to be about the average of the number of the input and output nodes.[19] Failure in learning can be attributed[49] to three main reasons:

- inadequate learning,
- inadequate number of hidden units, or
- presence of a stochastic rather than a deterministic relation between input and target in the training data, i.e., noisy training data.

1.5.5 ENHANCING CONVERGENCE RATE AND GENERALIZATION OF AN OPTIMIZATION MACHINE

While the steepest descent method used originally with the backpropagation algorithm, (Equation 1.48), can be an efficient method for obtaining the weights that minimize an error measure, error surfaces frequently possess properties that make this procedure of slow convergence. There are at least two reasons (correlated in a sense as will be seen below) for this slow rate of convergence.[56]

1. The magnitude of a gradient may be such that modifying a weight by a constant proportion, η as in Equation 1.48, of that gradient will yield too little reduction in the error measure. There are two cases for this situation. When the error surface is fairly smooth (or nearly flat), the gradient magnitude is small, and consequently the convergence is too slow. The other situation involves the case where the error curve is too wiggly. Even a small change in the weight space may result in 'overshooting,' which may produce a small reduction of the error measure. Oscillating over a local minimum can happen with this error function.
2. The second reason for the slow convergence is that the negative gradient may not point to the actual minima, as is usually the case. Figure 1.9 shows an example of an error function of the two parameters with the elliptic curves representing the contour of the error function. With the

given weight point $w(t)$ at time t, the negative gradient does not point to the real minima which are represented by a bullet in the center of the inner contour. Given the negative gradient, the magnitude in the direction of the major axis x_1 is too small, whereas the component in the minor direction x_2 is too large.

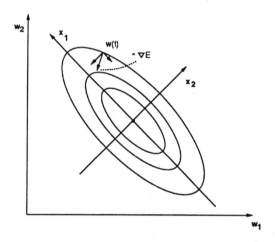

FIGURE 1.9   Error surface with contours over a two-dimensional weight space.

1.5.5.1   Suggestions for Improving the Convergence

Jacobs (1988)[56] summarized four heuristics proposed in the literature for increasing the rate of convergence:

1. Every parameter of the performance measure to be minimized should have its own individual learning rate, η_{ij}.
2. Every learning rate should be allowed to vary over time, $\eta_{ij}(k)$.
3. When the derivative of a parameter possesses the same sign for several consecutive time steps, the learning rate for that parameter should be increased.
4. When the sign of the derivative of a parameter alternates for several consecutive time steps, the learning rate for that parameter should be decreased.

Note that from Figure 1.9, by providing different learning rates for each parameter dimension, the current point in the weight space is not modified in the direction of the negative gradient, but toward the real minima.

Another cause for the slow convergence comes from the sigmoidal units $\phi()$'s that are used to impose the network with nonlinearity. The derivative of the nonlinear unit function has been shown to be in the form of Equation 1.55. The logistic units may become 'stuck' at a round value, either 0 or 1, since $\phi'(x) = y(1-y)$ (Equation 1.55) gives a very small value for an output $\simeq 0$ or 1:

$$\phi'(x) = y(1-y) \simeq 0 \quad \text{for } y \simeq 0 \text{ or } 1 \tag{1.59}$$

Unfortunately, any saturating unit function is bounded, resulting in the property: near the saturation points the derivative vanishes. With nonlinear units with the backpropagation learning and the general objective function $E = \|\mathbf{t} - \mathbf{y}\|^2$ giving the $\partial E/\partial w = y(1-y)$ the convergence of a network is known to be slow, as discussed earlier.

In the original work of Rumelhart, Hinton, and Williams[29] a 'momentum' term was added; that, is an exponential smoothing was applied to the correction term, so that

$$w_{ij} \leftarrow w_{ij} - \eta \left[(1-\alpha)\frac{\partial E}{\partial w_{ij}} + \alpha\left(\Delta w_{ij}\right) \right] \tag{1.60}$$

They also considered the 'on-line' version of Equation 1.60, that is

$$w_{ij} \leftarrow w_{ij} - \eta' y_i^p \delta_j^p + \alpha'\left(\Delta w_{ij}\right) \tag{1.61}$$

and updated the weights as each pattern was presented to the network.

1.5.5.2   Quick Prop

Some other interesting ideas to speed up the convergence have been introduced. *Quickprop*[57] used a second-order method, based loosely on Newton's method. Quickprop is based on two risky assumptions, (1) that the error vs. weight graph for each weight can be approximated by a parabola with one minimum value and (2) that the change in the slope of the error curve, as seen by each weight, is not affected by all the other weights that are changing at the same time.

Everything else proceeds as in standard backpropagation, but for each weight w_{ij} a set of information for the previous time update is retained to get a second order approximation. The steps to follow are (1) find the error derivative $S_{ij}(t-1) = \partial E(t-1)/\partial w_{ij}(t-1)$ and (2) update $\Delta w_{ij}(t-1) = w_{ij}(t) - w_{ij}(t-1)$. The computation for the next step size of a found direction according to the heuristics above is then given by:

$$\Delta w(t) = \frac{S(t)}{S(t-1) - S(t)} \Delta w(t-1) \tag{1.62}$$

where $S(t)$ and $S(t-1)$ are the current and previous values of $\partial E/\partial w$. This is a crude approximation to the optimal minima. The fraction portion η in each parameter w_{ij} is adaptively adjusted using the Equation 1.62.

To get around this pitfall, Fahlmann (1989)[57] suggested also using an offset in order for the *delta* (as in Equation 1.57) to be at least 0.1, i.e., $\phi'(x) = 0.1 + y(1-y)$.

1.5.5.3   Kullback-Leibler Distance

A more interesting treatment for the problem with the classical gradient descent method has been shown in the literature.[58-60] A relative (or cross) entropy of target **t** with respect to output **y** is defined and interpreted as Maximum A Posteriori (MAP) estimation for the optimal minima of the weight space,

$$E = \sum_p \sum_k \left[t_k^p \log \frac{t_k^p}{y_k^p} + \left(1 - t_k^p\right) \log \frac{1 - t_k^p}{1 - y_k^p} \right] \tag{1.63}$$

This entropy measure becomes the measure of 'maximum likelihood' if the targets t_k are $t_k \in \{0,1\}$,[20] and may be called the 'Kullback-Leibler' distance, one of the probabilistic distances.

The interpretation of the output vectors with this distance measure is that the output vector represents the conditional probability of target t, given the input pattern **x**. A binary random variable B_k associated with the kth output unit describes the presence $(B_k = 1)$ or absence $(B_k = 0)$ of the kth output attribute. For a given input pattern $\mathbf{x}^p$, the activity y^p reflects the conditional probabilities

$$P\left\{B_k = 1 | \mathbf{x}^p\right\} = y^p, \quad \text{and} \tag{1.64}$$

$$P\left\{B_k = 0 | \mathbf{x}^p\right\} = 1 - y^p. \tag{1.65}$$

With this distance measure the δ value in the generalized delta rule, (Equation 1.56), becomes simpler and linear with the error $(t^p - y_k)$:

$$\begin{aligned}
\delta_k &= \phi_k'(x_k) \frac{\partial E}{\partial y_k} \\
&= y_{jk}(1 - y_k) \left(\frac{-t^p}{y_k} + \frac{1 - t^p}{1 - y_k} \right) \\
&= y_k \left((1 - y_k) \frac{y_k - t^p}{y_k(1 - y_k)} \right) \\
&= y_k - t^p
\end{aligned} \tag{1.66}$$

Thus the error signal propagates towards the inner layers backwards and the pitfall problem (Equation 1.59) no longer exists for this distance measure.

1.5.5.4   Weight Decay

Another way to avoid saturation is to discourage large weights and hence large inputs:[20] ones with large deviations from the data set are used for training. One can modify the error function to obtain the regularization effects by adding an extra term, which penalizes the overfitting. Also the discouragement of unusual inputs (e.g., outlier patterns) works as *robust* learning. This generalization in learning is related to the *bias-variance* trade-off in the scatter plot smoothing.

A new error to be minimized is the sum of the squared error:

$$E' = E + \lambda \sum_{ij} w_{ij}^2 \tag{1.67}$$

where the λ is the *weight decay* parameter. The weight update rule, (Equation 1.48), turns out to be (with the penalty term)

$$w_{ij} \leftarrow w_{ij} - \eta \sum_p y_i^p \delta_j^p - 2\eta\lambda w_{ij} \tag{1.68}$$

This is the gradient (or steepest) descent learning method with a new error term.

Two effects from the weight decay can be realized. One is the *generalization* obtained by the shrinkage effect of the weight decay. This shrinkage method is the same idea as ridge regression in statistics, which may be written in a modified linear regression form as:

$$\left(X'X + \Lambda\right)\hat{\beta} = X'Y \tag{1.69}$$

where Λ is a non-negative diagonal matrix. This is motivated by a prior on β or as a penalty term or a device to avoid large parameter values in nearly collinear problems.[19] It is also known that weight decay helps the numerical stability of optimization algorithms, especially in avoiding almost flat regions in iterative methods, such as in Equation 1.48.

The extra penalty term in Equation 1.68, weight growing is equally discouraged; there is no discrimination of the weights by their hierarchical position in a multilayer network. With the help of Figure 1.7, the weights $\{w_{ij}\}$ relate the system as inputs $y_i = x_i$ and x_j, the input to the next layer units, but the weights $\{w_{jk}\}$ are between y_j and x_k. To give the same penalty for all the weights evenly, (Equation 1.68), the input vector $\mathbf{x}$ to the system should have the same range as the y_j's. Thus it is more sensible that the system inputs have the same range as the intermediate values y_j's, which is done by scaling so that the input $\{\mathbf{x}^p\}$ is in [0,1], approximately.

For the decay parameter λ, Ripley[28] suggested $\lambda \simeq 10^{-4} \sim 10^{-2}$ for the sum of the squares criterion (Equation 1.68), and $0.01 \sim 0.1$ for the entropy measure criterion, (Equation 1.63).

If regression and classification are to be considered in a unified frame, the distinguishing characteristic is in the interpretation and use of the response variable. Regression is a method of model fitting for the given data point pairs. Regression has the continuous response variable, representing outputs of the estimating function $\hat{f}(\cdot)$, and usually continuous in the region of the function $f(\cdot)$. One likes to find or estimate the underlying function that relates input and output pairs, $\{(\mathbf{x}_i, y_i)\}_1^N$, for many reasons. *Prediction* for future observations x_0, *inference* on the estimated function f, and *interpretation* of the function of covariate x_i are the principal objectives. Neural networks are a new surge in this regression paradigm, although research for regression purposes is not as active as it is for classification problems. Classification is meant to analyze different group data and to represent the group data well so that future observations could be classified as correctly as possible. The response variable can be considered as a categorical variable taking the value from a finite set of class labels.

The difference between regression and classification is whether the response variable is the continuous region of the function or the categorical variable for classification, respectively.

1.5.5.5   Regression Methods for Classification Purposes

The recent success and popularity of neural networks motivated some applied statisticians to look for similar methodologies in the statistical literature and to develop methods to use the existing nonparametric regression techniques[5] for classification. The classification problem is recast in the form of a regression problem. To establish a relationship between regression and classification, the two-class linear discriminant function can be shown to be the scalar (not a constant) multiple of the least square regression function in Section 1.5.6.

Generalization for multiple group settings is given in Section 1.5.7. A number of recently developed adaptive regression methods are studied. Those are Classification And Regression Tree (CART),[18] BRUTO,[61] and Multivariate Adaptive Regression Splines (MARS)[62] and incorporated with a bridging tool FDA (Flexible Discriminant Analysis)[5] for classification purposes.

1.5.6   TWO-GROUP REGRESSION AND LINEAR DISCRIMINANT FUNCTION

The linear discriminant function for two-group classification has been viewed by Fisher (1936) alternatively in a regression context. (See pp. 212–213 of Anderson (1984)[21]). The linear projector $W^{-1\prime}(\bar{\mathbf{x}}^{(2)} - \bar{\mathbf{x}}^{(1)})$ in the linear discriminant function $(\bar{\mathbf{x}}^{(2)} - \bar{\mathbf{x}}^{(1)})'W^{-1}\bar{\mathbf{x}}$ is actually a scalar multiple of the linear regression function.

A dummy variate is introduced for two class response values. Let the two variables be

$$y_i^{(1)} = \frac{n_2}{n_1 + n_2}, \quad i = 1, 2, \ldots, n_1, \tag{1.70}$$

$$y_i^{(2)} = \frac{-n_1}{n_1 + n_2}, \quad i = 1, 2, \ldots, n_2, \tag{1.71}$$

The regression function $\mathbf{b}'\mathbf{x}$ is obtained by minimizing the sum of squared residual (SSR)

$$\sum_{j=1}^{2} \sum_{i=1}^{n_j} \left[y_i^{(j)} - \mathbf{b}' \left(\mathbf{x}_i^{(j)} - \bar{\mathbf{x}} \right) \right]^2$$

where $\mathbf{x}_i^{(j)}$ is the ith observation from group j, $j = 1,2$ and $\bar{\mathbf{x}}$ is the overall mean of the training data.

The normal equations are obtained by taking the derivative of the SSR with respect to $\mathbf{b}$, the newly defined unknown coefficients of the two-group regression, and set it equal to zero:

$$\sum_{j=1}^{2} \sum_{i=1}^{n_j} \left(\mathbf{x}_i^{(j)} - \bar{\mathbf{x}} \right) \left(\mathbf{x}_i^{(j)} - \bar{\mathbf{x}} \right)' \mathbf{b} = \sum_{j=1}^{2} \sum_{i=1}^{n_j} y_i^{(j)} \left(\mathbf{x}_i^{(j)} - \bar{\mathbf{x}} \right)$$

$$= \frac{n_1 n_2}{n_1 + n_2} \left[\left(\bar{\mathbf{x}}^{(1)} - \bar{\mathbf{x}} \right) - \left(\bar{\mathbf{x}}^{(2)} - \bar{\mathbf{x}} \right) \right] \tag{1.72}$$

$$= \frac{n_1 n_2}{n_1 + n_2} \left(\bar{\mathbf{x}}^{(1)} - \bar{\mathbf{x}}^{(2)} \right) \tag{1.73}$$

The outer product in the LHS of Equation 1.72 is the total covariance of the predictor variables and can be decomposed in the form of within-covariance and between-covariance matrix combinations as

$$\sum_{j=1}^{2} \sum_{i=1}^{n_j} \left(\mathbf{x}_i^{(j)} - \bar{\mathbf{x}} \right) \left(\mathbf{x}_i^{(j)} - \bar{\mathbf{x}} \right)'$$

$$= \sum_{j=1}^{2} \sum_{i=1}^{n_j} \left(\mathbf{x}_i^{(j)} - \bar{\mathbf{x}}^{(j)} \right) \left(\mathbf{x}_i^{(j)} - \bar{\mathbf{x}}^{(j)} \right)'$$

$$+ n_1 \left(\bar{\mathbf{x}}^{(1)} - \bar{\mathbf{x}} \right) \left(\bar{\mathbf{x}}^{(1)} - \bar{\mathbf{x}} \right)' + n_2 \left(\bar{\mathbf{x}}^{(2)} - \bar{\mathbf{x}} \right) \left(\bar{\mathbf{x}}^{(2)} - \bar{\mathbf{x}} \right)' \tag{1.74}$$

$$= \sum_{j=1}^{2} \sum_{i=1}^{n_j} \left(\mathbf{x}_i^{(j)} - \bar{\mathbf{x}}^{(j)} \right) \left(\mathbf{x}_i^{(j)} - \bar{\mathbf{x}}^{(j)} \right)'$$

$$+ \frac{n_1 n_2}{n_1 + n_2} \left(\bar{\mathbf{x}}^{(1)} - \bar{\mathbf{x}}^{(2)} \right) \left(\bar{\mathbf{x}}^{(1)} - \bar{\mathbf{x}}^{(2)} \right)' \tag{1.75}$$

Thus Equation 1.72 is rewritten as

$$
\frac{n_1 n_2}{n_1 + n_2} \left(\overline{\mathbf{x}}^{(1)} - \overline{\mathbf{x}}^{(2)} \right)
$$

$$
= \left[\sum_{j=1}^{2} \sum_{i=1}^{n_j} \left(\mathbf{x}_i^{(j)} - \overline{\mathbf{x}}^{(j)} \right) \left(\mathbf{x}_i^{(j)} - \overline{\mathbf{x}}^{(j)} \right)' \right. \tag{1.76}
$$

$$
\left. + \frac{n_1 n_2}{n_1 + n_2} \left(\overline{\mathbf{x}}^{(1)} - \overline{\mathbf{x}}^{(2)} \right) \left(\overline{\mathbf{x}}^{(1)} - \overline{\mathbf{x}}^{(2)} \right)' \right] \mathbf{b}
$$

If we define the within-group SSP (sum of squares and products) as $\mathbf{W}$:

$$
W = \sum_{j=1}^{2} \sum_{i=1}^{n_j} \left(\mathbf{x}_{i=1}^{(j)} - \overline{\mathbf{x}}^{(j)} \right) \left(\mathbf{x}_{i=1}^{(j)} - \overline{\mathbf{x}}^{(j)} \right)',
$$

the normal Equation 1.76 has the form

$$
W\mathbf{b} = \frac{n_1 n_2}{n_1 + n_2} \left(\overline{\mathbf{x}}^{(1)} - \overline{\mathbf{x}}^{(2)} \right) - \frac{n_1 n_2}{n_1 + n_2} \left(\overline{\mathbf{x}}^{(1)} - \overline{\mathbf{x}}^{(2)} \right) \left(\overline{\mathbf{x}}^{(1)} - \overline{\mathbf{x}}^{(2)} \right)' \mathbf{b} \tag{1.77}
$$

$$
= \left(\overline{\mathbf{x}}^{(1)} - \overline{\mathbf{x}}^{(2)} \right) \left[\frac{n_1 n_2}{n_1 + n_2} - \frac{n_1 n_2}{n_1 + n_2} \left(\overline{\mathbf{x}}^{(1)} - \overline{\mathbf{x}}^{(2)} \right)' \mathbf{b} \right]. \tag{1.78}
$$

Since the whole bracket is a scalar, the solution $\mathbf{b}$ of Equation 1.78 is proportional to the projection vector $W^{-1} (\overline{\mathbf{x}}^{(1)} - \overline{\mathbf{x}}^{(2)})$ of the linear discriminant function.

1.5.7   MULTI-RESPONSE REGRESSION AND FLEXIBLE DISCRIMINANT ANALYSIS

Multiresponse linear/nonlinear regression can also be used for classification. The most simple and common way is to transform the categorical variable $j \in \{1,2,\ldots,J\}$ in the form of $(N \times J)$-matrix $Y_{N \times J}$ such that an element y_{ij} has a value of 1 in the jth column if the observation is in class j. The multiresponse multivariate regression is carried onto the predictors $\mathbf{x}$. A new observation $\mathbf{x}_0$ is fitted with the J fits and is classified by the class having the largest fitted value, i.e., Y_j.

Since we cannot expect the regression fit $\hat{y}_k = f_k(\mathbf{x}_0)$, the kth regression fit, to be in the region [0,1], the indicator matrix Y whose elements are either 0 or 1 is not a good way of introducing dummy response variables. *Optimal Scoring*, which will be studied in Section 1.5.8, transforms a categorical variable to real line R, such that linear regression of the transform is best regressed on the predictor variables $\mathbf{x}$.

1.5.7.1 Powerful Nonparametric Regression Methods for Classification Problems

Recently, Hastie, Buja, and Tibshirani[63] introduced a new treatment of regression methods to be used for classification problems. They showed that the discriminant analysis could be tackled via Optimal Canonical Correlation Analysis (CCA), especially its asymmetric version, Optimal Scoring (OS). The idea is based on the facts that CCA is equivalent to Linear Discriminant Analysis (LDA) and that the OS results to CCA, via various nonparametric regression methods.

Linear discriminant analysis in Section 1.4.2 of multi-group has been the traditional choice in classification and discriminant analysis. The robustness and the simplicity of LDA[64] in implementation and interpretation are responsible for its popularity. Recently a group of applied statisticians found and developed ways of using regression techniques for classification applications. Breiman and Ihaka (1984)[6] noticed that the regression approach to the classification problem can be extended from the two-group to a multi-group setting via *scaling* and ACE. This idea has been adapted by Hastie, Tibshirani, and Buja and was developed to render the Flexible Discriminant Analysis (FDA).[5]

The basic concept is that the LDA, CCA, and OS are equivalent. One can find the discriminant variates via either CCA or OS. Since this equivalence is so critical, some space is devoted here to the understanding of this property. The generalization of the LDA to nonlinear flexible discriminant analysis is due to the fact that an OS solution can be obtained by any linear/nonlinear regression method. This has the important consequence that we can simply use the tools for nonparametric regression to perform nonparametric discriminant analysis, which the authors termed as Flexible Discriminant Analysis (FDA). This section is a somewhat concise version of Section 3 of Hastie, Buja, and Tibshirani's unpublished paper.[63]

It is known that discriminant variates are the same as the so-called 'canonical variates,' which result from an associated canonical correlation analysis (CCA), and often the latter term is used interchangeably with discriminant variates. Somewhat less known is that an asymmetric version of canonical correlation analysis, here called optimal scoring (OS), well-known in correspondence analysis, can also yield a set of dimensions which coincide with those of LDA and CCA. Each of the three techniques (OS, CCA, LDA) to be discussed has an associated criterion and constraints under which the criterion is to be optimized. The equivalence of LDA, CCA, and OS follows as each of them are briefly described.

1.5.8 OPTIMAL SCORING

Optimal scoring is used to turn categorical variables into quantitative ones by assigning scores to classes (groups, categories). Suppose $\theta : \mathcal{J} \to \mathcal{R}$ is a function that assigns scores to the classes, such that the transformed class labels are optimally predicted by linear regression on X. This produces a one-dimensional separation between the classes. More generally, we can find K sets of independent scorings for the class labels, $\{\theta_1, \theta_2, ..., \theta_K\}$, and K corresponding linear maps $\eta_k(X) = X'\beta_k$,

$k = 1,2,\ldots, K$, chosen to be optimal for multiple regression in $\mathbb{R}^K$. Thus the OS problem is to find the two sets of unknown functions that minimize a certain criterion.

Let $(\mathbf{x}_i, g_i)$, $i = 1,2,\ldots, N$, be the training sample; then the scores $\{\theta_k(g)\}_1^K$ and the maps $\{\beta_k\}_1^K$ are chosen to minimize the average squared residual (ASR):

$$ASR = \frac{1}{N} \sum_{k=1}^{K} \sum_{i=1}^{N} \left(\theta_k(g_i) - \mathbf{x}_i' \beta_k \right)^2$$

In the criterion ASR above, $\theta(g)$ assigns a real number, θ_j, to the jth label of g, the categorical response variable. With the matrix notation, given a J-vector of such scores θ_k, a N-vector $Y\theta$ is a vector of scored training data which one may try to regress onto the predictor matrix H, the $N \times p$-matrix.

For simpler notational purposes, we proceed with a single solution only. The multiresponse multivariate regression can be thought of as simply the K duplicates for the single response multivariate regression. Thus a single solution pair (θ, β) is used in the following, instead of the series of solution (θ_k, β_k), $k = 1,2,\ldots, K$, to simplify the notation.

Definition: The Optimal Scoring problem is defined by the criterion

$$ASR(\theta, \beta_{OS}) = \min_{\beta} \left\{ \frac{1}{N} \left(\sum_{i=1}^{N} \left[\left(\theta(g_i) - h(\mathbf{x}_i)' \beta \right) \right]^2 \right) \right\} \tag{1.79}$$

$$= \min_{\beta} \frac{1}{N} \left\| Y\theta - H\beta \right\|^2 \tag{1.80}$$

which is to be minimized (or made stationary) under the constraint $N^{-1}\|Y\,\theta\|^2 = 1$ which is for a unique solution for θ.

A unified view for the three similar but equivalent techniques (OS, CCA, and LDA) can be conveniently achieved by rewriting the ASR in Equation 1.80 in a quadratic form:

$$ASR(\theta, \beta) = \theta' \Sigma_{11} \theta - 2\theta' \Sigma_{12} \beta + \beta' \Sigma_{22} \beta \tag{1.81}$$

where the matrices Σ are defined as:

- $\Sigma_{11} = \frac{1}{N} Y'Y$, a diagonal matrix with the class proportions $p_j = n_j/N$ in the diagonal,
- $\Sigma_{22} = \frac{1}{N} (H'H)$, the total covariance matrix of the predictor variables,
- $\Sigma_{12} = \frac{1}{N} (Y'H)$, $\Sigma_{21} = \Sigma_{12}'$

If all considered classes are in the sample, i.e., $n_j > 0$, Σ_{11} is invertible.

1.5.8.1 Partially Minimized ASR

If we assume that the score vector θ is fixed, the minimizing β for the OS problem is obtained by the least squares estimate of β:

$$\beta_{OS} = \left(H'H\right)^{-1} H'Y\theta = \Sigma_{22}^{-1}\Sigma_{21}\theta \qquad (1.82)$$

The linear regression of $Y\theta$ on to the design matrix H with the least square criterion gives the following results. From Equation 1.80 and Equation 1.82:

$$\min_{\beta} ASR(\theta,\beta) = \frac{1}{N}\left\|Y\theta\right\|^2 - \frac{1}{N}\left((Y\theta)'S'SY\theta\right) = 1 - \frac{1}{N}(Y\theta)'S(Y\theta) \qquad (1.83)$$

$$= 1 - \frac{1}{N}\theta'Y'SY\theta = 1 - \theta'\Sigma_{12}\Sigma_{22}^{-1}\Sigma_{21}\theta \qquad (1.84)$$

where $S = H\ (H'H)^{-1}H'$ denotes the 'hat' or 'smoother' matrix of the predictor matrix H, which is the result of the least square linear regression.

The same equation on the ASR (θ, β) has a matrix form as

$$\min_{\beta} ASR(\theta,\beta) = \frac{1}{N}\left\|Y\theta\right\|^2 - \frac{1}{N}\left((Y\theta)'S'SY\theta\right)$$

$$= \frac{1}{N}\left\|Y\theta\right\|^2 - \frac{1}{N}(SY\theta)'(SY\theta)$$

$$= \frac{1}{N}(Y\theta - SY\theta)'(Y\theta - SY\theta) \qquad (1.85)$$

$$= \frac{1}{N}\left\{(Y\theta)'(I-S)(I-S)Y\theta\right\}$$

$$= \frac{1}{N}\left\{\theta'Y'(I-P_H)Y\theta\right\}$$

with a new notation for the projection matrix, P_H, based on the predictor design matrix H for the least square linear regression

$$P_H = S = H\left(H'H\right)^{-1}H'$$

With the assumption of fixed θ we have reached the partially minimized ASR where the minimizing β was obtained via the least square linear regression. Now, we need to find the θ that transforms the indicator matrix to yield the scalings $Y\theta$

such that the linear regression yields the best fit to the new scalings. The question then is — Given Equation 1.85, what θ gives the least possible *ASR*?

It is the quadratic form of the symmetric matrix $Y'(I - P_H)Y$ that we like to look for the vector θ, that results in the minimum quadratic value. Minimizing θ for the whole matrix $Y'(I - P_H)Y$ is the same as maximizing θ for the matrix $Y\hat{Y} = YP_HY$, provided that the regression fit $Y\hat{Y} = P_HY$ is *shrunk*, which is a property of all linear smoothers.[65] The projection operation P_H is a linear smoother. Therefore, the minimizing θ in Equation 1.85 is the eigenvector corresponding to the largest eigenvalue of $Y\hat{Y} = YP_HY$.

This is the point at which nonlinear nonparametric regressions come into play for classification applications of regression. Direct calculation of the projector matrix P_H of the expanded predictor space $h(\mathbf{x})$, or spanned by the columns of the matrix H is possible, but the fact that any regression can calculate the fitted value $\hat{Y}$ allows various linear/nonlinear regressions to be used.

1.5.9 CANONICAL CORRELATION ANALYSIS

Canonical Correlation Analysis (CCA) seeks to identify and quantify the associations between two sets of variables. The correlation of two linear combinations of the two sets of variables is to be maximized.

Definition: The canonical correlation problem is defined by the criterion

$$COR(\theta_{CCA}, \beta_{CCA}) = \max_{\theta, \beta}\{\theta'\Sigma_{12}\beta\} \tag{1.86}$$

which is to be maximized under the constraints

$$\theta'\Sigma_{11}\beta = 1, \quad \text{and} \quad \beta'\Sigma_{22}\beta = 1. \tag{1.87}$$

The Σ's are the same as in the previous section for optimal scoring. The criteria of the optimal scoring $ASR(\theta,\beta)$ and canonical correlation analysis $COR(\theta,\beta)$ are related to each other by Equation 1.81 and the two CCA constraints:

$$ASR = 2 - 2\,COR$$

which means that the OS and the CCA differ only in the additional constraint on β through Equation 1.87.

The partially maximizing β_{CCA} with θ for both the OS and the CCA obtained by minimizing β_{OS} with the constraint of the β_{CCA} in Equation 1.87 is

$$\beta_{CCA} = \beta_{OS} \Big/ \sqrt{\beta'_{OS}\Sigma_{22}\beta_{OS}}. \tag{1.88}$$

The maximizer β_{CCA} representation in terms of the minimizer β_{OS} in the above equation (Equation 1.88) and the definition of the *CCA* (Equation 1.86) entails the identity in the fixed linear coefficients θ in the OS and CCA:

$$\max_{\substack{\theta'\Sigma_{11}\theta=1 \\ \beta'\Sigma_{22}\beta=1}} COR(\theta, \beta) = \theta'\Sigma_{12}\beta = \frac{\theta'\Sigma_{12}\beta_{OS}}{\sqrt{\beta'_{OS}\Sigma_{22}\beta_{OS}}}$$

$$= \left(\frac{\theta'\Sigma_{12}\beta_{OS}\beta_{OS}^{-1}\Sigma_{21}\theta}{\beta'_{OS}\Sigma_{22}\beta_{OS}}\right)^{\frac{1}{2}} \qquad (1.89)$$

$$= \left(\theta'\Sigma_{12}\beta_{OS}\beta'_{OS}\Sigma_{22}^{-1}\beta_{OS}^{'-1}\beta'\Sigma_{21}\theta\right)^{\frac{1}{2}}$$

$$= \left(\theta'\Sigma_{12}\Sigma_{22}^{-1}\Sigma_{21}\theta\right)^{\frac{1}{2}}$$

which verifies the identity of θ in that the minimizer in Equation 1.84 is the same as the one in the maximizer in Equation 1.86. With the identity of θ for both the OS and the CCA as just shown and the relationship between the β's (Equation 1.88) verifies that the OS is essentially the same as the CCA with the constraint on the β_{CCA}.

1.5.10   LINEAR DISCRIMINANT ANALYSIS

Linear Discriminant Analysis (LDA) is a standard tool for classification and dimension reduction purposes. The LDA is a special case of the Bayesian Classifier as in Section 1.4.2, where the group conditional distributions are assumed to be multivariate normal, have a common covariance matrix, and have different mean vectors for the different classes.

1.5.10.1   LDA Revisited

The optimizing problem of the multiclass data is to find the $K \leq J - 1$ linear combinations which separate the class means $\mathbf{m}_j$ as much as possible in the K dimensional subspace satisfying the constraint that the linear combinations are to be spherical, i.e., uncorrelated and with unit variance, with respect to Σ_w, the within-class covariance. The columns of the matrix U of LDA vectors $\mathbf{u}_k$ are the eigenvectors corresponding to the K largest eigenvalues of the matrix of $\Sigma_B^{-1}\Sigma_w$. The procedure for the LDA is first to sphere $\mathbf{x}$ with respect to the common within-groups covariance matrix, project these data onto the $J - 1$ dimensional subspace spanned by the J group mean vectors $\mathbf{m}_j$'s, and then classify the new discriminant covariate, $U\mathbf{x}_0$, vector to the class corresponding to the closest centroid.

Following the notations of the two sets of variables as in Section 1.5.8, the matrix M of mean vectors, Σ_B, and Σ_W have the following simple form with $P_Y = Y(Y^tY)^{-1}Y^t$ the projector onto a Y-column space:

- $M = \Sigma_{11}^{-1} \Sigma_{12}$, a $J \times p$-matrix whose rows are the class means $\mathbf{m}_j = \text{avg}\{\mathbf{h}_i;$ $i \in \text{Class } j\} : M = (\mathbf{m}_1, \mathbf{m}_2, \ldots, \mathbf{m}_J)'$

- $\Sigma_B = \frac{1}{N}(P_Y H)'(P_Y H) = \Sigma_{21} \Sigma_{11}^{-1} \Sigma_{12} = M'\Sigma_{11}M$

- $\Sigma_W = \frac{1}{N}[((I - P_Y)H)'(I - P_Y)H] = \Sigma_{22} - \Sigma_B$

The matrix M consists of rows of class mean vectors $\mathbf{m}_j$. The between-class covariance Σ_B is the covariance of H regressed onto Y, or, equivalently, the class-weighted covariance of the class means. The within-class covariance is the left of the subtraction of the Σ_B from the total covariance Σ_{22}.

The criterion of the linear discriminant problem is the maximization problem of the between-class variance under a constraint on the within-class variance.

Definition: The criterion of the linear discriminant problem to be maximized is the between-class variance:

$$BV\,AR(\beta_{LDA}) = \max_\beta \{\beta'\Sigma_B\beta\} \qquad (1.90)$$

with the constraint:

$$WV\,AR(\beta_{LDA}) = \beta'_{LDA}\Sigma_W\beta_{LDA} = 1 \qquad (1.91)$$

1.5.11 TRANSLATION OF OPTIMAL SCORING DIMENSIONS INTO DISCRIMINANT COORDINATES

It is convenient to use *CCA* as a link between *OS* and *LDA*. *CCA* is a generalized singular value problem for Σ_{12} with regard to the metrics given by Σ_{11} and Σ_{22}. Remember that it is the maximizing problem Equation 1.86 in which the generalized quadratic form is used, hence it is called the generalized singular value problem.

The associated singular value decomposition (SVD), essentially a collection of stationary solutions of the *CCA* problem, takes on the form:

$$\Sigma_{11}^{-1}\Sigma_{12}\Sigma_{22}^{-1} = \Theta D_\alpha B' \qquad (1.92)$$

$$\Theta'\Sigma_{11}\Theta = I_L \qquad (1.93)$$

$$B'\Sigma_{22}B = I_L \qquad (1.94)$$

where $L = \min(J,p)$, Θ is a $J \times L$ matrix whose columns θ_k are left-stationary vectors, B is a $p \times L$ matrix whose columns β_k are right-stationary vectors, and D_α is a diagonal matrix of size $L \times L$ with non-negative diagonal elements of α_k sorted in descending order.

A simple (non-generalized) SVD of a form $A = UDV'$ entails the trivial consequences:

$$A = UDV'$$

$$AV = UD$$

$$A'U = VD$$

$$U'AV = D$$

$$V'A'AV = D^2$$

$$U'AA'U = D^2$$

These are translated to the generalized SVD as follows. The left column is for the regular SVD and the right column for the generalized SVD.

$$A = UDV' \qquad \Sigma_{11}^{-1}\Sigma_{12}\Sigma_{22}^{-1} = \Theta D_\alpha B' \tag{1.95}$$

$$AV = UD \qquad \Sigma_{11}^{-1}\Sigma_{12} = \Theta D_\alpha B'\Sigma_{22}$$
$$\Sigma_{11}^{-1}\Sigma_{12}B = \Theta D_\alpha \tag{1.96}$$

$$A'U = VD \qquad \Sigma_{22}^{-1}\Sigma_{21}\Sigma_{11}^{-1} = B D_\alpha \Theta'$$
$$\Sigma_{22}^{-1}\Sigma_{21}\Theta = B D_\alpha \Theta_t \Sigma_{11}\Theta = B D_\alpha \tag{1.97}$$

$$U'AV = D \qquad \Theta'\Sigma_{11}^{-1}\Sigma_{12}\Sigma_{22}^{-1}B = D_\alpha$$
$$\Theta'\Sigma_{12}B = D_\alpha \tag{1.98}$$

$$V'A'AV = D^2 \qquad \Theta'\Sigma_{12}\Sigma_{22}^{-1}\Sigma_{21}\Theta = D_{\alpha^2} \tag{1.99}$$

$$U'AA'U = D^2 \qquad B'\Sigma_{21}\Sigma_{11}^{-1}\Sigma_{12}B = D_{\alpha^2} \tag{1.100}$$

In particular, Equation 1.98 implies $COR(\theta_k, \beta_k) = \alpha_k$.

As noted before from Equation 1.84 and Equation 1.89, the stationary θ vectors of OS and CCA are the same, while the β vectors of OS and CCA are related according to Equation 1.82 and Equation 1.97 by

$$B_{OS} = BD_\alpha^t, \tag{1.101}$$

B_{OS} being a matrix of OS-stationary column vectors $\beta_{OS,k}$. From Equation 1.84 and Equation 1.89 it follows that $ASR(\theta_k, \beta_k) = 1 - \alpha_k^2$.

To link CCA and LDA, we rewrite Equation 1.100 using the expression of the $\Sigma_B = \Sigma_{21} \Sigma_{11}^{-1} \Sigma_{12}$ as:

$$B' \Sigma_B B = D_{\alpha^2} \tag{1.102}$$

and

$$B' \Sigma_W B = B' \left(\Sigma_{22} - \Sigma_B \right) B \tag{1.103}$$

$$= I_L - D_{\alpha^2} = D_{1-\alpha^2} \tag{1.104}$$

These two equations, (Equation 1.102) and Equation 1.104 show that B diagonalize both Σ_B and Σ_W. If we define,

$$B_{LDA} = B D_{(1-\alpha^2)^{\frac{1}{2}}}$$

we get a matrix whose columns $\beta_{LDA,k}$ are stationary solutions of the LDA problem:

$$B'_{LDA} \Sigma_W \Sigma_{LDA} = I_L, \tag{1.105}$$

$$B'_{LDA} \Sigma_B B_{LDA} = D_{\alpha^2/(1-\alpha^2)} \tag{1.106}$$

Finally, the relation between the LDA and the OS solutions is given by

$$B_{LDA} = B_{OS} D_{[\alpha^2(1-\alpha^2)]^{-\frac{1}{2}}} \tag{1.107}$$

1.5.12 Linear Discriminant Analysis via Optimal Scoring

The minimization criterion, average squared residual (ASR), for a multi-response Optimal Scoring has the form

$$
\begin{aligned}
ASR &= \frac{1}{N} \sum_{k=1}^{K} \sum_{i=1}^{N} \left(\theta_k(g) - x_i' \beta_k \right)^2 \\
&= \frac{1}{N} \| Y\Theta - XB \|^2
\end{aligned}
\tag{1.108}
$$

with a constraint $N^{-1} \| Y\Theta \|^2 = 1$ for a unique solution Θ.

If Θ is fixed we get the transformed value $\Theta_{N \times K}^* = Y_{N \times J} \, \Theta_{J \times K}$.

With a new notation for the projection matrix, P_H, and smoothing operation S, based on the predictor design matrix H for the least square linear regression $P_H = S = H(H'H)^{-1}H'$, the partially minimizing ASR with the Θ^* fixed becomes

$$
\begin{aligned}
ASR(\Theta, B) &= \frac{1}{N}\|Y\Theta\|^2 - \frac{1}{N}\left((Y\Theta)'S'SY\Theta\right) \\
&= \frac{1}{N}\|Y\Theta\|^2 - \frac{1}{N}(SY\Theta)'(SY\Theta) \\
&= \frac{1}{N}(Y\Theta - SY\Theta)'(Y\Theta - SY\Theta) \qquad (1.109) \\
&= \frac{1}{N}\left\{(Y\Theta)^+(I-S)(I-S)Y\Theta\right\} \\
&= \frac{1}{N}\left\{\Theta'Y'(I-P_H)Y\Theta\right\}
\end{aligned}
$$

If we set the constraints on the Θ^* of zero mean and being unit variance and uncorrelated:

$$
\frac{1}{N}\sum_{i=1}^{N}\Theta_i^* = 0 \qquad \frac{1}{N}\Theta^{*'}\Theta^* = I_K
$$

the minimizing Θ is obtained from Equation 1.109 by the K largest eigenvectors Θ of $Y'P_HY$ with the constraint $\Theta'D_p\,\Theta = I_K$ and with $D_p = Y'Y/N$.

A direct approach for such optimal score Θ would be by explicitly building the project (or hat) matrix P_X and doing eigen analysis via Singular Value Decomposition,

$$
P_X = X\left(X'X\right)^{-1}X'
$$

$$
Y'P_XY = \Theta\Lambda\Theta'
$$

A more convenient approach avoids the explicit calculation P_X and takes advantage of the fact that P_X computes the linear regression $\hat{Y} = P_XY$. An algorithmic approach to compute the usual canonical variates by OS provides an equivalent procedure to get the LDA by OS.

1.5.12.1 LDA via OS

As the equivalence of OS and LDA from Equation 1.107 the algorithm for LDA via OS is:

1. *Initialization*: form $Y_{N\times J}$, the indicator matrix, whose index y_{ij} is 1 if the ith observation belongs to the jth group, otherwise is 0.
2. *Linear multivariate regression*: find the linear regression

$$\hat{Y} = P_X Y = SY$$

and by the linear least squares, set B such that

$$\hat{Y} = XB.$$

3. *Optimal scores*: find the eigenvector matrix Θ of rank $K \leq J$ matrix $Y'\hat{Y}$ via SVD

$$Y'\hat{Y} = \Theta \Lambda \Theta' \quad \text{with} \quad \Theta' D_p \Theta = I_J$$

4. *Update* the coefficient matrix of the linear combination matrix B obtained in step 2.

$$B \leftarrow B\Theta$$

The final coefficient matrix B_{OS} is, up to a diagonal scale matrix, the same as the LDA coefficient matrix B_{LDA} obtained from Equation 1.107.

$$B_{LDA} = B_{OS} D$$

where the diagonal matrix D has the elements

$$d_{kk} = \left[\alpha_k^2 \left(1 - \alpha_k^2\right)\right]^{-1/2}$$

and α_k is the kth element of the diagonal matrix Λ, in the spectral decomposition of the rank $K \leq J$ matrix $Y'\hat{Y}$ via SVD:

$$Y'\hat{Y} = \Theta' \Lambda \Theta$$

1.5.13 FLEXIBLE DISCRIMINANT ANALYSIS BY OPTIMAL SCORING

If we apply nonparametric regression $\hat{Y} = S(\hat{\lambda})Y$, in step 2 above, we can reduce the *flexibility* of the nonparametric regression into a classification problem. Here the smoothing parameter $\hat{\lambda}$ controls the fitness of the regression $\hat{Y}$ to Y, and is thus the control parameter.

The nonparametric multivariate regression in $\hat{Y} = S(\hat{\lambda})Y$ comes into play in two ways:[5]

- the regularization property by bias-variance control is obtained, and
- a model selection (i.e., variable selection) and interaction between variables may be exploited in the multivariate regression.

There exist many powerful nonparametric multivariate regression methods, and more are expected to be developed. The most recently developed are (1) Projection Pursuit Regression (PPR),[66] (2) Alternate Conditional Expectation (ACE),[67] (3) Additivity and Variance Stabilization (AVAS),[68] (4) Additive Model (AM),[63] (5) Multivariate Adaptive Regression Splines (MARS),[62] (6) π-method,[69] (7) Interaction spline method,[70] (8) Hinging-hyperplanes,[71] and (9) Neural networks.

The FDA by OS method is similar to the algorithmic LDA by OS of the previous Section 1.5.12. The steps to follow are

1. *Initialize*: Choose an initial score matrix Θ_0 satisfying the constraints

$$\Theta'_{K\times J} D_p \Theta_{J\times K} = I_K$$

and get the scoring matrix $\Theta_0^* = Y \Theta_0$. The Θ_0 may be obtained by a contrast matrix.*

2. *Multivariate nonparametric regression*: Fit a multi-response, adaptive nonparametric regression of Θ_0^* of X by one of the nonparametric regressions listed above.

$$\Theta_0^* = S\left(\hat{\lambda}\right)\Theta_0^* = \eta(\mathbf{x})$$

where $\eta(\mathbf{x})$ is the vector of fitted regression functions.

3. *Optimal scores*: Obtain the eigenvector matrix Φ of $\Theta_0^{*t} \hat{\Theta}_0^*$ and hence the optimal scores $\Theta_{J\times K} = \Theta_0 \Phi$.

4. *Update* the final model from step 2 using the optimal scores:

$$\eta(\mathbf{x}) \leftarrow \Theta' \eta(\mathbf{x})$$

It is worth noting step 3 in both procedures in order to distinguish the way of obtaining the optimal scores. For the first procedure for the LDA via OS, the indicator matrix Y is regressed on to X. But in the second procedure for FDA via OS, the transformed score data, $\Theta_0^* = Y \Theta_0$ are regressed onto X by any of the various nonparametric regression methods. The optimal score Θ is thus updated as $\Theta = \Theta_0 \Phi$.

For a J class problem, it is known from the discriminant analysis that the vector of canonical variates or functions $\eta(\mathbf{x})$ has at most $K = J - 1$ components. If $\bar{\eta}^j = \Sigma_{g_i=j} \eta(\mathbf{x}_i)/n_j$ denotes the fitted centroid of the jth class in this space of canonical variates, the discrimination rule has the form of a (weighted) nearest centroid rule:

---

* The contrast matrix is the $K - 1$ linear combinations of a factor variable with K levels. It is an encoding method of the factor variable such that the linear combination of the levels becomes linearly independent. There exist the Helmert, polynomial contrasts and others (see References 72, Ch.2).

$$\mathbf{x} \in j = \arg \min_{k} \left\{ \left\| D\big(\eta(\mathbf{x}) - \overline{\eta}^{k}\big) \right\|^{2} \right\} \tag{1.110}$$

D is the diagonal matrix of scale factors that convert optimally scaled fits to discriminant analysis variables.

1.6 COMPARISON OF EXPERIMENTAL RESULTS

In general, any pattern recognition system consists of two basic subsystems: feature extraction and classifier design. In this study, however, we are mainly interested in classifiers. There are many different classifiers from the simple and powerful non-parametric KNN rule to the recently popularized neural networks, as well as the newly developed multivariate regression methods. Eleven classifiers, which are all explained in Section 1.1, are experimented with the same data set obtained by Zernike Moments, a global feature extraction method.[73–75]

A new branch in the growing tree of the classifiers has been developed in applied statistics[6,5] and is by now popularized. It is based on the fact that Optimal Scoring (OS) is equivalent to the Linear Discriminant Analysis (LDA) (Equation 1.107) and the OS can be obtained by various regression techniques which are well researched in statistics (Section 1.5.11). The multivariate regression methods were used for classification, and the results were proven to be competitive to the classical statistical methods.

Table 1.1 describes the classifiers in a simple format with control parameters, learning and operation process. Details on the classifiers are given in Section 1.1.

The core part of the software for the classifiers used in the study has been obtained from contributed software. They are written mostly by originators or some active researchers in the area. The archive package "classif" is a collection contributed by B. Ripley and is maintained in the statlib@lib.stat.cmu.edu which is accessible by *anonymous ftp*. It can be found under "S" directory of the maintainer. This "classif" library also contains LDA, OLVQ1, KNN, and others that we did not experiment with.

Hastie and Tibshirani contributed the programs that are recently developed by themselves and A. Buja. The package "fda" contains the Flexible Discriminant Analysis (FDA), which is a way of using Optimal Scoring by nonparametric regression for classification problems. The library "fda" comes with POLYREG, BRUTO. MARS and BRUTO are the recently developed multivariate regression methods. MARS can also be obtained from the directory "general" in the same maintainer, statlib@lib.stat.cmu.edu. The CART and PPREG can also be found from the "S" directory of the same maintainer. These are also available in function type "tree ()" and "ppreg ()" from the commercial package Splus.*

The NNET neural networks written by Ripley are different from the original ones[29] in that he uses the modified Newton's optimization algorithm with BFGS algorithm (the most popular Hessian matrix update algorithm). The description is

---

* The commercial version of "S"[76] which is developed in AT&T Bell Lab. Splus is an extended version of "S" from Statistical Sciences, Inc. Seattle, WA., USA.

TABLE 1.1
The List of the Classifiers Used

Classifiers	Control Parameters	Learning	Operation
LDA	F, μ_i		$\arg\min_{i \sim G}$ $\|F(\mathbf{x} - \mu_i)\|^2$
OLVQ1	Codebook	find $\{\mathbf{m}_i\}_1^l$	find $d_i(\mathbf{x}, \mathbf{m}_i)$
KNN	$k=1,3,5$		$\arg\min_i$ $\{d_i(\mathbf{x}, \mathbf{x}_i)\}$
NNET	$h = 15$ $\lambda = 0.005$	Minimize $$\sum(\hat{y}_i - t_i)^2 + \lambda \sum W^2$$	$\arg\max_j \{P(j|\mathbf{x})\}$
CART		find $B_m(\mathbf{x}) =$ $$\Pi_{k=1}^{L_m} H\left[s_{km}\left(x_{v(k,m)} - t_{km}\right)\right]$$	$\mathcal{L}\left[\arg\max_m \{B_m|\mathbf{x}\}_1^M\right]$
LREG	$\deg = 1$	$P_x = X(X'X)^{-1}X'$	$\hat{y} = P_x \mathbf{y}$
POLY	$\deg = 2$	$P_H = H(H'H)^{-1}H'$	$\hat{y} = P_H \mathbf{y}$
PPREG	$\min = 9$ $\max = 15$	Minimize $$\sum\left(y_i - \sum\left(\beta_m \psi_m(\alpha_m^t \mathbf{x}_i)\right)\right)^2$$	$\hat{y} = \sum \beta_m \phi_m(\alpha^t \mathbf{x})$
BRUTO	$\cos t = 2.5$	Backfitting	$\hat{y} = \sum f_i(\mathbf{x}_i)$
MARS	$\cos t = 2$ $\deg = 1$	TURBO	$\hat{y} = \sum f_i(\mathbf{x}_i)$
Nnet	$h = 15$	Minimize $$\sum(\hat{\theta}(j) - \theta(j))^2 + \sum W^2$$	$\arg\min_j$ $\{(\theta(j) - \hat{\theta}(j))^2\}$

depicted in Figure 1.8. The NNET has been very reliable in experiments and yields a better convergence to better minima than any other software that has been tested for the feed-forward multilayer neural network study with backpropagation.

1.7 SYSTEM PERFORMANCE ASSESSMENT

In practice we are given a data set and required to design a system for a certain objective. The system is a realization of the function of an unknown input space $\mathcal{D}$.

If we know all the necessary characteristics of the input space D, it is fairly easy to design an optimal system for the objective, such as the Bayesian classification rule with class conditional distributions and a priori probabilities for the classes. We, however, usually do not know the underlying generating function that generates the sample we have at hand. Instead, from the sample we like to find the underlying generating function, i.e., the population distribution. This is the *inference* problem.

Let us say that the input space is fully described by a certain distribution function $F(\cdot)$. The system we are interested in can be represented as a functional θ that takes the population distribution F: $\theta(D) = \theta(F)$. The functional θ is known, but the distribution F is not. θ could be any statistic or a complicated error rate in a classification problem.

The distribution is usually estimated parametrically or non-parametrically, thus providing the input argument to the system functional $\theta()$ in order to estimate the system's functional of the real population distribution F. Thus we have an estimation for $\theta(F)$:

$$\theta(F) \simeq \theta(\hat{F}).$$

With this estimation strategy the next question is how accurate $\hat{\theta}$ is as the estimator of θ.

1.7.1   CLASSIFIER EVALUATION

Once we have designed a classifier, we like to know how accurately the system can do the job or quantify the quality of the system performance. Prediction error is the criterion that we like to employ to see how good the designed system is. For both regression and classification system design, the usual system performance measure is its prediction error. In the context of regression, prediction error refers to the expected squared difference between the response value and its prediction from the model

$$PE = E(y - \hat{y})^2 \tag{1.111}$$

The expectation operation refers to the repeated sampling from the true underlying population distribution.

Prediction error also arises in classification problems, where the response falls into one of J not ordered classes. The prediction error is commonly defined as the probability of an incorrect classification

$$PE = \text{Prob}\left(\hat{y} \neq y\right) \tag{1.112}$$

which is called misclassification rate.

How to assess the system performance is an important issue in order to better quantify the designed system in terms of a criterion, e.g., error rate.

1.7.1.1   Hold-Out Method

If the data set at hand is large, we may divide it in two parts; use one for training and hold out the other for testing, hence the name *hold-out* method. This is a popular method to assess the system's performance. In most cases the data are limited in size; thus a hold-out method is ad-hoc in the sense of which subset is held out for testing. The performance evaluation via this method depends on how the data are separated.

1.7.1.2   K-Fold Cross-Validation

A natural compromise to the hold-out above is the so-called *K*-fold cross-validation method. The given data are divided evenly into *K* parts. One or more of the *K* parts is used to test the designed system by the remaining parts of the data. An average among the results is called the *K*-fold cross-validation estimate of the true error rate. An extreme case results to the *leave-one-out* method, in which one observation, $(y_k, \mathbf{x}_k)$, is left out and the rest $N - 1$ cases, $\{ (y_i, \mathbf{x}_i)_{i \neq k} \}$ are used for training. The prediction error, *PE* (Equation 1.111) from the leave-one-out method is the average of the *N* errors

$$PE_{cv} = \frac{1}{N} \sum_{i=1}^{N} \left(y_i - \hat{f}^{-1}(\mathbf{x}_i) \right)^2 \tag{1.113}$$

where $\hat{f}^i(\mathbf{x}_i)$ is the estimation of the response of $f(\mathbf{x}_i)$ based on the system trained with the data in which the $\mathbf{x}_i$ is missing. In general, with a notation w_i being the index group in which the index *i* falls, the cross-validation has a form of prediction error in regression:

$$PE_{cv} = \frac{1}{N} \sum_{i=1}^{N} \left(y_i - \hat{y}^{-(w_i)}(\mathbf{x}_i) \right)^2$$

and in classification setting:

$$PE_{cv} = \frac{1}{N} \sum_{i=1}^{N} \left[y_i \neq \hat{y}^{-(w_i)}(\mathbf{x}_i) \right]. \tag{1.114}$$

Other than cross-validation for estimation some modification of the apparent error, the sum of squared residuals (*SSR*)

$$\frac{1}{N} \sum_{i=1}^{N} \left(y_i - \hat{f}(\mathbf{x}_i) \right)^2$$

has also been used [see Reference 77, Ch.17]; such as $SSR/(N-p)$, $SSR/(N-2p)$, and $C_p = SSR/N + 2p\hat\sigma^2/N$. Leaving these modifications of SSR aside (since they are beyond the scope of our interest) we like to use the *Bootstrap* estimate of prediction error, which is also used for the performance analysis of our classification system.

1.7.2 BOOTSTRAPPING METHOD FOR ESTIMATION

Bootstrapping is a method of nonparametric estimation of statistical errors, which are the bias and the standard error of an estimator. The nonparametric techniques known to date are the Bootstrap, the Jackknife, and the cross-validation. Nonparametric methods for testing the accuracy of an estimator have all some common desirable features: they require very little in the way of modeling, assumptions, or analysis, and can be applied in an automatic way to any statistics, no matter how complicated these are.[78]

In order to see what they are, a simple statistic, the sample mean $\bar{X}$, is employed to assess the accuracy of the estimation for the true mean μ. We consider the available data set as a random sample of size N from an unknown distribution F in the sense that it represents the population F relatively well. As shown in Figure 1.10, a random sample is drawn from an unknown probability distribution F,

$$X_1, X_2, \ldots, X_n \sim F \tag{1.115}$$

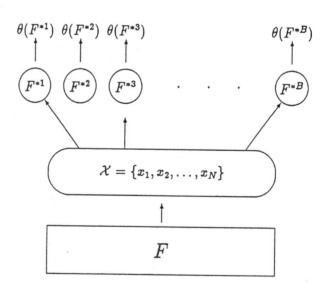

FIGURE 1.10　Illustration of the Bootstrap sampling.

With a sample from F, we compute the sample average $\bar{x} = \Sigma_1^N x_i/N$ as an estimate of the expectation of F, $E_F(X)$. For this special statistic (sample average), we can get more information about the estimator $\bar{x}$. The accuracy of the estimator is represented by the standard deviation of $\bar{x}$:

$$\hat{\sigma}(F;N,\bar{x}) = \left\{ \mathrm{var}\left(\bar{X}\right) \right\}^{1/2} = \left\{ \mathrm{var}(X)/N \right\}^{1/2}$$

$$= \left\{ \frac{1}{N(N-1)} \sum_{i=1}^{N} \left(x_i - \bar{x}\right)^2 \right\}^{1/2} \tag{1.116}$$

$$\approx \left(\frac{\mu_2(F)}{N} \right)^{1/2} \tag{1.117}$$

where $\mu_2(F)$ is the central moment of the population with distribution F. This standard error formula with the raw sample realization of Equation 1.115, does not extend to the other statistics, such as median, correlation, or prediction error. This is the point where computer methods, such as the resampling techniques for accuracy estimation, come into play.

1.7.2.1 Jackknife Estimation

Let $\bar{x}_{(i)}^*$ defined as

$$\bar{x}_{(i)} = \frac{1}{N-1} \sum_{j \neq i} x_j = \frac{N\bar{x} - x_i}{N-1} \tag{1.118}$$

with $N-1$ points, be the sample average of $N-1$ points for all $i = 1,2,\ldots,N$. Then the jackknife estimate of standard error is represented by

$$\hat{\sigma}_J(F;N,\bar{x}) = \left[\frac{N-1}{N} \sum_{i=1}^{N} \left(\bar{x}_{(i)} - \bar{x}_{(\cdot)}\right)^2 \right]^{1/2} \tag{1.119}$$

The $\bar{x}_{(\cdot)} = \sum_{i}^{N} \bar{x}_i/N$ is the average among the N $\bar{x}_{(i)}$'s. This can be proved to be equal to the standard error for the sample average of Equation 1.117 by substituting Equation 1.118 onto the Equation 1.119.

The jackknife standard error estimation of any statistic θ may have the form of Equation 1.119 to get the accuracy information of the estimator, $\hat{\theta}_J$. The advantage with the estimate of standard error for a statistic is to use Equation 1.119 where any statistic $\hat{\theta}_{(i)} = \hat{\theta}(X_1,\ldots,X_{i-1}, X_{i+1},\ldots, X_N)$ is replaced by $\bar{x}_{(i)}$ and $\hat{\theta}_{(\cdot)} = 1/N \sum_{i=1}^{N} \hat{\theta}_{(i)}$ for $\bar{x}_{(i)}$.

---

* Note the change of the notation in the deletion statistic from the usual superscript with negative sign, e.g., $\hat{f}^i(x_i)$ in Equation 1.113.

1.7.2.2   Bootstrap Method

Bootstrap generalizes Equation 1.117 in an apparently different way. Any statistic $\theta(F)$, which is a functional, requires the distribution F. But in practice F is not known and is difficult to estimate. An empirical distribution $\hat{F}$ from the given sample from an unknown distribution F is defined in a bootstrap setting by giving an equal probability mass $\frac{1}{N}$ to each of the values x_i, and draw a sample from the *empirical distribution* $\hat{F}$:

$$X_1^*, X_2^*, \ldots, X_N^* \sim \hat{F}$$

Each x_i^* is drawn independently *with replacement* and with equal probability from the set $\{x_1, x_2, \ldots, x_N\}$. Then the standard error of sample mean $\bar{X}^* = \Sigma_{i=1}^N X_i^*/N$ is given as

$$\sigma\left(F; N, \bar{x}^*\right) = \left(\frac{1}{N}\mu_2\left(\hat{F}\right)\right)^{1/2} = \left(\frac{1}{N}\sum_{i=1}^N \frac{1}{N}(x_i - \bar{x})^2\right)^{1/2}$$

$$= \left(\frac{1}{N^2}\sum_{i=1}^N (x_i - \bar{x})^2\right)^{1/2}$$

(1.120)

where $\mu_2(\cdot)$ is the second order central moment of a given distribution. Comparing this standard error for bootstrap sample average with Equation 1.117, we note that they are almost the same. Thus the jackknife (Equation 1.119) and the bootstrap (Equation 1.120) standard error for sample average (a simple statistic as an example) are shown to be nearly equal to Equation 1.117; a special statistic that is the sample average as an estimate for mean has an explicit form. Formulas like Equation 1.117 do not exist for most statistics.

This is where the computing intensive jackknife and bootstrap estimations are used. It turns out* that we can always numerically evaluate the bootstrap estimate for standard error $\hat{\sigma} = \sigma(\hat{F})$, without a simple expression like Equation 1.117.

1.8   ANALYSIS OF PREDICTION RATES FROM BOOTSTRAPPING ASSESSMENT

The boxplots in Figure 1.11 represent the E632 estimator superimposed by the distribution of the $B = 100$ bootstrap sample errors, $\hat{\theta}^*(b)$'s. The median value of the B error rates is replaced by the E632 estimate; thus the B bootstrap errors are shifted according to the E632 estimate. For ease of display and understanding the system performance, the recognition rates, i.e., $1 - \hat{\theta}$'s, are plotted.

---

* The proof can be found in Reference 77.

The *generality* issue of the designed system is related to its reliability in terms of standard error of the estimator for the prediction error. To make the analysis simpler, we assume the symmetry of the system performance of the classifiers in the boxplot figures. Then the standard error of the prediction rule by the mean (or median simply from the boxplots) is relatively approximated by the inter quartile range of the boxplots.

The mean value is the bootstrap sample estimate $\hat{\theta}$ of the true statistic $\theta = 1 - PE$. The standard deviation implicity represents the reliability of the estimate, i.e., standard error of the estimate. From the result of the classifiers considered in this study, (Figure 1.11), the 95% confidence interval of the estimate $\hat{\theta} = 0.955$ is given by

$$\hat{\theta} - \hat{\sigma} \times 1.645 \le \theta \le \hat{\theta} + \hat{\sigma} \times 1.645$$

$$0.941 \le \theta \le 0.96$$

where the multiple factor 1.645 is the 95% percentile point of the standard normal variate, $N(0,1)$.

The graphical display seems to reveal more for the comparison study of the classifiers and different treatments of the data. The boxplot display of a batch is a very simple and useful way to show the distribution of the sample. The Inter Quartile Range (IQR), which is the difference between the upper quartile and the lower quartile, is considered to be the robust estimation of the scalar multiple of the dispersion. The height of the box is the IQR. The median of the batch is represented by the line in the box. The whiskers represented by the dotted lines are extended up to the points in which the 1.5 times of the IQR contains. Outliers are represented by the individual dots to signify their existence. The boxplot, thus, displays the distribution very simply but well enough, especially when many different batches are to be compared.

The correct recognition rates from 11 classifiers are displayed with the boxplots for each data set obtained from the different treatments. Each boxplot shows the distribution of the recognition rate of the 100 systems designed by $B = 100$ bootstrap samples. The corresponding figures for the data are in Figure 1.11.

The results from the LDA and LREG (via linear regression) would have been the same due to the equivalence of the LDA and OS (Equation 1.107 and Equation 1.110) if the same bootstrap sample were used for both classifiers; the bootstrap samples used to train the classifiers are different for no reason!*

The best performance of the optimization machine with the feed-forward neural network structures can be observed (Figure 1.11). This is seen with the mean values for the estimation of the correct recognition error. Note that we do not consider the KNN classifier as a learning mechanism, so it is not of concern. It does not learn but performs by the exemplars; i.e., the computation in the operation phase is the largest, which is inappropriate in real-time processing applications.

---

* If the different classifiers were trained with the same B bootstrap samples, then the classification by the linear regression method and the LDA would have been the same.

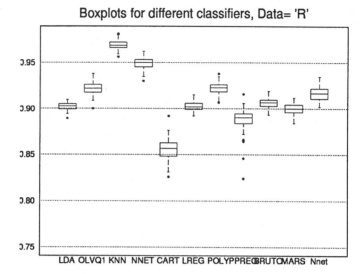

FIGURE 1.11   Boxplots for different classifiers for data set R. 100 bootstrap samples are used to assess each classifier.

REFERENCES

1. Lippmann, R. P., Pattern classification using neural networks, *IEEE Commun. Mag.*, 47, November 1989.
2. Lippmann, R. P., An introduction to computing with neural nets, *IEEE ASSP Mag.*, 4, April 1987.
3. Hush, D. and Horne, B., Progress in supervised neural networks, *IEEE Signal Process. Mag.*, 8, January 1993.
4. Weiss, S. M. and Kulikowski, C. A., *Computer Systems that Learn: Classification and Prediction Methods from Statistics, Neural Nets, Machine Learning, and Expert Systems*, Kaufman Publishers, San Francisco, CA,1991.
5. Hastie, T., Tibshirani, R., and Buja, A., Flexible discriminant analysis by optimal scoring. It can be obtained from /netlib/stat via ftp to netlib.att.com, February 1993.
6. Breiman, L. and Ihaka, R., Nonlinear Discriminant Analysis via Scaling and Ace. Technical report, Univ. California, Berkeley, 1984.
7. Härdle, W., *Smoothing Techniques With Implementation in S*, Springer-Verlag, Berlin, 1991.
8. Omohundro, S. M., Efficient algorithms with neural network behavior, *Complex Syst.*, 1; 273, 1987.
9. Huang, W. and Lippmann, R., Neural nets and traditional classifiers; in *Neural Information Processing Systems*, Anderson, D., Ed. American Institute of Physics, New York, 1986, 387.
10. Kohonen, T., The self-organizing map, *Proc. IEEE*, 78 (9); 1464, September 1990.
11. Carpenter, G. A. and Grossberg, S., Art2: Self-organization of stable category recognition codes for analog input patterns, *Appl. Opt.*, 26; 4919, 1987.
12. Gray, R. M., Vector quantization, *IEEE ASSP Mag.*, 1; 4, 1984.

13. Linde, Y., Buzo, A., and Gray, R. M., An algorithm for vector quantization, *IEEE Trans. Commun.*, COM-8; 84, 1980.

14. Powell, M. J. D., Radial Basis Functions for Multivariate Interpolation, Technical Report DAMPT 1985/NA12, Dept. of Appl Math. And Theor. Physics, Cambridge University., Cambridge, England, 1985.

15. Widrow, B., and Hoff, M., Adaptive switching circuits, in In 1960 IRE WESCON Convention Record, New York, NY, 1960, 96.

16. Quinlan, J. R., Induction of decision tree, *Machine Learning*, 1; 81, 1986.

17. Ripley, B. D., Neural networks and related methods for classification. PS file is available by anonymous ftp from markov.stats.ox.ac.uk (192.76.20.1) in directory pub/neural/papers.

18. Breiman, L., Friedman, J. H., Olshen, R. A., and Stone, C. J., *Classification and Regression Trees*, Wadsworth and Brooks/Cole, Monterey, CA, 1984.

19. Ripley, B. D., Statistical aspects of neural networks, in *Chaos and Networks: Statistical And Probabilistic Aspects* Barndorff-Nielsen, O. E., Cox, D. R., Jensen, J. L. and Kendall, S. S., Eds., Chapman & Hall, London, 1993.

20. Hinton, G. E., Connectionist learning procedures, *Artif. Intelligence*, 185, 1989.

21. Anderson, T. W., *An Introduction to Multivariate Statistical Analysis*, John Wiley & Sons, New York, 1984.

22. Fukunaga, K., *Introduction to Statistical Pattern Recognition*, 2nd ed., Academic Press, New York, 1990.

23. Johnson, R. A. and Wichern, D. W., *Applied Multivariate Statistical Analysis*, Prentice Hall, New York, 1998.

24. Kohonen, T., Kangas, J., Laaksonen, J., and Torkkola, K., Lvq-pak: The learning vector quantization program package. Technical report, Helsinki University of Technology, Laboratory of Computer and Information Science, 1992. lvq-pak is available for anonymous ftp user at the Internet site cochlea.hut.fi (130.233.168.48).

25. Wolpert, D., Alternative generalizers to neural nets. Abstracts of 1st Annual INNS Meeting, Boston, *Neural Netw.*, 1, 1988.

26. Farmer, J. D. and Sidorowich, J. J., Exploiting Chaos to Predict the future and Reduce Noise. Technical report, Los Alamos National Laboratory, Los Alamos, New Mexico, 1988.

27. Stanfill, C. and Waltz, D., Toward memory-based reasoning, *Commun. ACM*, 29 (12); I:213, 1986.

28. Ripley B. D., Neural networks and flexible regression and discrimination. PS file is available by anonymous ftp from markov.stats.ox.ac.uk (192.76.20.1) in directory pub/neural/papers.

29. Rumelhart, D. E., Hinton, G. E., and Williams, R. J., Learning internal representations by error backpropagation, in *Parallel Distributed Processing: Explorations in the Microstructure of Cognition, I Foundations*, Rumelhart, D.E., McClelland, J.L., and the PDP Research Group, Eds., MIT Press, 1986, chap. 8.

30. Minsky, M. and Papert, S., *Perceptrons: An Introduction to Computational Geometry*, MIT Press, Cambridge, MA, 1969.

31. Rosenblatt, R., *Principles of Neurodynamics*, Spartan Books, New York, 1959.

32. Bryson, A. E. and Ho, Y. C., *Applied Optimal Controls*, Bleisdell, New York, 1969.

33. Werbos, P. J., Beyond Regression: New Tools for Prediction and Analysis in the Behavior Sciences. Ph.D. thesis, Harvard University, Cambridge, MA, 1974.

34. Parker, D. B., Learning-Logic. Technical Report TR-47, Center for Comp. Res. in Econ. and Man., MIT, Cambridge, MA, April 1985.

35. Widrow, B., Generalization and information storage in networks of adaline 'neurons', in *Self-Organizing Systems*, Yovitz, M., Jacobi, G., and Goldstein, G., Eds., Spartan Books, Washington, DC, 1962, 435.

36. Widrow, B. and Lehr, M., 30 years of adaptive neural networks: perceptron, madaline, and backpropagation, *Proc. IEEE*, 78 (9); 1415, September 1990.

37. Zahner, D. and Micheli-Tzanakou, E., Alopex and backpropagation *Supervised and Unsupervised Pattern Recognition: Feature Extraction and Computational Intelligence*, CRC Press, Boca Raton, Fl, 1999, Chap. 2.

38. Micheli-Tzanakou, E., Neural networks in biomedical signal processing, in *The Biomedical Engineering Handbook*, Bronjino, J., Ed., CRC Press Inc., Boca Raton, Fl, 1995, 917.

39. Zahner, D. and Micheli-Tzanakou, E., Artificial neural networks: definitions, methods and applications, in *The Biomedical Engineering Handbook*, Bronjino, J., Ed., CRC Press Inc., Boca Raton, Fl, 1995, 2689.

40. Micheli-Tzanakou, E., Uyeda, E., Sharma, A., Ramanujan, K. S., and Dong, J., Face recognition: comparison of neural networks algorithms, *Simulation*, 64; 37, 1995.

41. Rumelhart, D. E., Hinton, G. E., and Williams, R. J., Learning representations by backpropagation errors, *Nature*, 323; 533, 1986.

42. Peressini, A. L., Sullivan, F. E., and Uhl J. J. Jr., *The Mathematics of Nonlinear Programming*, Springer-Verlag, Berlin, 1988.

43. Kirkpatrick, S., Gelatt, C. D., Jr., and Vecchi, M. P., Optimization by simulated annealing, *Science*, 220 (4598): 671, May 1983.

44. Hinton, G. E. and Sejnowski, T. J., Learning and relearning in boltzmann machines, in *Parallel Distributed Processing: Explorations in the Microstructure of Cognition, I. Foundations*, Rumelhart, D.E., McClelland, J.L. and the PDP Research Group, Eds., MIT Press, Cambridge, MA, Chap. 7.

45. Harth E. and Tzanakou E., Alopex: A stochastic method for determining visual receptive fields, *Vision Res.*, 14, 1475, 1974.

46. Unnikrishnan, K. P. and Venugopal, K. P., Alopex: A correlation-based learning algorithm for feed-forward and recurrent neural networks, *Neural Computation*, June 1994.

47. Engel, J., Teaching feed-forward neural networks by simulated annealing, *Complex Syst.*, 2; 641, 1988.

48. Hopfield, J. J. and Tank, D. W., Neural computation of decisions in optimization problems, *Biol. Cybern.*, 52; 141, 1985.

49. Hornik, K., Stichcombe, M., and White, H., Multilayer feedforward networks are universal approximators, *Neural Netw.*, 2; 359, 1989.

50. Kolmogorov, A. N., On the representation of continuous functions of many variables by superposition of continuous functions of one variable and addition, *Doklady Akad. Nauk SSR*, 114; 953, 1957.

51. Le Cun, Y., Medeles Connexionists de l'apprentissage, Ph.D. thesis, Universite Pierre et Marie Curie, Paris, 1987.

52. Lapedes, A. and Farber, R., How Neural Networks Work, Technical report, Los Alamos National Laboratory, Los Alamos, NM, 1988.

53. Gallant, A. R. and White, J., There exists a neural network that does not make avoidable mistables, in *IEEE Second Int. Conf. Neural Networks*, SOS Printing, San Diego, 1988, 657.

54. Hecht-Nielsen, R., Theory of the backpropagation neural network, in *Proc. Int, Joint Conf. Neural Networks*, SOS Printing, San Diego, 1989, I: 593.

55. Baum, E. and Haussler, D., What size net gives valid generalization? *Neural Comput.*, 1; 151, 1989.

56. Jacobs, R. A., Increased rates of convergence through learning rate adaptation, *Neural Netw.*, 1; 295, 1988.

57. Touretzky, D., Hinton, D., and Sejnowski, T., Eds., *Faster-learning Variations of Backpropagation: An Empirical Study*, Morgan Kaufmann, San Mateo, CA, 1989.

58. Solla, S. A., Levin, E., and Fleisher, M., Accelerated learning in layered neural networks, *Complex Syst.*, 2; 625, 1988.

59. Golden, R. M., A unified framework for connectionist systems, *Biol. Cybern.*, 59; 109, 1988.

60. van Ooyen, A. and Niehhuis, B., Improving the convergence of the backpropagation algorithm, *Neural Netw.* 5; 465, 1992.

61. Hastie, T., Discussion in flexible parsimonious smoothing and additive modeling, *Technometrics*, 31 (1); 23, 1989.

62. Friedman, J. H., Multivariate adaptive regression splines, *Ann. Stat.*, 19 (1); 1, 1991.

63. Hastie, T., Buja A., and Tibshirani, R., Penalized discriminant analysis. can be obtained from /netlib/stat via ftp to netlib.att.com., July 1993.

64. Gnanadesikan, R. and Kettenring, J., Discriminant analysis and clustering, *Stat. Sci.*, 4 (1); 34, 1989.

65. Buja, A., Hastie, T., and Tibshirani, R., Linear smoothers and additive models, *Ann. Stat.*, 17 (2); 453, 1989.

66. Friedman, J. H. and Stuetzle, W., Projection pursuit regression, *J. Am. Stat. Assoc.*, 76 (376); 817, December 1981.

67. Breiman, L. and Friedman, J. H., Estimating optimal transformations for multiple regression and correlation, *J. Am. Stat. Assoc.*, 80 (391); 580, September 1985.

68. Tibshirani, R., Estimation optimal transformations for regression via addivity and variance stabilization, *J. Am. Stat. Assoc.*, 83; 394,1988.

69. Breiman, L., The π-method for estimation multivariate functions from noisy data, *Technometrics*, 33 (2); 125, 1991.

70. Wahba, G., *Spline Models for Observational Data*, SIAM, Philadelphia, 1990.

71. Breiman, L., Hinging Hyperplanes for Regression, Classification and Function Approximation, Technical Report 324, Univ. California, Berkeley, 1991.

72. Chambers, J. M. and Hastie, T. J., Statistical models, in *Statistical Models in S*, Chambers, J. M. and Hastie, Trevor J., Eds. Wadsworth & Brooks, Pacific Grove, CA, 1991.

73. Teague, M. R., Image analysis via the general theory of moments, *J. Opt. Soc. Am.*, 70 (8); 920, August 1980.

74. Khotanzad, A. and Hong, Y. H., Invariant image recognition by zernlike moments, *IEEE Trans. Pattern Anal. Machine Intelligence*, 12 (5); 489, May 1990.

75. Chung, W., A Strategy for Visual Pattern Recognition, Ph.D. thesis, Electrical and Computer Engineering, Rutgers University, The State University of New Jersey, 1994.

76. Becker, R. A., Chambers, J. M., and Wilks, A. R., *The New S Language*, Wadsworth, Pacific Grove, CA, 1988.

77. Efron, B. and Tibshirani, R. J., *An Introduction to the Bootstrap*, Chapman & Hall, London, 1993.

78. Efron, B. and Gong, G., a leisurely look at the bootstrap, the jackknife, and cross-validation, *Am. Statistician*, 37 (1); 36, February 1983.

79. Efron, B. and Tibshirani, R. J., Bootstrap methods for standard errors, confidence intervals, and other measures of statistical accuracy, *Stat. Sci.* 1 (1); 54, 1986.

80. Efron, B., Estimating the error rate of a prediction rule: improvement on cross-validation, *J. Am. Stat. Assoc.*, 78 (382); 316, June 1983.
81. Jain, A. K., Dubes, R. C., and Chen, C. C., Bootstrap techniques for error estimation, *IEEE Trans. Pattern Anal. Machine Intelligence*, 9 (5); 628, September 1987.

2 Artificial Neural Networks: Definitions, Methods, Applications

Daniel A. Zahner and
Evangelia Micheli-Tzanakou

2.1 INTRODUCTION

The potential of achieving a great deal of processing power by wiring together a large number of very simple and somewhat primitive devices has captured the imagination of scientists and engineers for many years. In recent years, the possibility of implementing such systems by means of electro-optical devices and in very large scale integrations has resulted in increased research activities.

Artificial neural networks (ANNs) or simply neural networks (NNs) are made of interconnected devices called neurons (also called neurodes, nodes, neural units or simply units). Loosely inspired by the makeup of the nervous system, these interconnected devices look at patterns of data and learn to classify them. NNs have been used in a wide variety of signal processing and pattern recognition applications and have been successfully applied in such diverse fields as speech processing,[1-4] handwritten character recognition,[5-7] time series prediction,[8-9] data compression,[10] feature extraction,[11] and pattern recognition in general.[12] Their attractiveness lies in the relative simplicity with which the networks can be designed for a specific problem, along with their ability to perform nonlinear data processing.

As the neuron is the building block of a brain, a neural unit is the building block of a neural network. Although the two are far from being the same or from performing the same functions, they still possess similarities that are remarkably important. NNs consist of a large number of interconnected units that give them the ability to process information in a highly parallel way. The brain, as well, is a massively parallel machine as has long been recognized. As each of the 10^{11} neurons of the human brain integrates incoming information from all other neurons directly or indirectly connected to it, an artificial neuron sums all inputs to it and creates an output that is carrying information to other neurons. The connection from one neuron's dendrites or cell body to another neuron's processes is called a synapse. The strength by which two neurons influence each other is called a synaptic weight. In a NN all neurons are connected to all other neurons by synaptic weights that can have seemingly arbitrary values, but in reality, these weights show the effect of a stimulus on the neural network and the ability or lack of it to recognize that stimulus.

0-8493-2278-2/00/$0.00+$.50
© 2000 by CRC Press LLC

In the biological brain, two types of processes exist, static and dynamic. Static brain conditions are those that do not involve any memory processing, while dynamic processes involve memory processing and changes through time. Similarly, NNs can be distinguished as static or dynamic, the former being those that do not involve any previous memory and only depend on current inputs, the latter having memory and being described by differential equations that express changes in the dynamics of the system through time.

All NNs have certain architectures, and all consist of several layers of neuronal arrangements. The most widely used architecture is that of the perceptron first described in 1958 by Rosenblatt.[13] In the sections that follow we will build on this architecture but not necessarily on the original assumptions of Rosenblatt, the validity of which has been disputed by others.[27]

Since there are many names in the literature that express the same thing and usually create a lot of confusion for the reader, we will define the terms to be used and use them throughout the chapter. Terminology is a big concern for those involved in the field and for organizations such as IEEE. A standards committee has been formed to address issues such as nomenclature and paradigms. In this book, whenever possible, we will try to conform to the terms and definitions already in existence.

Some methods for training and testing of NNs will be described in detail, although many others will be left out due to lack of space, but references will be provided for the interested reader. A small number of applications will be given as examples, since many more are discussed in other chapters of this book, and it will be redundant to repeat them here.

2.2 DEFINITIONS

Neural Nets (NNs) go by many other names, such as connectionists models, neuromorphic systems, and parallel distributed systems, as well as artificial NNs, which distinguishes them from the biological ones. They contain many densely interconnected elements called neurons or nodes, which are nothing more than computational elements nonlinear in nature. A single node acts like an integrator of its weighted inputs. Once the result is found, it is passed to other nodes via connections that are called synapses. Each node is characterized by a parameter that is called threshold or offset and by the kind of nonlinearity through which the sum of all the inputs is passed. Typical nonlinearities are the hardlimiter, the ramp (threshold logic element), and the widely used sigmoid.

The simplest NN is the single layer perceptron[13,14] which is a simple net that can decide whether an input belongs to one of two possible classes. Figure 2.1 is a schematic representation of a simple one-neuron perceptron, the output of which is passed through a nonlinearity called an activation function. This activation function is of different types, the most popular being a sigmoidal logistic function.

Figure 2.2 is a schematic representation of some activation functions, such as the hardlimiter (or step), the threshold logic (or ramp), a linear, and a sigmoid. The neuron of Figure 2.1, receives many inputs, I_i, each weighted by a weight W_i ($i = 1, 2....N$). These inputs are then summed. The sum is then passed through the activation function, f, and an output, y, is calculated only if a certain threshold is exceeded.

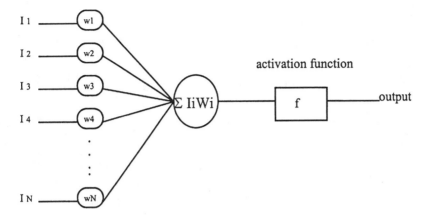

FIGURE 2.1 Artificial neuron.

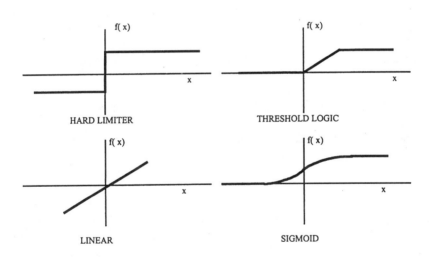

FIGURE 2.2 Typical activation functions.

Complex artificial neurons may include temporal dependencies and more complex mathematical operations than summation.[15] While each node has a simple function, their combined behavior becomes remarkably complex when organized in a highly parallel manner.

NNs are specified by their processing element characteristics, the network topology, and the training or learning rules they follow in order to adapt the weights, W_i. Network topology falls into two broad classes, feed-forward (nonrecursive) and feedback (recursive) NNs.[16] Nonrecursive NNs offer the advantage of simplicity of implementation and analysis. For static mappings a nonrecursive network is all one needs to specify any static condition. Adding feedback expands the network's range of behavior since now its output depends upon both the current input and network states. But one has to pay a price, longer times for teaching the NN.

Obviously the scheme of Figure 2.1 is quite simple and inadequate in solving problems. A multilayer perceptron (MLP) is the next choice. A number of inputs are now connected to a number of nodes at a second layer called the hidden layer. The outputs of the second layer may connect to a third layer and so on, until they connect to the output layer. In this representation, every input is connected to every node in the next layer and the outputs of one hidden layer are connected to the nodes of the next hidden layer and so on. More details on multilayer perceptrons can be found in Chapter 12.

Artificial neural networks usually operate in one of two modes. Initially there exists a training phase where the interconnection strengths are adjusted until the network has a desired output. Only after training does the network become operational, i.e., capable of performing the task it was designed and trained to do. The training phase can be either supervised or unsupervised. In supervised learning, there exists information about the correct or desired output for each input training pattern presented.[20] The original perceptron and backpropagation are examples of supervised learning. In this type of learning the NN is trained on a training set consisting of vector pairs. One of these vectors is used as input to the network; the other is used as the desired or target output. During training the weights of the NN are adjusted in such a way as to minimize the error between the target and the computed output of the network. This process might take a large number of iterations to converge, especially because some training algorithms (such as backpropagation) might converge to local minima instead of the global one. If the training process is successful, the network is capable of performing the desired mapping.

In unsupervised learning, no *a priori* information exists, and training is based only on the properties of the patterns. Sometimes this is also called self-organization.[20] Training depends on statistical regularities that the network extracts from the training set and represents as weight values. Applications of unsupervised learning have been limited. However, hybrid systems of unsupervised learning combined with other techniques produce useful results.[21-23] Unsupervised learning is highly dependent on the training data, and information about the proper classification is often lacking.[21] For this reason, most neural network training is supervised.

2.3 TRAINING ALGORITHMS

After McCulloch and Pitts[24] demonstrated, in 1943, the computational power of neuron-like networks, much effort was given to developing networks that could learn. In 1949, Donald Hebb proposed the strengthening of connections between presynaptic and post-synaptic units when both were active simultaneously.[25] This idea of modifying the connection weights as a method of learning is present in most learning models used today. The next major advancement in neural networks was by Frank Rosenblatt.[13,14] In 1960, Widrow and Hoff proposed a model, called the Adaptive Linear Element (ADALINE), which learns by modifying variable connection strengths, minimizing the square of the error in successive iterations.[26] This error correction scheme is now known as the Least Mean Square(LMS) algorithm, and it has found widespread use in digital signal processing.

There was great interest in neural network computation until Minsky and Papert published a book in 1969 criticizing the perceptron. This book contained a mathematical analysis of perceptron-like networks, pointing out many of their limitations. It was shown that the single layer perceptron was incapable of performing the XOR mapping. The single layer perceptron was severely limited in its capabilities. For linear activation functions, multilayer networks were no different from single layer models. Minsky and Papert pointed out that multilayer networks with nonlinear activation functions could perform complex mappings. However the lack of any training algorithms for multiple layer networks made their use impossible. It was not until the discovery of multilayer learning algorithms that interest in neural networks resurfaced. The most widely used training algorithm is the backpropagation algorithm, as already mentioned in the introduction.

Another algorithm used for multilayer perceptron training is the ALOPEX algorithm. ALOPEX was originally used for visual receptive field mapping by Tzanakou and Harth in 1973[28–30] and has since been applied to a wide variety of optimization problems. These two algorithms are explained in detail below.

2.3.1 BACKPROPAGATION ALGORITHM

The backpropagation algorithm is a learning scheme in which the error is back-propagated layer by layer and used to update the weights. The algorithm is a gradient descent method that minimizes the error between the desired outputs and the actual outputs calculated by the MLP. Let

$$E_p = \frac{1}{2} \sum_{i=1}^{N} (T_i - Y_i)^2 \tag{2.1}$$

be the error associated with template p. N is the number of output neurons in the MLP, T_i is the target or desired output for neuron i and Y_i is the output of neuron i calculated by the MLP. Let $E = \Sigma E_p$ be the total measure of error. The gradient descent method updates an arbitrary weight, w, in the network by the following rule:

$$w(n+1) = w(n) + \Delta w(n) \tag{2.2}$$

where

$$\Delta w(n) \alpha - \eta \frac{\partial E}{\partial w(n)} \tag{2.3}$$

where n denotes the iteration number and η is a scaling constant. Thus, the gradient descent method requires the calculation of the derivatives $\dfrac{\partial E}{\partial w(n)}$ for each weight, w, in the network. For an arbitrary hidden layer neuron, its output, H_j, is a nonlinear function f of the weighted sum of all its inputs (net_j).

$$H_j = f\left(net_j\right) \tag{2.4}$$

where f is the activation function. The most commonly used activation function is the sigmoid function given by

$$f(x) = \frac{1}{1 + e^{-x}} \tag{2.5}$$

Using the chain rule, we can write

$$\frac{\partial E}{\partial w_{ij}} = \frac{\partial E}{\partial net_j} \cdot \frac{\partial net_j}{\partial w_{ij}} \tag{2.6}$$

and since

$$net_j = \sum_{j=1}^{n} w_{ij} I_i \tag{2.7}$$

we have

$$\frac{\partial net_j}{\partial w_{ij}} = I_i \tag{2.8}$$

Thus Equation 2.6 becomes

$$\frac{\partial E}{\partial w_{ij}} = \frac{\partial E}{\partial net_j} \cdot I_i \tag{2.9}$$

$$\frac{\partial E}{\partial net_j} = \sum_{k=1}^{m} \frac{\partial E}{\partial net_k} \cdot \frac{\partial net_k}{\partial H_j} \cdot \frac{\partial H_j}{\partial net_j} \tag{2.10}$$

recalling that

$$net_k = \sum_{j=1}^{n} w_{ik} H_j \tag{2.11}$$

it follows that

$$\frac{\partial net_k}{\partial H_j} = w_{jk} \tag{2.12}$$

also

$$\frac{\partial H_j}{\partial net_j} = f'\left(net_j\right) \tag{2.13}$$

Therefore,

$$\frac{\partial E}{\partial net_j} = f'\left(net_j\right) \cdot \sum_{k=1}^{n} \frac{\partial E}{\partial net_k} \cdot w_{jk} \tag{2.14}$$

Assuming f to be the sigmoid function of Equation 2.5, then

$$f'\left(net_j\right) = Y_i\left(1 - Y_i\right) \tag{2.15}$$

Equation 2.14 gives the unique relation that allows the backpropagation of the error to all hidden layers. For the output layer

$$\frac{\partial E}{\partial net_j} = \frac{\partial E}{\partial H_j} \cdot f'\left(net_j\right) \tag{2.16}$$

$$\frac{\partial E}{\partial H_j} = -\left(T_i - Y_i\right) \tag{2.17}$$

In summary, then, first the output Y_i for all the neurons in the network is calculated. The error derivative needed for the gradient descent update rule of Equation 2.2 is calculated from

$$\frac{\partial E}{\partial w} = \frac{\partial E}{\partial net} \cdot \frac{\partial net}{\partial w} \tag{2.18}$$

If j is an output neuron, then

$$\frac{\partial E}{\partial net_j} = -\left(T_i - Y_i\right) \cdot Y_i\left(1 - Y_i\right) \tag{2.19}$$

If j is a hidden neuron, then the error derivative is backpropagated by using Equations 2.14 and 2.15. Substituting, we get

$$\frac{\partial E}{\partial net_j} = Y_i\left(1 - Y_i\right) \cdot \sum_{k=1}^{m} \frac{\partial E}{\partial net_k} \cdot w_{jk} \tag{2.20}$$

Finally the weights are updated, as in Equation 2.2.

There are many modifications to the basic algorithm that have been proposed to speed the convergence of the system. Convergence is defined as a reduction in the overall error below a minimum threshold. It is the point at which the network is said to be fully trained. One method[31] used is the inclusion of a momentum term in the update equation such that

$$w(n+1) = w(n) - \eta \frac{\partial E}{\partial w(n)} + \alpha \Delta w(n) \qquad (2.21)$$

η is the learning rate and is taken to be 0.25. α is a constant momentum term which determines the effect of past weight changes on the direction of current weight movements.

Another approach used to speed the convergence of backpropagation is the introduction of random noise.[32] It has been shown that while inaccuracies resulting from digital quantization are detrimental to the algorithm's convergence, analog perturbations actually help improve convergence time.

One of these variations is the modification by Fahlman,[33] called the *quickprop,* that uses second derivative information without calculating the Hessian needed in the straight backpropagation algorithm. It requires saving a copy of the previous gradient vector, as well as the previous weight change. Computation of the weight changes uses only information associated with the weight being updated:

$$\Delta w = \nabla w_{ij}(n) / \left[\nabla w_{ij}(n-1) - \nabla w_{ij}(n) \right] / \Delta w_{ij}(n-1) \qquad (2.22)$$

where $\Delta w_{ij}(n)$ is the gradient vector component associated with the weight w_{ij} at iteration n. This algorithm assumes that the error surface is parabolic, concave upward around the minimum, and that the slope change of the weight $\Delta w_{ij}(n)$ is independent of all other changes in weights. There are obviously problems with these assumptions, but Fahlman suggests a "maximum growth factor" μ in order to limit the rate of increase of the step size, namely that if $\Delta w_{ij}(n) > \mu \Delta w_{ij}(n-1)$ then $\Delta w_{ij}(n) = \mu \Delta w_{ij}(n-1)$. Fahlman also used a hyperbolic arctangent function to the output error associated with each neuron in the output layer. This function is almost linear for small errors, but it blows up for large positive or large negative errors. Quickprop is an attempt to reduce the number of iterations needed by straight backpropagation, and it succeeded in doing so by a factor of 5, but this factor is problem dependent. This method also required several trials before the parameters were set to acceptable values.

Backpropagation has achieved widespread use as a training algorithm for neural networks. Its ability to train multilayer networks has led to a resurgence of interest in the field. Backpropagation has been used successfully in applications such as adaptive control of dynamical systems and in many general neural network applications. Dynamical systems require monitoring of time in ways that monitor the past. In fact, the biological brain performs in an admirable way just because it has access to and uses values of different variables from previous instances. Backpropagation through time is another extension of the original algorithm proposed by Werbos in 1990[34] and has been previously applied in the "Truck Backer-Upper" by Nguyen and Widrow.[35] In this problem a sequence of decisions must be made without an immediate indication of how effective these steps are. No indication of performance exists until the track hits the wall. Backpropagation through time solves the problem, but it has its own inadequacies and performance difficulties. Despite its tremendous effect on neural networks, the algorithm is not without its problems. Some of the problems have been discussed above. In addition, the complexity of the algorithm makes hardware implementations of it very difficult.

2.3.2 THE ALOPEX ALGORITHM

The ALOPEX process is an optimization procedure that has been demonstrated successfully in a wide variety of applications. Originally developed for receptive field mapping in the visual pathway of frogs, ALOPEX's usefulness and its flexible form have increased the scope of its applications to a wide range of optimization problems. Since its development by Tzanakou and Harth in 1973,[28] ALOPEX has been applied to real-time noise reduction,[36] pattern recognition,[37] adaptive control systems,[38] and multilayer neural network training to name a few.

Optimization procedures, in general, attempt to maximize or minimize a function $F(\ )$. The function $F(\ )$ is called the cost function, and its value depends on many parameters or variables. When the number of parameters is large, finding the set $(x_1, x_2, \dots x_N)$ that corresponds to the optimal (maximal or minimal) solution is exceedingly difficult. If N were small, then one could perform an exhaustive search of the entire parameter space, in order to find the "best" solution. As N increases, intelligent algorithms are needed to quickly locate the solution. Only an exhaustive search can guarantee that a global optimum is found; however, near-optimal solutions are acceptable because of the tremendous speed improvement over exhaustive search methods.

Backpropagation, described earlier, being a gradient descent method often gets stuck in local extrema of the cost function. The local stopping points often represent unsatisfactory convergence points. Techniques have been developed to avoid the problem of local extrema, with simulated annealing[39] being the most common. Simulated annealing incorporates random noise, which acts to dislodge the process from local extremes. Crucial to the convergence of the process is that the random noise be reduced as the system approaches the global optimum. If the noise is too large, the system will never converge and can be dislodged mistakenly from the global solution.

ALOPEX is another process which incorporates a stochastic element to avoid local extremes in search of the global optimum of the cost function. The cost function or response is problem-dependent and is generally a function of a large number of parameters. ALOPEX iteratively updates all parameters simultaneously based on the cross-correlation of local changes, ΔX_i, and the global response change ΔR, plus an additive noise. The cross-correlation term $\Delta X_i \Delta R$ helps the process move in a direction that improves the response. Table 2.1 shows how this can be used to find a global maximum of R.

TABLE 2.1

	ΔX		ΔR	$\Delta X\,\Delta R$
$X\uparrow$	+	$R\uparrow$	+	+
$X\uparrow$	+	$R\downarrow$	−	−
$X\downarrow$	−	$R\uparrow$	+	−
$X\downarrow$	−	$R\downarrow$	−	+

All parameters X_i are changed simultaneously at each iteration according to

$$X_i(n) = X_i(n-1) + \gamma\, \Delta X_i(n)\Delta R(n) + r_i(n) \tag{2.23}$$

The basic concept is that this cross-correlation provides a direction of movement for the next iteration. For example, take the case where $X_i\downarrow$ and $R\uparrow$. This means that the parameter X_i decreased in the previous iteration, and the response increased for that iteration. The product $\Delta X_i\Delta R$ is a negative number, and thus X_i would be decreased again in the next iteration. This makes perfect sense since a decrease in X_i produced a higher response; if you are looking for the global maximum, then X_i should be decreased again. Once X_i is decreased and R also decreases, then $\Delta X_i\Delta R$ is now positive and X_i increases.

These movements are only tendencies, since the process includes a random component that will act to move the weights unpredictably, avoiding local extrema of the response. The stochastic element of the algorithm helps it to avoid local extrema at the expense of slightly longer convergence or learning period.

The general ALOPEX updating Equation 2.23 is explained as follows. $X_i(n)$ are the parameters to be updated, n is the iteration number, and $R(\ )$ is the cost function, of which the "best" solution in terms of X_i is sought. Gamma, γ, is a scaling constant, $r_i(n)$ is a random number from a Gaussian distribution whose mean and standard deviation are varied, and $\Delta X_i(n)$ and $\Delta R(n)$ are found by:

$$\Delta X_i(n) = X_i(n-1) - X_i(n-2) \tag{2.24}$$

$$\Delta R(n) = R(n-1) - R(n-2) \tag{2.25}$$

the calculation of $R(\ )$ is problem dependent and can be easily modified to fit many applications. A detailed description of the response calculation can be found in other chapters. This flexibility was demonstrated in the early studies of Harth and Tzanakou.[29] In mapping receptive fields, no *a priori* knowledge or assumptions were made about the calculation of the cost function, instead a "response" was measured. By using action potentials as a measure of the response[28,29,40,41] receptive fields could be determined by using the ALOPEX process to iteratively modify the stimulus pattern until it produced the largest response.

It should be stated that due to its stochastic nature, efficient convergence depends on the proper control of both the additive noise and the gain factor γ. Initially all parameters X_i are random, the additive noise has a Gaussian distribution with mean 0, and standard deviation, σ, initially large. The standard deviation, σ, decreases as the process converges to ensure a stable stopping point. Conversely, gamma, γ, increases with iterations. As the process converges, ΔR becomes smaller and smaller, and an increase in gamma is needed to compensate for this.

Additional constraints include a maximal change permitted for X_i, for one iteration. This bounded step size prevents the algorithm from drastic changes form one iteration to the next. These drastic changes often lead to long periods of oscillation, during which the algorithm fails to converge.

2.3.3 MULTILAYER PERCEPTRON (MLP) NETWORK TRAINING WITH ALOPEX

A MLP can also be trained for pattern recognition using ALOPEX. A response is calculated for the jth input pattern based on the observed and desired output

$$R_j(n) = O^{des}{}_k - \left(O^{obs}{}_k(n) - O^{des}{}_k\right)^2 \tag{2.26}$$

Where $O^{obs}{}_k$ and $O^{des}{}_k$ are vectors corresponding to O_k for all k. The total response for iteration n, is the sum of all the individual template responses, $R_j(n)$.

$$R(n) = \sum_{j=1}^{m} R_j(n) \tag{2.27}$$

In Equation 2.27 m is the number of templates used as inputs. ALOPEX iteratively updates the weights using both the global response information and local weight histories, according to the following:

$$W_{ij}(n) = r_i(n) + \gamma \Delta W_{ij}(n)\Delta R(n) + W_{ij}(n-1) \tag{2.28}$$

$$W_{jk}(n) = r_i(n) + \gamma \Delta W_{jk}(n)\Delta R(n) + W_{ik}(n-1) \tag{2.29}$$

where γ is an arbitrary scaling factor, $r_i(n)$ is an additive Gaussian noise, ΔW represents the local weight change and ΔR represents the global response information. These values are calculated by:

$$\Delta W_{ij}(n) = W_{ij}(n-1) - W_{ij}(n-2) \tag{2.30}$$

$$\Delta W_{jk}(n) = W_{jk}(n-1) - W_{jk}(n-2) \qquad (2.31)$$

$$\Delta R(n) = R(n-1) - R(n-2) \qquad (2.32)$$

Besides its universality to a wide variety of optimization procedures, the nature of the ALOPEX algorithm makes it suitable for VLSI implementation. ALOPEX is a biologically influenced optimization procedure that uses a single value global response feedback, to guide weight movements toward their optimum. This single value feedback, as opposed to the extensive error propagation schemes of other neural network training algorithms, makes ALOPEX suitable for fast VLSI implementation.

Recently, a digital VLSI approach to implementing the ALOPEX algorithm was undertaken by Pandya and Venugopal.[66] Results of their study indicated that ALOPEX could be implemented using a Single Instruction Multiple Data (SIMD) architecture. A simulation of the design was carried out, in software, and good convergence for a 4x4 processor array was demonstrated.

In our laboratory, an analog VLSI chip was designed to implement the ALOPEX algorithm. By making full use of the algorithm's tolerance to noise, an analog design was chosen. As discussed earlier, analog designs offer larger and faster implementations than those of digital designs. More details are given in Chapter 12.

2.4   SOME APPLICATIONS

2.4.1   EXPERT SYSTEMS AND NEURAL NETWORKS

Computer-based diagnosis is an increasingly used method that tries to improve the quality of health care. Systems that depend on artificial intelligence (AI), such as knowledge-based systems or expert systems, as well as hybrid systems such as the above combined with other techniques, like NNs, are coming into play. Systems of that sort have been developed extensively in the last ten years with the hope that medical diagnosis and therefore medical care will improve dramatically. Hatzilyger-oudis et al.[42] are developing such a system with three main components; a user interface, a database management system, and an expert system for the diagnosis of bone diseases. Each rule of the knowledge representation part is an Adaline unit that has as inputs the conditions of the rule. Each condition is assigned a significance factor corresponding to the weight of the input to the Adaline unit, and each rule is assigned a number, called a bias factor, that corresponds to the weight of the bias input of the unit. The output is calculated as the weighted sum of the inputs filtered by a threshold function.

Hudson et al.[43] developed a NN for symbolic processing. The network has four layers. A separate decision function is used for layer three and a threshold for each node in the same layer. If the value of the decision function exceeds the corresponding threshold value, a certain symbol is produced. If the value of the decision function does not exceed the threshold, then a different symbol is produced. The so generated symbols of adjacent nodes are combined at layer four according to a well-structured grammar. A grammar provides the rules by which these symbols are combined.[44] The addition of a symbolic processing layer enhances the NN in a number of ways. It is, for instance, possible to supplement a network that is purely diagnostic, with a level which recommends further actions, or to add additional connections or nodes in order to more closely simulate the nervous system.

With increasing network complexity, parameter variance increases, and the network prediction becomes less reliable. This difficulty can be overcome if some prior knowledge can be incorporated into the NN to bias it.[45] In medical applications in particular, rules can either be given by experts or can be extracted from existing solutions to the problem. In many cases the network is required to make reasonable predictions before it has gone through any sufficient training data, relying only on *a priori* knowledge. The better this knowledge is initially, the better the performance and the shorter the training.[46,47]

2.4.2 APPLICATIONS IN MAMMOGRAPHY

One of the leading causes of death of women in America is breast cancer. Mammography has been proven to be an effective diagnostic procedure for early detection of breast cancer. An important sign in its detection is the identification on the mammograms of microcalcifications, especially when they form clusters. Chan et al.[48] have developed a computer-aided diagnosis (CAD) scheme based on filtering and feature extracting methods. In order to improve on the false positives, Zhang et al.[49] applied an artificial NN which is shift invariant. They evaluated the performance of the NN by the "jack-knife" method[50] and receiver operating characteristic analysis.[51,52] A shift invariant NN is a feed-forward NN with local, spatially invariant interconnections similar to those of the neocognitron[53] but without the lateral interconnections. BP was also used for training for individual microcalcifications and a cross-validation technique was employed in order to avoid overtraining. In this technique the data set is divided into two sets, one used for training and the other for validating the predetermined intervals. The training of the network is terminated just before the performance of the network for the validating set decreases. The shift-invariant NN was proven to be much better in dropping the false positive classifications by almost 55% over previously used NNs.

In another study, Zheng et al.[54] used a multistage NN for detection of microcalcification clusters with almost 100% success and only one false positive per image. The multistate NN consists of more than one NN connected in series. The first stage is called the "detail network," with inputs the pixel values of the original image, while the second network, the "feature network" gets as inputs the output from the first stage and a set of features extracted from the original image. This approach has higher sensitivity of classification and a lower false positive detection than the previous reports.

Another approach was used by Floyd et al.[55] where radiologists read the mammograms and came up with a list of eight findings, which were used as features for a NN. The results from biopsies were taken as the truth of diagnosis. For indeterminate cases, as classified by radiologists, the NN had a performance index of 0.86, which is quite high.

Downes[56] used similar techniques to identify stellate lesions. He used texture quantification via fractal analysis methods instead of using the raw data. In mammograms, specific textures are usually indicative of malignancy. The method used for calculating the fractal dimension of digitized images was based upon the relationship between the fractal dimension and the power spectral density.

Giger et al.[57] aligned the mammograms of left and right breasts and used a subtraction technique to find initial candidate masses. Various features were then extracted and used in conjunction with NNs in order to reduce false positives resulting from bilateral subtraction. Receiver operating characteristic (ROC) analysis was applied to evaluate the output of the NN. The methods used were evaluated using pathologically confirmed cases. This scheme yielded a sensitivity of 95% at an average of 2.5 false positive detections per image.

2.4.3 CHROMOSOME AND GENETIC SEQUENCES CLASSIFICATION

Several clinical disorders are related to chromosome abnormalities that are difficult to identify accurately and also classify the individual chromosome. Automated systems can greatly help human capabilities in dealing with some of the problems involved. One way to deal with this problem is the use of NNs. Several studies have already been done toward enhancing the ability of an automated computerized system to analyze chromosome identification.[58] One such study by Sweeney and Musavi[59] analyzed the metaphase of chromosome spreads employing probablistic NNs (PNNs), which have been used as alternatives to various classification problems. Firstly introduced by Specht,[60,61] PNNs are combinations of a kernel-based estimator for estimation of probability densities and the Bayes rule for classification decision. The estimation with the highest value specifies the correct class. Thus, training of PNNs means to find appropriate kernel functions, usually taken to be Gaussian densities, and therefore the problem is reduced to the selection of a scalar parameter, namely the standard deviation, of the Gaussian. A way to improve the accuracy of a PNN for chromosome classification is to use the knowledge that there can be a maximum of only two chromosomes assigned to each class. This knowledge can be easily incorporated into the NN. Similar or better results were obtained to the classical BP-trained NN.

A hybrid symbolic/NN machine learning algorithm was introduced by Noordewier et al.[62] for the recognition of genetic sequences. The system uses a knowledge base of hierarchically structured rules to form an artificial NN in order to improve the knowledge base. They used this system in recognizing genes in DNA sequences. The learning curve of this system was compared to that of a randomly initialized, fully connected two-layer NN. The knowledge-based NN learned much faster than the other one, but the error of the randomly initialized NN was slightly lower (5.5 vs. 6.4%). Methods have also been devised to investigate what the NN has learned by an automatic translation into symbolic rules of trained NN initialized by the knowledge-based method.[63]

Medical axis transform (MAT) based features as inputs to a NN have been used in studying human chromosome classification.[64] Prenatal analysis, genetic syndrome diagnosis, and others make this research very important. Human chromosome classification based on NN requires no *a priori* knowledge or even assumptions on the data. MAT is a widely used method for transformations of elongated objects and requires less storage and time while preserving the topological properties of the object. MAT also allows for a transformation from a 2D image to a 1D representation of it. The so obtained features are then fed as inputs to a two-layer feed-forward NN trained by BP, with almost perfect results in classifying chromosomes. An optimization on an MLP was also done.[65]

REFERENCES

1. Mueller, P. and Lazzaro, J., Real time speech recognition, in *Neural Networks for Computing*, Dember, J. Ed., American Inst. of Physics, New York, 1986, 321–326.
2. Bourland, H. and Morgan, N., A continuous speech recognition system embedding a multilayer perceptron into HMM, in *Advances in Neural Information Processing Systems, 2*, Touretzky, D., Ed., Morgan Kauffman, San Mateo, CA, 1990, 186.
3. Bridle, J. S. and Cox, S. J., RecNorm: simultaneous normalization and classification applied to speech recognition, in *Advances in Neural Information Processing Systems, 3*, Lippmann, R. P., Moody, J. E., and Touretzky, D. S., Eds., Morgan Kauffman, San Mateo, CA, 1991, 234.
4. Lee, S. and Lippman, R. P., Practical characteristics of neural networks and conventional classifiers on artificial speech problems, in *Advances in Neural Information Processing Systems, 2*, Touretzky, D. S., Ed., Morgan Kauffmann, San Mateo, CA, 1990, 168.
5. Fukushima, K., Neocognition: a neural network model for a mechanism of visual pattern recognition, IEEE Trans. on Systems, Man Cybernetics, SMC-13(5), 1983, 826.
6. Dasey, T. J. and Micheli-Tzanakou, E., An unsupervised system for the classification of handwritten digits, comparison with backpropagation training, *Handbook of Industrial Electronics*, Irwin, D., Ed., 1994.
7. LeCun, Y., Boser, B., Denker, J. S., Henderson, D., Howard, R. E., Hubbard, W., and Jackel, L. D., Backpropagation applied to handwritten zip code recognition, *Neural Comput.*, 1(4), 541, 1989.
8. Hakim, N., Kaufman, J. J., Cerf, G., and Medows, H. E., A discrete time neural network model for system identification, *Proc. of IJCNN*, 90(Vol. 3), 593, 1990.
9. Hesh, D., Abdallah, C., and Horne, B., Recursive neural networks for signal processing and control, in *Proc. First IEEE-SP Workshop on Neural Networks for Signal Processing*, Princeton, NJ, 1991, 523.
10. Cottrell, G. W., Munro, P. N., and Zipser, D., Image compression by backpropagation, a demonstration of extensional programming, in *Advances in Cognitive Science*, Vol. 2, Ablex Publ., Norwood, NY, 1989, 208.
11. Oja, E. and Lampinen, J., Unsupervised learning for feature extraction, in *Computational Intelligence Imitating Life*, Zurada, J. M., Marks, R. J., II, and Robinson, C. J., Eds., IEEE Press, New York, 1994.
12. Fogelman Soulie, F., Integrating neural networks for real world applications, in *Computational Intelligence Imitating Life*, Zurada, J. M., Marks, R. J., II, and Robinson, C. J., Eds., IEEE Press, New York, 1994.
13. Rosenblatt, F., The perceptron: a probabilistic model for information storage and organization in the brain, *Psychol. Rev.*, 65, 386, 1958.

14. Rosenblatt, F., *Principles of Neurodynamics*, Spartan Books, New York, 1962.
15. Lippman, R. P., An introduction to computing with neural nets, *IEEE, ASSP Mag.*, 4, 1987.
16. Moore, K., Artificial neural networks: weighing the different ways to systemize thinking, *IEEE Potentials*, 23, 1992.
17. Huang, S. and Huang, Y., Bounds on the number of hidden neurons in multilayer perceptrons, *IEEE Trans. Neural Networks*, 2(1), 47, 1991.
18. Kung, S. Y., Hwang, J., and Sun, S., Efficient modeling for multilayer feedforward neural nets, Proc. IEEE Conf. on Acoustics, Speech Signal Processing, New York, 1988, 2160.
19. Mirchandani, G., On hidden nodes for neural nets, *IEEE Trans. Circuits and Systems*, 36(5), 661, 1989.
20. Kohonen, T., *Self-Organization and Associative Memory*, Springer-Verlag, New York, 1988.
21. Hecht-Nielsen, R., Counterbackpropagation networks, *Proc. of the IEEE First Int. Conf. on Neural Networks*, Vol. 2, 1987, 19.
22. Dasey, T. J. and Micheli-Tzanakou, E., The unsupervised alternative to pattern recognition I: classification of handwritten digit, *Proc. 3rd Workshop on Neural Networks*, Auburn, AL, 1992, 228.
23. Dasey, T. J. and Micheli-Tzanakou, E., The unsupervised alternative to pattern recognition II: detection of multiple sclerosis with the visual evoked potential, *Proc. 3rd Workshop on Neural Networks*, Auburn, AL., 1992, 234.
24. McCulloch, W. C. and Pitts, W., A logical calculus of the ideas imminent in nervous activity, *Bull. Math. Biophys.*, 5, 115, 1943.
25. Hebb, D., *The Organization of Behavior*, Wiley, New York, 1949.
26. Widrow, B. and Lehr, M. A., 30 Years of adaptive neural networks: perceptron, Madaline, and backpropagation, *Proc. IEEE*, 78(9), 1415, 1990.
27. Minsky, M. and Papert, S., *Perceptrons: An Introduction to Computational Geometry*, M.I.T. Press, Cambridge, MA, 1969.
28. Tzanakou, E. and Harth, E., Determination of visual receptive fields by stochastic methods, *Biophys. J.*, 15, (42a), 1973.
29. Harth, E. and Tzanakou, E., Alopex: a stochastic method for determining visual receptive fields, *Vis., Res.*, 14, 1475, 1974.
30. Tzanakou, E., Michalak, R., and Harth, E., The ALOPEX process: visual receptive fields by response feedback, *Biol. Cybern.*, 35, 161, 1979.
31. Rumelhart, D. E. and McClelland, J. L., Eds., *Parallel Distributed Processing*, M.I.T. Press, Cambridge, MA, 1986.
32. Holstrom, L. and Koistinen, P., Using additive noise in backpropagation training, *IEEE Trans. Neural Networks*, 3(1), 24, 1992.
33. Fahlmann, S. E., Faster learning variations of backpropagation: an emprical study, in *Proc. of the Connectionist Models Summer School*, Touretzky, D., Hinton, G., and Sejnowski, T., Eds., Morgan Kaufmann, San Mateo, CA, 1988.
34. Werbos, P. J., Backpropagation through time: what it does and how to do it, *Proc. IEEE*, 78(30), 1550, 1990.
35. Nguyen, D. and Widrow, B., The truck backer-upper: an example of self-learning in neural networks, *Proc. Int. Joint Conf. on Neural Networks*, Vol. II, IEEE Press, New York, 1989, 357.
36. Ciaccio, E. and Tzanakou, E., The ALOPEX process: Application to real-time reduction of motion artifact, Ann. Int. Conf. of IEEE EMBS, Vol. 12, no. 3, 1990, 1417.

37. Dasey, T. J. and Micheli-Tzanakou, E., A pattern recognition application of the Alopex process with hexagonal arrays, Int. Joint Conf., on Neural Networks, Vol. II, 1990, 119.

38. Venugopal, K., Pandya, A., and Sudhakar, R., ALOPEX algorithm for adaptive control of dynamical systems, *Proc. of IJCNN*, Vol. II, 1992, 875.

39. Kirkpatrick, S., Gelatt, C. D., and Vecchi, M. P., Optimization by simulated annealing, *Science*, 220, 671, 1983.

40. Micheli-Tzanakou, E., Non-linear characteristics in the frog's visual system, *Biol. Cybern.*, 51, 53, 1984.

41. Micheli-Tzanakou, E., Visual receptive fields and clustering, *Behav. Res. Meth. Instrument.*, 15(6), 553, 1983.

42. Hatzilygeroudis, I., Vassilakos, P. J., and Tsakalidis, A., An intelligent medical system for diagnosis of bone diseases, *Proc. Int. Conf. on Med. Physics and Biom. Eng.*, Vol. 1, Cyprus, 1994, 148.

43. Hudson, D. L., Cohen, M. E., and Deedwania, P. C., A neural network for symbolic processing, *Proc. 15th Ann. Int. Conf. of the IEEE/EMBS*, Vol. 1, 1993, 248.

44. Hoperoft, J. E. and Ullman, J. D., *Formal Languages and their Relation to Automata*, Addison-Wesley, Reading, MA, 1969.

45. Roscheisen, M., Hofmann, R., and Tresp, V., Neural control for running mills: incorporating domain theories to overcome data deficiency, in *Advances in Neural Information Processing Systems 4*, Morgan Kauffman, San Mateo, CA, 1992.

46. Towell, G. G., Shavlik, J. W., and Noordemier, M. O., Refinement of approximately correct domain theories by knowledge-based neural networks, *Proc. 8th Nat. Conf. on Artif. Intelligence*, 1990, 861.

47. Tresp, V., Hollatz, J., and Ahmad, S., Network structuring and training using rule-based knowledge, in *Advances in Neural Information Processing Systems 5*, Morgan Kauffman, San Mateo, CA, 1994, 871.

48. Chan, H.-P., Doi, K., Vyborny, C. J. et al., Improvement in radiologists' detection of clustered microcalcifications on mammograms. The potential of computer aided diagnosis, *Inv. Radiol.*, 25, 1102, 1990.

49. Zhang, W., Giger M. L., Nishihara, R, M., and Doi, K., Application of a shift-invariant artificial neural network for detection of breast carcinoma in digital mammograms, *Proc. World Congr. on Neural Networks*, Vol. I, 1994, 45.

50. Fukunaga, K., *Introduction to Statistical Pattern Recognition*, 2nd ed., Acad. Press, New York, 1990.

51. Metz, C.E., Current problems in ROC analysis, *Proc. Chest Imaging Conf.*, Madison, WI, 1988, 315.

52. Metz, C.E., Some practical issues of experimental design and data analysis in radiological ROC studies, *Inv. Radiol.*, 24, 234, 1989.

53. Fukushima, K., Miyake, S., and Ito, T., Neocognitron: a neural network model for a mechanism of visual pattern recognition, IEEE Trans. on Systems, Man, and Cybernetics, SMC-13, 1983, 826.

54. Zheng, B., Qian, W., and Clarke, L. P., Artificial neural network for pattern recognition in mammography, *Proc. World Congr. on Neural Networks*, Vol. I, 1994, 57.

55. Floyd, C. E., Jr., Yun, A. J., Lo, J. Y., Tourassi, G., Sullivan, D. C., and Kornguth, P. J., Prediction of breast cancer malignancy for difficult cases using an artificial neural network, *Proc. World Congr. on Neural Networks*, Vol. I, 1994, 127.

56. Downes, P., Neural network recognition of multiple mammographic lesions, *Proc. World Congr. on Neural Networks*, Vol. I, 1994, 133.

57. Giger, M. L., Lu, P., Huo, Z., and Zhang, W., Application of artificial neural networks to the task of merging feature data in computer-aided diagnosis schemes, *Proc. World Congr. on Neural Networks*, Vol. I, 1994, 43.

58. Piper, J., Granunn, E., Rutovitz, D., and Ruttledge, H., Automation of chromosome analysis, *Signal Proc.*, 2(3), 109, 1990.

59. Sweeney, W. P., Jr. and Musavi, M. T., Application of neural networks for chromosome classification, *Proc. 15th Annu. Int. Conf. of the IEEE/EMBS*, Vol. 1, 1994, 239.

60. Specht, D., Probabilistic neural networks for classification, mapping, or associative memory, IEEE Int. Conf. on Neural Networks, San Diego, CA, 1988.

61. Specht, D., Probabilistic neural networks, *Neural Net.*, 3, 109, 1990.

62. Noordewier, M. O., Towell, G. G., and Shavlik, J. W., Training knowledge-based neural networks to recognize genes in DNA sequences, *Adv. Neural Info. Proc. Sys.*, 3, 530, 1993.

63. Towell, G. G., Graven, M., and Shavlik, J. W., Automated interpretation of knowledge-based neural networks, Technical Report, Univ. of Wisc. Computer Sci. Dept., Madison, WI, 1991.

64. Lerner, B., Rosenberg, B., Levistein, M., Guterman, H., Dinstein, I., and Romem, Y., Medical axis transform based features and a neural network of human chromosome classification, *Proc. World Congr. on Neural Networks*, Vol. 3, 1994, 173.

65. Lerner, B., Guterman, H., Dinstein, J., and Romem, Y., Learning curves and optimization of a multilayer perceptron neural network for chromosome classification, *Proc. World Congr. on Neural Networks*, Vol. 3, 1994, 248.

66. Pandya, A.S. and Vinugopal, P., A Stochastic Parallel Algorithm for Supervised Learning in Neural Networks, IEEE Trans. Inf. Sys., E77-D No. 4, 1994, 376.

3 A System for Handwritten Digit Recognition

Woogon Chung and Evangelia Micheli-Tzanakou

3.1 INTRODUCTION

Visual pattern recognition has long been an interesting problem, both from the application and technical aspects. We hope to design a system that understands characters, words, and even sentences. Handwritten digit recognition is one of the most challenging problems. Its applications are extensive—automatic document processing, banking systems, etc. Depending on the writer's environment, the writing style differs, and this causes the difficulty in the system design, even though the fundamental assumption in writing communications is that differences between characters are more significant than differences among the same character.

The handwritten digit recognition has a long history, and many researchers have proposed different models.[1-6] These are mostly *model-based*. The developed model is usually specific to the given data set, and its applicability for a different data set is rather restricted. These methods find local properties, or primitives, e.g., arcs, lines, starting/end points, and the rules that combine the individual properties, from the skeletonized images. Painstaking processes to find and tune the properties are some of the difficulties and variabilities of the resulting systems.

A simple and important image pattern analysis (of Arabic numerals, for example) is carried out to demonstrate that a simple *model-free* strategy, via global moments with proper statistical analysis, renders a quite acceptable result. The moment calculation for features is model-free, since no other information of the data set than the group label is required in order to design the pattern recognition system. All the groups of data are treated the same way to extract the global features, while the model-based methods are required to describe each different digit by a certain list of properties.

3.2 PREPROCESSING OF HANDWRITTEN DIGIT IMAGES

The images are passed through a sequence of preprocessing steps before the Zernike moments calculation, a global feature extraction method which will be described in Section 3.3. A block diagram for the sequence of preprocessing procedures and the intermediate results of digit images are shown in Figure 3.1.

0-8493-2278-2/00/$0.00+$.50
© 2000 by CRC Press LLC

I

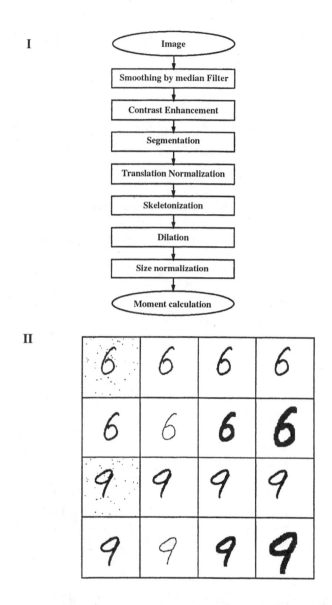

II

FIGURE 3.1   (I) Sequence of the preprocessing. (II) Two original images (9,6) and their preprocessed results. Starting with the original images, the results of 'smoothing,' 'contrast enhancement,' 'thresholding,' 'centering,' 'skeletonization,' 'dialization,' and 'size normalization' are presented from left to right and top to bottom of the figure.

The objectives and the methods for each preprocessing are described in the following paragraphs. The major objective in the preprocessing stage of the pattern recognition system is getting unique features from the same group of patterns.

Noise due to acquisition or transmission is reduced by a *smoothing* operation with neighboring pixel values which generally is low-pass filtering. Smoothing substitutes the value of the pixel in the center of a window with the average of pixels in the window. Such an operation has the effect of suppression of the distortions in the gray values caused by sensor noise or transmission errors. Edges in an object are typical changes in the gray levels. Thus, *smoothing* and *edge detection* are contradictory. In image analysis, however, one likes to smooth without distorting the edges.

Median filtering, which is a nonlinear operation, is well known for noise removal while preserving the edges[7] rendering solution to this contradiction. Since the binary noise (i.e., shot noise) is the noise type to be removed, we apply *median* filtering to our data. The pixels in the window (usually 3×3-matrix) are sorted, and a robust median value is chosen to replace the pixel value. Since the binary noise, like the shot noise, completely changes the gray level value, it is very unlikely to be the median value in the window. Thus, the median of the pixel values in the window is used to estimate its gray level value.

Due to variations in the acquisition systems, e.g., cameras and scanners, reflection angle, etc., recorded pixel values are not exactly what objects really are. Thus the smoothed images are further processed for gray-scale modifications to enhance constrast.

The contrast of an image in a given gray level range can be increased by stretching the range of gray levels in the image. The brightest and the darkest pixel values are found, and they are assigned to white and black, i.e., 255 and 0 in an 8-bit representation. This is an affine transformation taking the acquisition value and changing it to the full gray levels. Some benefits from the contrast enhancement (usually known as histogram equalization) are

- the elimination of the irregular acquisition effects, and
- the enhancement of contrast.

The enhanced contrast not only helps in viewing but also in building more confidence in finding the threshold in order to separate an object from the background. Segmentation of an image into parts is an important stage in image analysis. It uses clustering of pixels by their values. An ideal clustering would result in homogeneity in the distribution of pixels in a cluster, thus segmenting the images into parts by their pixel values.

In digit recognition we have only one object to be segmented from the background. For this purpose, simply taking the midpoint as the threshold of the gray level in the histogram will result in good binary images.

Another preprocessing step is done for the varied positions of the centroids of the digits, as seen in Figure 3.2. This translational variance of the images is interpreted as the camera movement in a direction perpendicular to the optical axis. The centroid of an image $f(x, y)$ is given by

$$\bar{x} = M_{1,0} / M_{0,0}, \ \bar{y} = M_{0,1} / M_{0,0}$$

where

$$M_{p,q} = \sum\sum x^p y^q f(x,y)$$

is the $(p + q)$-th order moment. The image is translated to the center of the frame by moving the centroid to that point.

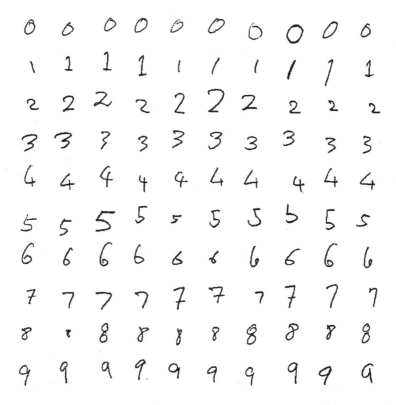

FIGURE 3.2   Some digits from the training data. Five people are involved in writing digits on a grid and of one inch square. We assume that the digits are well separated, that is *interaction* and *occlusion* problems are solved already. Different sizes and widths of writing styles are notable.

Depending upon the writing instruments and the writer's habits, stroke widths are different, as can be seen in the sample digits of Figure 3.2. *Skeletonization** is used in order to find an approximation to the medial axis of planar objects.

The basic requirements in the skeletonization algorithms are end-point preservation and pixel connectivity.[8,9] The algorithm used for our study is that of Zhang and Suen.[9] Eight neighbor pixel values, either 0 or 1, are usually compared, and a

---

* Some other terms, like shrinking and thinning, appear in the literature and are used interchangeably.[8,9]

decision is made as to whether to delete the center pixel or not. The eight neighbors are denoted as $(p_2, p_3,\ldots,p_9)$, as shown in the Figure 3.3(a). Using the eight neighbor values, we test for four conditions in order to decide for the removal of the center pixel, p_1.

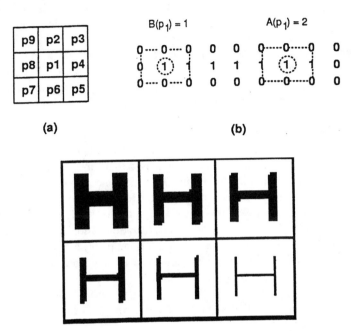

(a)   **(b)**

FIGURE 3.3 (I) (a)Neighboring pixels and (b)preventing end-points and middle points from deletion. (II) A series of skeletonized patterns next to the original pattern. Starting from upper left, original pattern, 1st, 2nd, 3rd, 4th, and 5th (the last one) are displayed. As the procedure goes, it peels off the boundary points and an opposite corner point; then it does the same from the opposite direction. In the first peeling-off, all the N/W boundary points and a S/E corner point are deleted.

The algorithm works in two directions. The conditions for the two directions are

$$2 \le B(p_1) \le 6 \qquad\qquad 2 \le B(p_1) \le 6 \tag{3.1}$$

$$A(p_1 = 1) \qquad\qquad A(p_1) = 1 \tag{3.2}$$

$$p_2 * p_4 * p_6 = 0 \qquad\qquad p_2 * p_4 * p_8 = 0 \tag{3.3}$$

$$p_4 * p_6 * p_8 = 0 \qquad\qquad p_2 * p_6 * p_8 = 0 \tag{3.4}$$

where the first two conditions of the second set are the same as the ones in the first set of conditions. $B(p_1)$ is the sum of all the eight neighboring pixels, that is, $B(p_1) = p_2 + p_3 + \cdots + p_9$, and $A(p_1)$ represents the number of the (0, 1) patterns around the neighboring pixels (Figure 3.3b).

The conditions of Equation 3.3 and Equation 3.4 in the first set above are satisfied when $p_4 = 0$ or $p_6 = 0$ or ($p_2 = 0$ and $p_8 = 0$). So point p_1, which has been removed, might be an East/South boundary point or a North-West corner point. This set of conditions is valid for East/South boundary point or North-West corner point deletion. The conditions of Equation 3.1 and Equation 3.2 protect the end-points from being deleted (Figure 3.3): the first loop at the left end-point has $B(p_1) = 1$ which does *not* meet the condition of Equation 3.1, and the second loop shows that $A(p_1) = 2$, meaning the middle point cannot be deleted. A set of skeletonized patterns and the original is also displayed in Figure 3.3. Note that the procedures take turns in both directions as the algorithm passes the two subiterations with the corresponding conditions.

After segmentation by thresholding, the binary images are skeletonized to obtain the invariance of the stroke width that resulted from different writing styles and writing instruments. For global moment calculation a dilation process is desired. Pen path-width standardization by dilation is proven to be important for that purpose. (This will be indirectly seen later in Figure 3.6, where the reconstruction of patterns is progressively done for some font images. In the reconstruction, the narrow strokes are less prominent compared to the wider width parts of the fonts).

Another reason for the path-width standardization is that the moment values obtained from the skeletonized images (width of one pixel) are more vulnerable to perturbation by a little change in the location of the skeletonized pixels (Figure 3.4). Therefore, a certain width in a given image size is desired in order

- to stabilize the moment values against the variation of the skeletonized patterns and
- to build tighter clusters in the same group and larger separations between the clusters of the different classes.

Nonlinear morphological processing, as opposed to the linear processing (e.g., convolution) achieves certain effects such as dilation, erosion, opening, closing, and boundary extraction.[7,10,11]

Let $\mathcal{F}$ be the set of all the pixels of the matrix which are not zero and $\mathcal{M}$ the set of the non-zero mask pixels. With $\mathcal{M}_p$ we denote the mask shifted or centered on this reference point to the pixel p.

Dilation is defined with a set operation as follows:

$$\mathcal{F} \oplus M = \left\{ p : M_p \cap \mathcal{F} \neq \varnothing \right\}$$

that is, the dilation operation produces the points on which the mask $\mathcal{M}$ and the image $\mathcal{F}$ have at least one non-zero pixel in common. Erosion is defined as

$$\mathcal{F} \ominus M = \left\{ p : M_p \subseteq \mathcal{F} \right\}$$

that is, the erosion produces the points for which the mask is a subset of the original image. These are equivalent to the regular binary operations for dilation and erosion, respectivley:

$$f'_{xy} = \bigvee_{k=-K}^{K} \bigvee_{k=-K}^{K} \left(M_{k,l} \wedge f_{x-k,y-l} \right) \tag{3.5}$$

and

$$f'_{xy} = \overline{\bigvee_{k=-K}^{K} \bigvee_{k=-K}^{K} M_{k,l} \wedge \overline{f_{x-k,y-l}}} \tag{3.6}$$

where the $\vee$ and $\wedge$ denote the logical [OR] and [AND] operations, respectively. The binary image f is convolved with a symmetric $(2K + 1) \times (2K + 1)$ mask M. The erosion has to be done as shown in Equation 3.6 since the all-zero mask $\mathcal{M}$ would have no meaning in a binary [AND] operation. In other words, the erosion operation is done by first dilating with the background and then inverting the result to get the erosion effect.

3.2.1 Optimal Size of the Mask for Dilation

The intuition for the dilation operation is justified via a simulation to find an optimal dilation matrix of size, $2K + 1$. The strategy is that given a size of the image frame, find the size of the dilation matrix of size $2K + 1$ which gives a larger separation between group means (or higher confidence in order to reject the null hypothesis of MANOVA model), in comparing J population mean vectors. The MANOVA model and the modified Wilks' statistic (or Bartlett statistic)[12] is used to measure the separation. Leaving the details to Reference 13, we introduce its definition as well as results from a simulation study.

3.2.2 Bartlett Statistic

This is the modified Wilks' lambda statistic, given by

$$\Lambda^* = \frac{|WSSP|}{|BSSP + WSSP|} \tag{3.7}$$

where the WSSP and BSSP are the "within" and "between" sums of squares and cross-products. A simple modification results to the Bartlett statistic, provided that the null hypothesis (i.e., same group means) is true and $N = \sum_{j=1}^{J} n_j$ is large:

$$-\left(N - 1 - \frac{(p+J)}{2} \right) \ln \left(\frac{|WSSP|}{|BSSP + WSSP|} \right) > \chi^2_{p(J-1)}(\alpha) \tag{3.8}$$

where $\chi^2_{p(J-1)}$ (α) is the upper (100α)th percentile of a chi-square distribution with $p(J - 1)$ degrees of freedom and J is the number of classes while p represents the dimensionality of the covariate.

The size of the digital images used in this study is about 128×128 because the moment approximation by digital calculation requires high resolutions. This fact is partly studied for lower moment invariants[14] requiring the image size to be larger than 60×60 pixels. For the higher order moments, a higher resolution may be required. With the image size fixed (129×129), Bartlett statistics (or modified Wilks' lambda Λ^*) are calculated for different dilation matrix size, $2K + 1$, as in Equation 3.5.

For the simulation study, an image pattern 'A' is preprocessed in the same way except for the size of dilation. The skeletonized image is dilated with dilation matrix sizes $2K + 1 = 1, 3, 5, 7, 9, 11, 13, 15$. A set of Zernike moments are obtained for different dilation sizes, and the Bartlett statistics (Equation 3.8) are calculated and plotted against the size of the dilation matrix $2K + 1$ (Figure 3.4). The null hypothesis (that is, all the mean vectors are the same) test is obviously rejected in all K values at the significance level $\alpha = 0.01$.

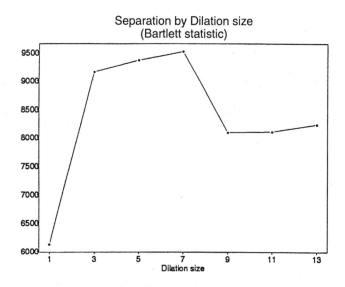

FIGURE 3.4   Bartlett statistic against dilation matrix size. Dilation increases the statistic as K increases and starts decreasing after size $2K + 1 = 7$ with the image frame of size 129×129.

From Figure 3.4, the statistic with $2K + 1 = 7$ is the highest. In fact results using size 7 look the best (Figure 3.1) for an image of size around 129×129, which is the size we have chosen.

It is worth noting the assumption made on the statistics. The statistics of Equation 3.8 assume that the error term follows the multinormal distribution $\in_{i,j} \sim N(0, \Sigma)$ in the one-way classification model

$$X_{ij} = \mu_X + \mu_j + \in_{ij}$$

where $i = 1, 2, ..., n_j$ and $j = 1, 2, ..., J$. μ_X is an overall mean and μ_j represents the jth treatment effect (or jth group mean) with $\sum_{j=1}^{J} n_j \mu_j = 0$.

Furthermore, the statistic does not necessarily measure the separation between multigroup mean vectors where $J > 2$. For example, with a scalar statistic in a two dimensional three group setting, a large statistic may result also from the case that any two mean vectors are unacceptably close, but the other mean vector is far from the two. However, the more ideal separation among the groups is in the case when the three mean vectors are equilateral in distance. The Bartlett statistic in this sense gives little insight on how well the mean vectors are separated; however, it still gives some feeling about the separation.

After the translation invariance has been obtained by the translation standardization stage in Figure 3.1, size standardization follows. The radius of an image function $f(x, y)$ can be defined[15] as

$$r = \left(\mu_{20} + \mu_{02} \right)^{1/2} \tag{3.9}$$

where μ_{20} and μ_{02} are the moments of order 2 after the centralization and represent the variance in x- and y-directions of the ellipsoidal approximation of the image. In the stage of size standardization the desired radius r^s, after normalization, is fixed to be 60% of one-half the smaller side of the image frame:

$$r^s = 0.6 * \min \left\{ ncol/2, nrow/2 \right\} \tag{3.10}$$

where $ncol \times nrow$ is the size of the image frame. All the object pixels are scaled in such a way that the radius r^s of the scaled object becomes the prescribed value. The 60% restriction can be thought of as a control parameter that contains all the scaled objects inside the frame. This prevents the scaled objects from spilling outside the frame, and it corresponds to the coordinate normalization in the Zernike moment calculation, which will be treated in Section 3.3. It should be noted that the radius in Equation 3.9 is neither the principal axis length a nor the secondary principal axis length b of an ellipsoid approximation of the image function $f(x, y)$, but that it is directly related to a and b; the area of an ellipse of parameters (a, b) is equal to πab. Digits such as '1' have a larger major principal axis but smaller secondary principal axis, whereas the digit '0' and '4' give relatively equal principal and secondary principal axes a and b. The effect of the size normalization with the control constant 0.6 in Equation 3.10 is shown in Figure 3.1.

3.3 ZERNIKE MOMENTS (ZM) FOR CHARACTERIZATION OF IMAGE PATTERNS

The complex Zernike moments of order n with repetition l are defined as

$$A_{n,l} = \frac{n+1}{\pi} \int_0^{2\pi} \int_0^\infty \left[V_{nl}(r,\theta) \right]^* f(r\cos\theta, r\sin\theta) \, r \, dr \, d\theta \tag{3.11}$$

where $n = 0, 1, 2, \cdots, \infty$ and l takes on positive and negative integer values such that

$$n - |l| = even, \quad |l| \le n. \tag{3.12}$$

The Zernike polynomials[16] given by

$$V_{nl}(r \cos \theta, r \sin \theta) = R_{nl}(r)\exp(il\theta) \tag{3.13}$$

are a complete set of complex-valued orthogonal functions on a unit disk $x^2 + y^2 \le 1$:

$$\int_0^{2\pi} \int_0^1 \left[V_{nl}(r,\theta)\right]^* V_{mk}(r,\theta)\, r\, dr\, d\theta = \frac{\pi}{n+1}\delta_{mn}\delta_{kl} \tag{3.14}$$

In Figure 3.5 the luminance of gray images represents the real part of the polynomials which are in [—1, 1] and 256 gray levels are assigned to the discrete level of the polynomials. The periodicity in Equation 3.13 being equal to $2\pi/l$ related the polynomial image to an l-fold symmetric range.

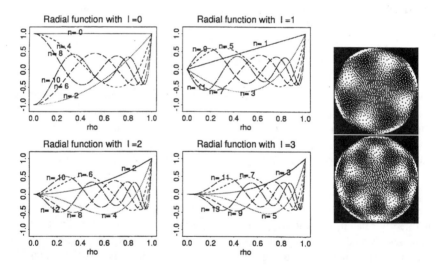

FIGURE 3.5   Radial and Zernike polynomials $R_{nl}(r)$ for different orders for a given azimuthal repetition l. Two real parts of Zernike polynomials with $(n, l) = (6, 4)$ and $(n, l) = (9, 5)$ are also shown.

The real-valued radial polynomial shown in Figure 3.5 and represented by Equation 3.13 satisfies the following condition:

$$\int_0^1 R_{nl}(r)R_{mk}(r)\, r\, dr = \frac{1}{2(n+1)}\delta_{mn} \tag{3.15}$$

and is defined as

$$R_{nl}(r) = \sum_{s=0}^{(n-|l|)/2} (-1)^s \frac{(n-s)!}{s!\left(\frac{n+|l|}{2}-s\right)!\left(\frac{n-|l|}{2}-s\right)!} r^{n-2s}$$

$$= \sum_{\substack{k=|l| \\ n-k=even}} B_{nl|l|k} r^k$$

(3.16)

where the $B_{nl|l|k}$ is the new expression (by changing the variable) for the coefficient part of the radial polynomial:

$$B_{nl|l|k} = (-1)^{\frac{n-k}{2}} \frac{\left(n-\frac{n+k}{2}\right)!}{\left(\frac{n-k}{2}\right)!\left(\frac{k+l}{2}\right)!\left(\frac{k-l}{2}\right)!}$$

The orthogonality of the Zernike polynomials enables a given $f(x, y)$ to be expressed in terms of the polynomials

$$f(x,y) = \sum_{n=0}^{\infty} \sum_{\substack{|l|\leq n \\ n-|l|=even}} A_{nl} V_{nl}(x,y)$$

(3.17)

where the Zernike moments A_{nl} are computed over the unit disk $x^2 + y^2 \leq 1$:

$$A_{nl} = \frac{n+1}{\pi} \iint_{x^2+y^2\leq1} \left[V_{nl}(r,\theta)\right]^* f(x,y) \, dxdy$$

$$= \left[A_{n,-l}\right]^*$$

This is obtained simply by the orthogonality property of the Zernike polynomials in Equation 3.13. The second equal sign holds because $f(x, y)$ is real, and the radial polynomials satisfy $R_{n,l} = R_{n,-l}$. $A_{n,l}$ can be interpreted as the projection, correlation, or proximity of a given image onto each complex valued polynomial. Thus the set of Zernike moments is the collection of the projections of a given image onto the set of the Zernike polynomials with order n and azimuthal repetition l.

In practice, we cannot have an infinite limit in the summation of Equation 3.17. Instead the finite order of N is used:

$$\hat{f}(x,y) = \sum_{n=0}^{N} \sum_{\substack{|l|\leq n \\ n-|l|=even}} A_{nl} V_{nl}(x,y).$$

(3.18)

This approximation with the finite order N is the optimal among all the other representations of $f(x, y)$ expressed by moments due to the orthogonality property.

The Zernike moments can be represented by the regular geometric moments (GM) by expressing the terms r^k in Equation 3.16 and $\exp(-il\theta)$ in Equation 3.13 in terms of x and y:

$$r^k = \left(x^2 + y^2\right)^{\frac{k}{2}}$$

$$\exp(-\theta) = \left(x^2 + y^2\right)^{-\frac{1}{2}}(x - iy)$$

$$\exp(-il\theta) = \exp(-i\theta)^l = (\cos\theta - i\sin\theta)^l \qquad (3.19)$$

$$= \left(x^2 + y^2\right)^{-\frac{1}{2}}(x - iy)^l$$

$$= \left(x^2 + y^2\right)^{-\frac{1}{2}}\sum_{m=0}^{l}\binom{l}{m}(-i)^m x^{l-m} y^m$$

The resulting expression for the A_{nl} is

$$A_{nl} = \left(\frac{n+1}{\pi}\right)\iint R_{nl}(r)\exp(-il\theta)f(x, y)dxdy$$

$$= \frac{n+1}{\pi}\sum_{\substack{k=|l| \\ n-k=even}}^{q}\sum_{j=0}^{|l|}\sum_{m=0}^{|l|} \qquad (3.20)$$

$$w^m\binom{q}{j}\binom{|l|}{m}B_{n|l|k}M_{k-(2j+m),2j+m}$$

where $w = -i, +i$ for $l > 0$, $l \leq 0$, respectively, and $q = \frac{1}{2}(k - |l|)$.

3.3.1 RECONSTRUCTION BY ZERNIKE MOMENTS

In designing a pattern recognition system, one should be concerned with what constitutes the feature elements. What is the best set (if any at all) of the possible features for the classification purpose? How does one get it? A trade-off is to be made between representability and complexity of the system that resulted from the selected set of the global features.

The order of the ZM to be included can be found by the reconstruction process. Due to the orthogonality of the Zernike polynomials (Equation 3.14), we are able to reconstruct the image $\hat{f}(x, y)$ by its finite order representation (Equation 3.18) of the original image $f(x, y)$. In order to illustrate the reconstruction process and to find the optimal order to be used, we revisit Equation 3.18 and simplify it in terms of real-valued functions.[17]

$$\hat{f}(x,y) = \sum_{n=0}^{N}\sum_{l<0} A_{nl}V_{nl}(\rho,\theta) + \sum_{n=0}^{N}\sum_{l\geq0} A_{nl}V_{nl}(\rho,\theta)$$

$$= \sum_{n=0}^{N}\sum_{l>0} A_{n,-l}V_{n,-l}(\rho,\theta) + \sum_{n=0}^{N}\sum_{l\geq0} A_{nl}V_{nl}(\rho,\theta)$$

$$= \left[\sum_{n=0}^{N}\sum_{l>0}\left[A_{nl}^{*}V_{nl}^{*}(\rho,\theta) + A_{nl}V_{nl}(\rho,\theta)\right]\right]$$

$$+ \sum_{n=0}^{N} A_{n0}V_{n0}(\rho,\theta)$$

$$= \sum_{n=0}^{N}\left[\sum_{l>0}\left(C_{nl}\cos l\theta + S_{nl}\sin l\theta\right)R_{nl}(\rho) + \frac{C_{n0}}{2}R_{n0}(\rho)\right]$$

with

$$C_{nl} = 2\,\mathrm{Re}\left(A_{nl}\right) = \frac{2(n+1)}{\pi}\iint_{x^2+y^2\leq1} f(x,y)$$

$$R_{nl}(\rho)\cos l\theta\, dx\, dy$$

$$S_{nl} = 2\,\mathrm{Im}\left(A_{nl}\right) = \frac{-2(n+1)}{\pi}\iint_{x^2+y^2\leq1} f(x,y)$$

$$R_{nl}(\rho)\sin l\theta\, dx\, dy$$

In Section 3.3 the azimuthal index l is limited by the condition

$$n - l = \text{even} \quad \text{and} \quad n \geq l \tag{3.21}$$

Two digits of **times-bold 14** font were reconstructed from the ZM. The reconstruction is done up to a certain high order, say 15; the order up to 15 renders a total of 72 moments:

$$72 = \sum_{n=0}^{15}\left\{\left\lfloor\frac{n}{2}\right\rfloor + 1\right\}$$

Figure 3.6 shows the original image and its reconstruction by ZM. It is evident that lower order ZMs capture gross shape information and that the more fine structures are filled in by higher order moments. Each digit consists of 16 small frames, which are the original, top-left, and its reconstruction in the direction from left to right and top down for orders 1 to 15. Most of the digits are well reconstructed by

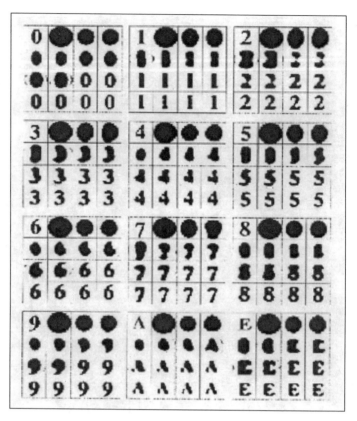

FIGURE 3.6   Reconstruction via ZM. The original image and the reconstruction by 1st to 15th orders of moment show the effects of the orders in the reconstruction.

order around 11 ~ 15, except the digit '4'. We conjecture that the handwritten digits with various writing styles need orders up to 15 for the reconstruction to be close enough to the original images. The possible redundant variables included by higher moments will be removed via PCA (see Section 3.4).

Order 15 was chosen to be the cut-off point for our handwritten digit data through visual inspection of Figure 3.6. In this way we have resolved the question of how large the feature set should be.

3.3.2 FEATURES FROM ZERNIKE MOMENTS

The advantage of ZMs for pattern recognition has been reported in terms of noise immunity, discrimination power[18] and image representation ability, noise insensitivity, and information relevance.[19] These are considered a basic theoretical support for ZM. A simulation study that supports the theoretical work can be found in Reference 20. The application of ZM for pattern recognition is also in favor of the ZMs compared to others.[21]

Functions of the Zernike moments, called Zernike Moment Invariants (ZMIs), are introduced in order to get the rotational invariance from different orders m and azimuthal indices h for a given order n and l. Teague[22] introduced a form of rotational invariance

$$ZMI_{n0} = A_{n0}; \quad ZMI_{nl} = |A_{nl}|^2 \tag{3.22}$$

$$ZMI_{nz} = \left[A_{nl}^*\left(A_{mh}\right)^p\right] \pm \left[A_{nl}^*\left(A_{mh}\right)^p\right]^* \tag{3.23}$$

where the integers m, n, h, l and positive integer p are constraints such as

$$m = \text{any integer} \quad p = \frac{l}{h} \text{ with } l \bmod h = 0$$

$$h \leq l \qquad z = p + l + h \text{ for index}$$

The first two invariants in Equation 3.22 are called primary invariants and the third in Equation 3.23 secondary invariants. The number of the primary invariants for a given order n is $\lfloor \frac{n}{2} \rfloor + 1$, due to the constraint $n - |l| = $ even in Equation 3.12 of the ZM definition. The secondary invariants are found by forcing the exponential term to be 1, thus to become independent of the angle θ,

$$\left[A_{nl}^*\left(A_{mh}\right)^p\right] + \left[A_{nl}^*\left(A_{mh}\right)^p\right]^*$$

$$= R_{nl}(r)\exp(-jl\theta)R_{mh}^p(r)\exp(jph\theta)$$

$$+ R_{nl}(r)\exp(jl\theta)R_{mh}^p(r)\exp(-jph\theta) \tag{3.24}$$

$$= R_{nl}(r)R_{mh}^p(r)\left[\exp\{j(ph-l)\theta\} + \exp\{-j(ph-l)\theta\}\right]$$

$$= R_{nl}(r)R_{mh}^p(r) \cdot \cos(ph-l)\theta$$

with the constraint on p, h, and l ensuring the cos() term to be one, thus resulting in $R_{nl}(r)\,R_{mh}^p(r)$ being independent of the angle θ. Since there is no restriction on the order m of the secondary invariant, we could have an infinite number of invariants by varying m while satisfying Equation 3.23. However by the definition of the functional independence of the invariants, only $n + 1$ number of invariants are functionally independent. The moment invariants are functionally independent if the invariants can be solved for the moments which form them.[21] $n + 1$ is the number of the independent moments from the definition of ZM (Equation 3.11) and its constraints on the indices (Equation 3.12).

Another set of Zernike moment invariants has been introduced recently[21] The idea is the same as that of Teague's in Equation 3.22 and Equation 3.23, and is given by

$$ZMI'_{n0} = A_{n0}; \quad ZMI'_{nl} = |A_{nl}| \tag{3.25}$$

$$ZMI'_{nz} = \left[A^*_{mh}\left(A_{nl}\right)^p \right] \pm \left[A^*_{mh}\left(A_{nl}\right)^p \right]^* \tag{3.26}$$

where

$$m \leq n \quad p = \frac{h}{l} \text{ with } 0 \leq p \leq 1$$

$$h \leq l \quad z = \frac{l}{h} \text{ for index}$$

The difference of this formulation from the original ones (Equation 3.22 and Equation 3.23) is that the modulus values are taken instead of their squares, and the constraints on the indices are rational power multiplications rather than integer power. The first constraint $m \leq n$ ensures that only combinations of moments of orders lower than n are used to form secondary invariants. The factor p ranges between 0 and 1. This constraint tends to decrease the magnitudes of the secondary invariants since p decreases as l increases. This magnitude decreasing property of the new invariants ZMI' (Equation 3.26) is desirable and was not present in the original ZMI of Equation 3.23.

The secondary parts of the ZMI and ZMI' (Equation 3.23 and Equation 3.26) are the additional $(\lfloor n/2 \rfloor)$ rotational invariant values that are obtained from the power multiplication of the higher order moments or lower order moments, respectively.

As shown in the ZMI and ZMI' the rotational invariance is obtained in various ways by forcing the phase information of complex-valued ZMs to be one. Using only radial information means that all the points of a circle of radius r, in the complex domain, are the same. In addition, in digit recognition, 180-degree rotation conflict digits such as 9 and 6 are not taken care of.

Khotanzad and Hong[17] used the modulus value of the complex-valued ZM, the primary invariant, to eliminate the rotational problem. Their argument is based on the fact that the ZM for a rotated image $f'(x, y)$ due to rotation by θ, results to a simple phase shift:

$$
\begin{aligned}
A^r_{nl} &= \frac{n+1}{\pi} \int_{\phi=0}^{2\pi} \int_{r=0}^{1} f(r, \phi - \theta) R_{nl}(r) \exp(-il\phi) r\, dr\, d\phi \\
&= \frac{n+1}{\pi} \int_{\phi^*=0}^{2\pi} \int_{r=0}^{1} f(r, \phi^*) R_{nl}(r) \exp\left(-il(\phi^* + \theta)\right) r\, dr\, d\phi^* \\
&= A_{nl} \exp(-il\theta)
\end{aligned}
\tag{3.27}
$$

The original function $f(x, y)$ and the rotated one $f'(x, y)$ result to the same modulus value:

$$\left| A_{nl} \right| = \left| A^r_{nl} \right|$$

As a remedy to this problem we have included skewness information into the modulus value of all the variables used (up to order 15). The skewness of a two-dimensional function $f(x, y)$ is obtained for each variable x and y. The skewness for an image function $f(x, y)$ is given by

$$S_x = \frac{\mu_{3,0}}{\left(\mu_{2,0}\right)^{3/2}} \tag{3.28}$$

$$S_y = \frac{\mu_{0,3}}{\left(\mu_{0,2}\right)^{3/2}} \tag{3.29}$$

Two more new variables for the skewness information are added to the modulus values of the ZM order from 2 to 15. The 0th and 1st order are deleted since the image has been preprocessed to be size standardized and to be centered by the centroid. The new moment moduli with the skewness values added are now not only rotation invariant but also free of the 180-degree rotation conflict.

Section 3.5 includes the results from both the modulus values of ZM called 'V' and the modulus values of ZM with skewness information added, called 'V1'.

An argument is developed here to justify the use of only the real components of the ZM. The 180-degree rotation conflict problem is taken care of by the third order moments $\mu_{0,3}$ and $\mu_{3,0}$ of Equation 3.28 and Equation 3.29. This skewness information is contained in the real part of the phase components of the lower orders of ZM ($A_{3,1}$ and $A_{2,1}$). We call 'R' the real part of the ZM. The number of the real part of the ZM for a given order n is $\left\lfloor \frac{n}{2} \right\rfloor + 1$ and is obtained with $m =$ even from Equation 3.20. That is, the real part of ZM is given by

$$
\begin{aligned}
C_{nl} &= 2Re[A_{nl}] = 2\frac{n+1}{\pi} \iint_{x^2+y^2 \le 1} R_{nl}(r)\cos(l\theta)f(x,y)\, dx\, dy \\
&= 2\frac{n+1}{\pi} \sum_{\substack{k=|l| \\ n-k=even}} \sum_{j=0}^{q} \sum_{\substack{m=0 \\ m=even}}^{|l|} (-1)^m \binom{q}{j}\binom{|l|}{m} B_{n|l|k} M_{k-(2j+m),2j+m}
\end{aligned}
\tag{3.30}
$$

with $q = \frac{1}{2}(k - |l|)$.

The rotational invariance by the modulus operation of ZM or moment invariants has been successful with the patterns that have no 180-degree rotation conflict, such as printed English alphabets, the aerial views of the four Great Lakes, aircraft recognition tasks, etc.

The circular symmetry property of the Zernike polynomials seems to handle the rotational variance of the patterns well. The Zernike polynomial $V_{nl}(r, \theta)$ is circularly symmetric in periods of $2\pi/l$ (Equation 3.13) and has a wedge shape implying the rotational variance of patterns.

If the patterns from a group vary within a certain orientation range (as is the case with handwritten digits), the modulus operation or the ZMI costs too much for the rotational invariance. The range of the modulus operation of the complex-valued ZM is only the distance, represented by a radius in a complex domain. The real part of ZM, however, has a range twice as large as that of the modulus value; it explains more than the radius does.

The modulus value (called 'V') or squared modulus of the complex-valued ZM is the primary part of the Zernike moment invariants [ZMI] (Equation 3.22 and Equation 3.25). The secondary part of the ZMI shown in Equation 3.23 and Equation 3.26 is not included in our features because the secondary invariants are simply the power multiplication that adds another $(\lfloor n/2 \rfloor)$ number of the orientation independent values. Instead, we have followed the strategy of including the primary invariants of all the moments that have been included by the reconstruction process in finding the finite number of moments for the given patterns.

3.4   DIMENSIONALITY REDUCTION

The subject of dimensionality reduction in pattern recognition is concerned with mathematical tools for reducing the size of the features. The most revealing facts with dimensionality reduction are discussed in reference 23 and summarized below:

- Reduction of the physical system complexity is as required by feasibility limitations of either a technical or economical nature.
- It ensures the reliability of the decision making procedure by removing the redundant and irrelevant information which has a derogatory effect on the classification process.
- More importantly, the dimensionality is strongly related to the size of the sample used for training: as the dimensionality increases, the size of the training required grows exponentially. Neural networks, however, train well regardless of the dimensionality, except that the networks require more time to learn and result to poor convergence.

Two stages are employed for this purpose: Principal Component Analysis (PCA) is followed by Discriminant Analysis (DA), both of which are eigen analyses on the covariance-type matrices. These eigen analyses can be interpreted as finding $p < q$ directions on which the projections of the data result to some interesting properties, such as large variance or separation among the group means under a set of constraints.

3.4.1   PRINCIPAL COMPONENT ANALYSIS

Principal component analysis of a multivariate random sample can be viewed as finding an axis optimizing a criterion in a geometric sense. Illustration with projection of simulated two-dimensional data points is shown in Figure 3.7. Pearson (1901) looked for a new axis on which the projection gives the *least sum of squares* of d_i. Hotelling (1933) was interested in finding a new axis on which the *maximum variance* of the projection values is obtained (see reference 24). Even though the approaches are different and opposite, the resulting axis from the two different approaches is the same. The optimal axis for minimal sum of the squares of d_i's is the same as the one with the axis in which maximum variance of z_i is obtained.

$$\max_{P_Y} \sum \frac{z_i^2}{N-1} \Leftrightarrow \min_{P_Y} \sum d_i^2 \qquad (3.31)$$

where P_Y is the projection operator defined by a projection axis.

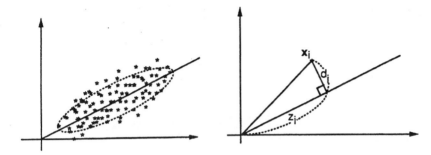

FIGURE 3.7   Illustration of projection of a vector point y_i onto the principal axis. z_i represents the projected value of x_i onto the axis and d_i the error component of the projection. $z_i^2 + d_i^2 = const$ confirms the equivalence of the two motivations for finding the optimal axis.

The idea of the PCA is to find a rotational transformation (i.e., an orthogonal transformation) matrix $R_{q \times q}$ such that the sample variances of the new rotated variables are in decreasing order of magnitude.[24] Thus the first principal component is such that the projections of the given points onto it have maximum variance among all possible linear coordinates; the second principal component has maximum variance subject to being orthogonal to the first; and so on.

The PCA is done on the sample version of total covariance matrix $T_{q \times q}$ of the handwritten data matrix, $Y_{N \times q}$, where the dimensionality $q = 70$ and the size $N = \Sigma_{j=1}^{J} n_j = 1000$, and where $J = 10$ is the number of classes.

The lower curve of the plot in Figure 3.8 is called *scree plot* and represents the variance information contained in the new derived variates. The upper curve represents the accumulated version of the lower scree curve, which is the total variance of the newly obtained variables from the first to the corresponding variable indices. The 95% and 99% of the accumulated variance are indicated by the two broken lines. The 99% explanation of the variance information is obtained by the first 35 newly obtained variables.

Since the dimensionality of the original data $q = 70$ was too large, we reduced the dimensionality using PCA of the total covariance. With the new data set, which is supposed to be uncorrelated (or less correlated), we are ready to do more statistical treatment in order to find multidimensional outliers for robust analysis and reduce the *heteroscedacity* (and as a by-product enhance the multinormality, if possible at all).

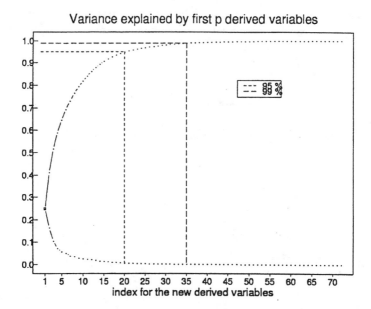

FIGURE 3.8   PCA on the sample total covariance matrix of the handwritten data set 'R.'

A strategy we follow for such large dimensionality is a two-step dimensionality reduction. First principal component analysis on the total sample covariance matrix, T, is carried out. Then discriminant analysis follows, in order to reduce the dimensionality even further to $J — 1$.

Even though the PCA is well known to be sensitive to outliers,[25,26] we argue that the whole data set is preserved, as much as we want, in a lower dimensional space, provided that the explanation of the variance information is over, say, 99%. The whole data set as a single batch from the different clusters of different classes is decorrelated via principal component analysis. Now the lower $p = 35$ dimensional space is processed by discriminant analysis for further dimensionality reduction.

3.4.2 DISCRIMINANT ANALYSIS

Suppose that we wish to find a linear transformation matrix F, which maximizes some distance criterion d defined over a sample of random vectors in a new transform space. Two interesting pairwise distance measures are the *intraset* and the *interset* distances.[27] The intraset distance, or averaged within-class distance, between the kth variable of all pattern vectors in one class, averaged over all classes is

$$d_W^{(k)} = \frac{1}{2}\sum_{i=1}^{J} P(w_i)\frac{1}{n_i^2}\sum_{j=1}^{n_i}\sum_{l=1}^{n_i}\mathbf{f}_k'\big(\mathbf{y}_{ij}-\mathbf{y}_{il}\big)\big(\mathbf{y}_{ij}-\mathbf{y}_{il}\big)'\mathbf{f}_k \qquad (3.32)$$

where n_i is the number of vectors $\mathbf{y} \in w_i$ and $\mathbf{f}_k$ is the kth column of the transformation matrix F.

The interset distance, or between-class distance, of the kth direction in the new transform space is defined as

$$d_B^{(k)} = \sum_{i=2}^{J} P(w_i) \sum_{h=1}^{i-1} P(w_h) \frac{1}{n_i n_h} \sum_{j=1}^{n_i} \sum_{l=1}^{n_h} \mathbf{f}_k' \left(\mathbf{y}_{ij} - \mathbf{y}_{hl} \right) \left(\mathbf{y}_{ij} - \mathbf{y}_{hl} \right)' \mathbf{f}_k \qquad (3.33)$$

The first two summation indices hold for $N(N-1)/2$ interpoint distances.

These averaged distance measures are expressed in terms of sample within-groups covariance matrix W and between-groups matrix B,[28] defined as:

$$W = \frac{1}{N-J} \sum_{i=1}^{J} \sum_{j=1}^{n_i} \left(\mathbf{y}_{ij} - \bar{\mathbf{y}}_i \right) \left(\mathbf{y}_{ij} - \bar{\mathbf{y}}_i \right)'$$

$$B = \frac{1}{J-1} \sum_{i=1}^{J} n_i \left(\bar{\mathbf{y}}_i - \bar{\mathbf{y}} \right) \left(\bar{\mathbf{y}}_i - \bar{\mathbf{y}} \right)'$$

respectively, where $N = \Sigma_{i=1}^{J} n_i$.

Using the definition of W and B, the distance measures $d_W^{(k)}$ (Equation 3.32) and $d_B^{(k)}$ (Equation 3.33) can be written, in terms of W and B, as follows:

$$d_W^{(k)} = \mathbf{f}_k' W \mathbf{f}_k$$

$$d_B^{(k)} = \mathbf{f}_k' B \mathbf{f}_k$$

Now, we are interested in maximizing a distance measure $d_B^{(k)} = \mathbf{f}_k' B \mathbf{f}_k$ with respect to the transformation vector $\mathbf{f}_k$ subject to a constraint, e.g., holding a distance measure, (i.e., $d_W^{(k)} = \mathbf{f}_k' W \mathbf{f}$) constant. The constraints are usually chosen to be irrelevant for the maximization of $d_B^{(k)}$ while guaranteeing a unique solution $\mathbf{f}_k$, i.e., $\mathbf{f}' W \mathbf{f} = 1$. The solution for this kind of optimization problem can be obtained by the method of Lagrange multipliers. Maximization of $d_B^{(k)}$, subject to $d_W^{(k)}$ constant, has the form

$$\max_{\mathbf{f}_k} \left\{ J = d_B^{(k)} - \lambda \left(d_W^{(k)} - \text{const} \right) = \mathbf{f}_k' B \mathbf{f}_k - \lambda \left(\mathbf{f}_k' W \mathbf{f}_k - \text{const} \right) \right\}$$

Setting the first derivative of J with respect to $\mathbf{f}_k$ equal to zero yields

$$\frac{\partial J}{\partial \mathbf{f}_k} = B \mathbf{f}_k - \lambda W \mathbf{f}_k = 0$$

$$(B - \lambda W) \mathbf{f}_k = 0$$

If we premultiply the above by W^{-1}, it results in an eigenvalue problem, i.e.,

$$\left(W^{-1}B - \lambda I\right)\mathbf{f}_k = 0 \tag{3.34}$$

The traditional disCRIMinant COORDinate system (or CRIMCOORD) is interpreted as finding functions that maximize the quadratic forms:

$$d_B^{(k)} = \mathbf{f}_k^t B\mathbf{f}_k,$$

with respect to $\mathbf{f}_k$, subject to the constraint of

$$d_W^{(k)} = \mathbf{f}_k^t W\mathbf{f}_k = 1 \tag{3.35}$$

resulting to the solution of Equation 3.34.

Two consecutive linear transformations by R (via PCA) followed by F (via DA) are represented by a linear transformation matrix FR of dimension $J - 1 \times q$, for example, 9×70 for our data set. Figure 3.9 shows two-dimensional projections of 30 randomly selected patterns from each group on the first five discriminate variates (CRIMCOORD) with corresponding digit representation. Remarkably, some distinction of the digits is clear from the figures, implying that the discriminant variates discriminate among the different groups.

3.5 ANALYSIS OF PREDICTION ERROR RATES FROM BOOTSTRAPPING ASSESSMENT

Prediction error is usually a good measure of the performance of pattern recognition systems. In practice, a random sample, called training data, from an unknown population described by distribution F is given. Any statistic $\theta(F)$ requires distribution F, but in practice, the F is not known and is difficult to estimate. An empirical distribution $\hat{F}$ from the given sample from an unknown distribution F is defined in a bootstrap setting by giving an equal probability mass $1/N$ on each of the values x_i. A bootstrap sample is a random sample from the empirical distribution

$$X_1^*, X_2^*, \ldots, X_N^* \sim \hat{F}$$

Each x_i^* is drawn independently *with replacement* and with equal probability from the sample, i.e., training data:

$$\mathcal{X} = \left\{x_i\right\}_1^N - \left\{(v_i, y_i)\right\}_1^N$$

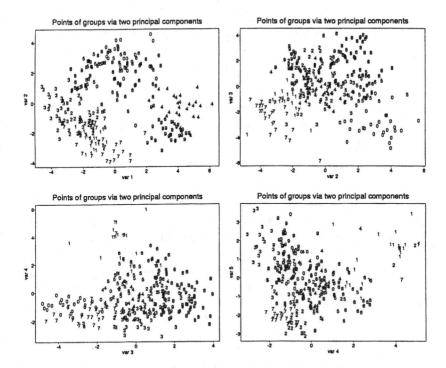

FIGURE 3.9   Two-dimensional projections of the handwritten data with the first five discriminant variates.

Standard error and bias estimation using Bootstrap resampling techniques can be found from the references.[29,30] Here we introduce the algorithms for estimation of the standard error and the bias for prediction error estimation, leaving the technical details in the references above.

The Monte Carlo bootstrapping algorithm proceeds in three steps:

1. using a random number generator, independently draw a large number 50 $\leq B \leq$ 200 of bootstrap samples, $\{F^{*b}\}_{b=1}^{B}$,
2. for each bootstrap sample F^{*b}, evaluate the statistic of interest, $\hat{\theta}^*(b) = \hat{\theta}(F^{*b})$ for $b \in \{1, 2,..., B\}$ from the training data X,
3. calculate the sample standard deviation of $\hat{\theta}^*(b)$ values

$$\hat{\sigma}_B = \left(\frac{1}{B-1} \sum_{b=1}^{B} \left\{ \hat{\theta}^*(b) - \hat{\theta}^*(\cdot) \right\}^2 \right)^{1/2},$$

$$\hat{\theta}^*(\cdot) = \frac{1}{B} \sum_{b=1}^{B} \hat{\theta}^*(b)$$

(3.36)

Standard errors are crude but useful measures of statistical accuracy.[31] An approximated confidence interval for an unknown parameter θ is given by

$$\theta \in \hat{\theta} \pm \hat{\sigma} z^{(\alpha)} \tag{3.37}$$

where $z^{(\alpha)}$ is the $100 \cdot \alpha$ percentile point of a standard normal variate, e.g., $z^{(0.95)} = 1.64485$. The standard error approximation (Equation 3.37) for a confidence interval bears the assumption that

$$\frac{\hat{\theta} - \theta}{\hat{\sigma}} \sim N(0,1)$$

Bias about an estimator $\hat{\theta}$ is the next to be considered. Bootstrap bias estimation is an estimation of the *optimistic bias op* resulting from using the same training data for prediction, e.g., via the resubstitution method. One way to estimate the system performance from the given sample is to correct the *apparent error rate* (or resubstitution error rate) by the estimation of the optimistic (or positive) bias. The optimistic bias is defined as

$$op(X;F) = \theta - \theta_{app}$$

where θ is the true error rate for the unknown distribution F and θ_{app} for the apparent error rate. Since we do not know the bias $op(X, F)$, the bootstrap estimate of the bias, op_{boot} is found instead and the optimistic θ_{app} is corrected by adding the estimated bias

$$\hat{\theta} = \theta_{app} + op_{boot} \tag{3.38}$$

Let $\eta(v, X)$ be a decision rule based on the training set X and let $Q[y_i, \eta(v_i, X)]$ be an indication of misclassification of v_i by $\eta()$:

$$Q[y_i, \eta(v_i, X)] = \begin{cases} 1 & \eta(v_i, X) \neq y_i, \\ 0 & \text{otherwise} \end{cases}$$

Thus $Q[\cdot] = 1$ indicates the misclassification of a training observation from the system designed by the training data.

The bootstrap procedure for estimating the bias, op_{boot}, follows:

1. Select $50 \leq B \leq 200$ bootstrap samples from the empirical distribution $\hat{F}$.
2. From each bootstrap sample compute the bias w_b

$$w_b = \sum_{i=1}^{N} \left(\frac{1}{N} - P_i^{*b} \right) Q[y_i, \eta(v_i, F^{*b})]$$

with P_i^{*b} indicating the proportion of the bootstrap sample on x_i, i.e.,

$$P_i^{*b} = \text{Cardinality of } \left\{ j \middle| x_j^{*b} = x_i \right\} / N$$

and $\eta(v_i, F^{*b})$ being the prediction of v_i from the system trained by F^{*b}.
3. Repeat step 2 to get $\{w_1, w_2, \cdots, w_B\}$.

Then the bootstrap bias op_{boot} is estimated by

$$op_{boot} = \frac{1}{B} \sum_{b=1}^{B} w_b$$

and thus, the bootstrap error estimate $\hat{\theta}$ (Equation 3.38) is obtained.

E0 prediction error estimation is equivalent to counting the number of patterns that are not included in the bootstrap samples and normalizing the misclassification count of the samples[32] by the total number of the training patterns not selected in the bootstrap samples. Thus, E0 uses the testing set, which is asymptotically 36.8% of the original training, according to the argument that follows. In a typical bootstrap sample, about 63% of the original observations are likely to be chosen. This is easily seen since the probability that an observation does not belong to a bootstrap sample is

$$(1 - 1/N)^N = 1/e.$$

Thus an observation x_i will be in the bootstrap sample with about $1 - 1/e = 63.2\%$ chances.

Let $A_b = \{i \mid P_i^{*b} = 0\}$ denote the index set of training patterns that do not appear in the F^{*b}, then the prediction error θ_0 estimated by the E0 estimator is defined by

$$\theta_0 = \frac{\sum_{b=1}^{B} \sum_{i \in A_b} Q\left[y_i, \eta\left(v_i, F^{*b}\right) \right]}{\sum_{b=1}^{B} \text{Cardinality of } \{A_b\}}.$$

This E0 estimator is a form of cross-validation in that the testing data have not been used in training. The difference from the cross-validation is that the E0 separates the training and the testing data randomly while the cross-validation selects the testing pattern sequentially such that all the training patterns are used for testing.

The testing patterns used in the apparent error rate obtained by the resubstitution method are too close or the distance is 'zero' from the training patterns, while the test patterns for E0 estimator are 'too far' from the training set. From that the asymptotic probability argument that a pattern will not be included in a bootstrap sample is 0.368, the weighted average of θ_{app} and θ_0 involves patterns at the 'right' distance from the training set in estimating the error rate:[32]

$$\theta_{632} = 0.368 * \theta_{app} + 0.632 * \theta_0 \qquad (3.39)$$

The E632 was shown to be optimal in terms of least variance and bias from a comparison study for various estimators[32,33] among cross-validation, ordinary bootstrap bias correction (Equation 3.38) and E632 (Equation 3.39). We used the E632 prediction error as a standard performance measure. (The bootstrap package bootstrap.funs* contains various resampling techniques and is available via *anonymous* ftp to statlib@lib.stat.cmu.edu.)

The boxplots in Figure 3.10 represent the E632 estimator superimposed to the distribution of the $B = 100$ bootstrap sample errors, $\hat{\theta}^*(b)$'s (Equation 3.36). The median value of the B error rates is replaced by the E632 estimate; thus the B bootstrap errors are shifted according to the E632 estimate. For ease of display and understanding the system performance, the recognition rates, $1 - \hat{\theta}'s$, are plotted.

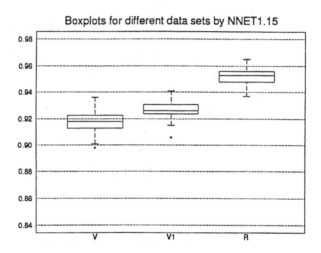

FIGURE 3.10 Boxplots from nnet with hidden layer size = 15 for data sets of V, V1, R. E632 prediction error rate and $B = 100$ bootstrap samples are used.

The height of the box is the inter-quartile range, which is the difference between the upper quartile and the lower quartile, and is considered to be a robust estimation of the scalar multiple of the dispersion. The median of the batch is represented by the line in the box. The bars, represented by the vertical dotted lines, are extended up to the points 1.5 times of the inter-quartile range. Outliers are represented by the individual dots to signify their existence. The boxplots display the distribution very simply but well enough, especially when many different batches are to be compared.

The correct recognition rates from three-layer feed-forward neural networks with the Broyden, Flecher, Goldfarb and Shannon (BFGS) algorithm (in reference 34) are displayed in the boxplots for each data set obtained from the different treatment

---

* The bootstrap was contributed by Efron and Tibshirani.

in Section 3.3.2. The classifiers* used in this study can be obtained via *anonymous* ftp to statlib@lib.stat.cmu.edu. Each boxplot shows the distribution of the recognition rate of 100 systems designed by 100 bootstrap samples.

3.6 SUMMARY

A simple model-free feature extraction by the two-dimensional Zernike polynomials was shown to be a powerful pattern recognition system (via the correct recognition rate $\gtrsim$ 95% via the E632 prediction error measure) for handwritten digits. The images are preprocessed before ZM calculations take place, and the dimensionality of the feature vectors is reduced by PCA followed by DA.

For the 180-degree rotation conflict data, addition of the skewness variables improves the performance of the system. Simply by taking the real (or imaginary) part of the complex-valued Zernike Moments, one obtains more information than what is lost by the rotational invariance operation for the rotational variance of the patterns, which is inherent in handwritten digit data. The rotational variance of the patterns seems to be observed by the *wedge* type Zernike polynomials.

The addition of skewness information (V1) to the modulus value (V) of the complex-valued ZM improves generally the correct recognition rate by 2–3%, while the real part (R) yields generally 3–4% improvement over the modulus value (V). The wedge shape of the polynomial also possesses an important property that the variation, at around the outer region of the patterns, results to less variance than the one from the Cartesian coordinates, such as the regular moments and their invariants.

ACKNOWLEDGMENTS

The authors wish to thank Dr. R. Gnanadesikan for insightful discussions and Mr. G. Kontaxakis for help with the final version of the manuscript.

REFERENCES

1. Tappert, C. C., Suen, C. Y., and Wakahara, T., The state of the art in on-line handwriting recognition, *IEEE Trans. Pattern Anal. Machine Intelligence*, 12(8), 787, 1990.
2. Le Cun, Y., Jacket, L. D., Boser, B., Denker, J. S., Graf, H. P., Guyon, I., Henderson, D., Howard, R. E., and Hubbard, W., Handwritten digit recognition: applications of neural network chips and automatic learning, *IEEE Commun. Mag.*, 41, November 1989.
3. Suen, C. Y., Berthod, M., and Mori, S., Automatic recognition of handprinted characters—the state of the art, *Proc. IEEE*, 68(4), 469, April 1980.
4. Mantas, J., An overview of character recognition, *Patt. Recog.*, 19(6), 425, 1986.
5. Bitchell, B. T. and Gillies, A. M., A model-based computer vision system for recognizing handwritten zip codes. *Mach. Vision Applic.*, 2, 231, 1989.

---

* The package nnet is contributed by Ripley.

6. Wang, C. H. and Srihari, S. N., A framework for object recognition in a visually complex environment and its application to locating address blocks on mail pieces, *Int. J. Comput. Vision*, 2, 125, 1988.

7. Jähne, B., *Digital Image Processing: Concepts, Algorithms and Scientific Applications*, Springer-Verlag, Berlin, 1991.

8. Pavlidis, T., A thinning algorithm for discrete binary images, *Comput. Graphics Image Process.*, 13, 142, 1980.

9. Zhang, T. Y. and Suen, C. Y., A fast parallel algorithm for thinning digital patterns, *Commun. ACM*, 27(3), 236, 1984.

10. Haralilck, R. M. and Shapiro, L. G., *Computer and Robot Vision*, Vol. 1. Addison-Wesley, Reading, MA, 1992.

11. Serra, J., *Image Analysis and Mathematical Morphology*. Academic Press, New York, 1982.

12. Johnson, R. A. and Wichern, D. W., *Applied Multivariate Statistical Analysis*, Prentice-Hall, Englewood Cliffs, NJ, 1988.

13. Chung, W., A Strategy for Visual Pattern Recognition, PhD thesis, Electrical and Computer Engineering, Rutgers University, The State University of New Jersey, 1994.

14. Teh, C-H. and Chin, R. T., On digital approximation of moment invariants, *Comput. Vision, Graphics, Image Process.* 33, 318, 1986.

15. Dudani, S. A., Breeding, K. J., and McGhee, R. B., Aircraft identification by moment invariants, *IEEE Trans. Comput.*, 26(1), 39, January 1977.

16. Bhatia, A. B. and Wolf, E., On the circle polynomials of Zernike and related polynomials orthogonal sets, *Proc. Camb. Phil. Soc.*, 50, 40, 1954.

17. Khotanzad, A. and Hong, Y. H. Invariant image recognition by Zernike moments, *IEEE Trans. Patt. Anal. Mach. Intell.*, 12(5), 489, May 1990.

18. Abu-Mostafa, Y. S. and Psaltis, D., Recognitive aspects of moment invariants, *IEEE Trans. Patt. Anal. Mach. Intell.*, 6(6), 698, November 1984.

19. Teh, C-H. and Chin, R. T., On image analysis by the methods of moments, *IEEE Trans. Patt. Anal. Mach. Intell.*, 10(4), 496, July 1988.

20. Chung, W. and Micheli-Tzanakou, E., A simulation study for different moment sets, in *Document Recognition*, IS&E/SPIE 1994 International Symposium on Electronic Imaging, February 1994, 378.

21. Belkasim, S. O., Shridhar, M., and Ahmadi, M., Pattern recognition with moment invariants: a comparative study and new results, *Patt. Recog.*, 24(12), 1117, 1991.

22. Teague, M. R., Image analysis via the general theory of moments, *J. Opt. Soc. Am.*, 70(8), 920, August 1980.

23. Kittler, J., Feature selection and extraction, in *Handbook of Pattern Recognition and Image Processing*, Academic Press, New York, 1986, Chap. 3, 59.

24. Gnanadesikan, R., *Methods for Statistical Data Analysis of Multivariate Observations*, John Wiley & Sons, New York, 1977.

25. Ammann, L. P., Robust singular value decompositions: a new approach to projection pursuit, *J. Am. Stat. Assoc.*, 88(422), 505, 1993.

26. Devlin, S., Gnanadesikan, R., and Kettenring, J., Robust estimation of dispersion matrices and principal components, *J. Am. Stat. Assoc.*, 76, 354, 1981.

27. Kittler, J., Mathematical methods of feature selection in pattern recognition, *Int. J. Man-Machine Stud.*, 7, 609, 1975.

28. Kittler, J. and Young, P. C., A new approach to feature selection based on the Karhunen-Loeve expansion, *Patt. Recogn.*, 5, 335, 1973.

29. Efron, B. and Gong, G., A leisurely look at the bootstrap, the jackknife, and cross-validation, *Am. Stat.*, 37(1), 36, February 1983.

30. Efron, B. and Tibshirani, R. J., *An Introduction to the Bootstrap.* Chapman & Hall, London, 1993.
31. Efron, B. and Tibshirani, R. J., Bootstrap methods for standard errors, confidence intervals, and other measures of statistical accuracy, *Stat. Sci.*, 1(1), 54, 1986.
32. Efron, B., Estimating the error rate of a prediction rule: improvement on cross-validation, *J. Am. Stat. Assoc.*, 78(382), 316, June 1983.
33. Jain, A. K., Dubes, R. C., and Chen, C. C., Bootstrap techniques for error estimation, *IEEE Trans. Patt. Anal. Mach. Intell.*, 9(5), 628, September 1987.
34. Peressini, A. L., Sullivan, F. E., and Uhl, J. J., Jr., *The Mathematics of Nonlinear Programming*, Springer-Verlag, Berlin, 1988.

4 Other Types of Feature Extraction Methods

Evangelia Micheli-Tzanakou, Ahmet Ademoglu, and Cynthia Enderwick

4.1 INTRODUCTION

If a signal contains frequency components emerging and vanishing in certain time intervals, then a time as well as a frequency localization is required. The traditional method proposed for such an analysis is the Short Time Fourier Transform (STFT) or Gabor Transform.[5] The STFT enables the time localization of a certain sinusoidal frequency but with an inherent limitation of the Heisenberg's uncertainty principle, which states that resolution in time and frequency cannot be arbitrarily small, because their product is lower bounded by

$$\Delta t \Delta f \geq \frac{1}{4\pi} \qquad (4.1)$$

In order to overcome the resolution limitation of the STFT, a decomposition of square integrable signals $L^2(R)$ has recently been developed under the name of wavelets.[1,10] These families of functions $h_{a,b}$

$$h_{a,b}(t) = |a|^{\frac{1}{2}} h(t-b)a, b \in \_, a \neq 0 \qquad (4.2)$$

are generated from a single function $h(t)$ by the operation of dilations and translations. The wavelet transform of a continuous signal $x(t)$ can be defined as

$$CWT_x(b,a) = < x(t), |a|^{-1/2} h^*((t-b)/a) > = |a|^{-1/2} \int x(t) h^*((t-b)/a) dt \qquad (4.3)$$

where * represents the complex conjugation and where <> represents the inner product. Equation 4.3 is interpreted as a multiresolution decomposition of the signal into a set of frequency channels having the same bandwidth in a logarithmic scale (i.e., constant Q or constant relative bandwidth frequency analysis by octave band filters). For the STFT, the phase space is uniformly sampled, whereas in wavelet transform the sampling in frequency is logarithmic, which enables one to analyze higher frequencies in shorter windows and lower frequencies in longer windows in time.

4.2 WAVELETS

As mentioned in the introduction, wavelet analysis has practically become a ubiquitous tool in image and signal processing. Two basic properties, space and frequency localization and multiresolution analysis, make this a very attractive tool in signal and image analysis. Unlike the complex sine and cosine basis functions of the Fourier transform, the basis functions of the wavelet transform are localized in both space and frequency.

Taking the wavelet transform of an image involves convolving a pair of filters, one high pass and one low pass, with the image. This is followed by decimation by two and repeated for as many octaves as desired. This algorithm is depicted in Figure 4.1, which shows a one-octave decomposition of an image into four components: low pass rows, low pass columns (LP-LP); high pass rows, low pass columns (HP-LP); low pass rows, high pass columns (LP-HP); and high pass rows, high pass columns (HP-HP). These will later be referred to as components 0 through 3, respectively. For the purposes of decomposing images, subsequent octaves were created by transforming the LP-LP component of the previous octave.

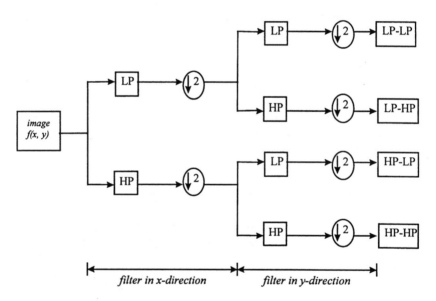

FIGURE 4.1   Wavelet transform algorithm — sub-band decomposition of one octave. HP = high-pass, LP = low-pass, ↓ 2 represents decimation by 2.

The result of the wavelet transform on a test image is shown in Figure 4.2. Figure 4.2a shows the original image Lena.bmp. Figure 4.2b shows the first octave wavelet decomposition using a 4-tap Daubechies filter bank[1] (see also Press et al.[15]) enhanced to show the high-pass components. From left to right, top to bottom is LP-LP (component 0), HP-LP (component 1), LP-HP (component 2), and HP-HP (component 3). Notice that component 1 accentuates the vertically oriented details,

FIGURE 4.2 The wavelet transform of Lena.bmp. Note that (b) has been enhanced to accentuate the detail coefficients.

component 2 accentuates the horizontally oriented details, and component 3 accentuates the 45° and 135° diagonal details.

While there exist many practical wavelet filters applicable to a wide variety of problems, the filter used in our analysis is the Daubechies 8-tap filter. These filter coefficients are

High-pass coefficients = {0.010597, 0.032883, –0.030841, –0.187035,
0.027984, 0.630881, –0.714847, 0.230378}
Low-pass coefficients = {0.230378, 0.714847, 0.630881, –0.027984,
–0.187035, 0.030841, 0.032883, –0.010597}

4.2.1 DISCRETE WAVELET SERIES

Although various ways of discretizing time-scale parameters are possible, the conventional scheme is the so-called dyadic grid sampling, where time remains continuous but time-scale parameters are sampled by choosing $a = 2^i$ and $b = k2^i$ $i,k \ e \ Z$. The wavelets in this case are given by

$$h_{ik}(t) = 2^{-\frac{i}{2}} h\left(2^{-i} t - k\right). \tag{4.4}$$

A wavelet series decomposes a signal $x(t)$ onto a basis of continuous-time wavelets or the so-called synthesis wavelets $\alpha_{i,k}(t)$ as shown

$$x(t) = \sum_{i \in Z} \sum_{k \in Z} C_{i,k} \alpha_{i,k}(t) \tag{4.5}$$

The wavelet coefficients are defined as

$$C_{i,k} = \int x(t) h_{i,k}^*(t) dt \tag{4.6}$$

The signal decomposition may be done by using orthogonal wavelets,[1] in which case the analysis and synthesis wavelets are identical.

4.2.2 Discrete Wavelet Transform (DWT)

The DWT is very close to a wavelet series, but in contrast, it applies to discrete-time signals $x[n]$. It achieves a multiresolution decomposition of $x[n]$ on I octaves labeled by $i = 1,\ldots,I$, given by

$$x[n] = \sum_{i=1}^{\infty} \sum_{k \in Z} a_{i,k} g_i \left[n - 2^i k \right] + \sum_{k \in Z} b_{I,k} h_I \left[n - 2^I k \right] \qquad (4.7)$$

The DWT computes wavelet coefficients $a_{i,k}$ for $i = 1,\ldots,I$ and scaling coefficients $b_{i,k}$, which are given by

$$DWT\left\{ x[n]; 2^i, k2^i \right\} = a_{i,k} = \sum_n x[n] g_i^* \left[n - 2^i k \right] \qquad (4.8)$$

and

$$b_{I,k} = \sum_n x[n] h_I^* \left[n - 2^I k \right] \qquad (4.9)$$

where the $g_i[n - 2^i k]$'s are the discrete wavelets and the $h_I[n - 2^I k]$ are the scaling sequences.

4.2.3 Spline Wavelet Transform

The attractiveness of the Gabor representation of a signal comes from its optimal time-frequency localization.[5] However, the use of fixed window size, redundancy, and nonorthogonality are the major limitations of the Gabor analysis. The use of B-spline wavelets is shown to have near optimal time-frequency localization by Unser et al.[25,26] Although they are not orthogonal as the Battle/Lemarie polynomial spline wavelets used by Mallat[10] which are exponentially decaying, they are semiorthogonal and have a compact support.

The B-Spline Wavelet Transform is used to construct a sequence of embedded polynomial spline function spaces $\{S_{(i)}^n \; i \in Z\}$ of order n such that $S_{(i)}^n \supset S_{(i+1)}^n$ for $i \in Z$ where Z is the set of integers.[26] $S_{(i)}^n$ is the subset of functions in $L_2(R)$ that are of class C^{n-1}, i.e., continuous functions with continuous derivatives up to order $(n-1)$ and are equal to a polynomial of degree n in intervals $[k2^i,(k+1)2^i]$ with $k \in Z$. Hence

$$S_i^n = \phi_i^n(x) = \sum_{k=-\infty}^{\infty} c_{(i)}(k) \beta_{2^i}^n \left(x - 2^i k \right) \qquad (4.10)$$

where

$$\beta_{2^i}^n(x) = \beta^n\left(x/2^i\right)$$ (4.11)

The basis function $b^n(x)$ is the B-spline of order n. For the spline basis functions of order n the wavelet sequence q^n is

$$q^n(k-1) = (-1)^k b^{2n+1}(k)^* p^n(k)$$ (4.12)

where $b^{2n+1}(k) = \beta^{2n+1}(k)$, and $p^n(k)$, the scaling sequence, is the binomial kernel

$$p^n(k) = 1/2^n \binom{n+k}{k}, \quad k = 0, \ldots, n+$$ (4.13)

When the wavelet and the scaling sequences are determined, the wavelet function $\underline{b}_2^n(x)$ can be constructed by a scaling function $\beta^n(x)$ by solving a two-scale equation (a dilation equation)

$$\underline{\beta}_2^n(x) = \underline{\beta}^n(x/2) = \sum_{k=-\infty}^{+\infty} q^n(k)\beta^n(x-k)$$ (4.14)

Given a function $\phi^n(x)$, we can obtain the B-spline representation at the finest resolution level that is defined as level (0) using

$$\phi_{(0)}(x) = \sum_{k=-\infty}^{\infty} c_{(0)}(k)\beta^n(x-k)$$ (4.15)

The essence of the wavelet transform is to decompose the above expression using basis functions that are expanded by a factor of two

$$\phi_n^{(0)}(x) = \sum_{k=-\infty}^{+\infty} d_{(1)}(k)\underline{\beta}_2^n(x-2k) + \sum_{k=-\infty}^{+\infty} c_{(1)}(k)\beta_2^n(x-2k)$$ (4.16)

The B-spline wavelet is a polynomial compact support with the property that $\underline{\beta}_2^n \perp \beta_2^n(x-2k)\ k \in Z$.[25] This means that the first term of the right-hand side of Equation 4.16 is the projection of $\phi_{(0)}^n$ on $S_{(1)}^n$ and the second term represents the residual error.

The decomposition can be implemented iteratively up to a level I, which yields the wavelet representation

$$\phi_n^{(0)}(x) = \sum_{i=1}^{I} \sum_{k=-\infty}^{+\infty} d_{(i)}(k)\beta_{2^i}^n\left(x - 2^i k\right) + \sum_{k=-\infty}^{+\infty} c_I(k)\beta_{2^I}^n\left(x - 2^I k\right)$$ (4.17)

where

$$\underline{\beta}_{2^i}^n(x) = \underline{\beta}_{2^i}^n\left(x/2^{i-1}\right)$$ (4.18)

The coefficients $\{d_{(1)},\dots,d_{(l)}\}$ are the so-called wavelet coefficients ordered from fine to coarse while the sequence $\{c_{(l)}\}$ characterizes the lower resolution signal at level (l).

4.2.4 THE DISCRETE B-SPLINE WAVELET TRANSFORM

It is possible to take a wavelet function which may be regarded as a band-pass filter and represent it as a combination of a low-pass and a high-pass filter. In this case, the wavelet analysis becomes a multiresolution analysis. The low-pass and the high-pass filters for n^{th} order spline wavelet multiresolution decomposition may be computed as

$$h(k) = 1/2\left[b^{2n+1}\right]^{-1} \uparrow_2 *b^{2n+1} * p^n(k)$$ (4.19a)

$$g(k+1) = 1/2\left[b^{2n+1}\right]^{-1} \uparrow_2 *(-1)^k p^n(k)$$ (4.19b)

where $\uparrow_2$ indicates up-sampling by 2.

4.2.5 DESIGN OF QUADRATIC SPLINE WAVELETS

For the analysis of the waveforms used in this study, the quadratic spline wavelets ($n = 2$) are designed particularly for their antisymmetric property which conforms to the morphological character of the signal. The quadratic spline and wavelet functions are shown in Figures 4.3 and 4.4. The low-pass $h(n)$ and the high-pass $g(n)$ filter kernels for the quadratic spline wavelet are

$$h(k) = \frac{1}{2}\left[\left(b^5\right)^{-1}\right](k) \uparrow_2 *b^5(k) * p^2(k)$$ (4.20)

$$g(k+1) = \frac{1}{2}\left[b^5(k)\right]^{-1} \uparrow_2 *(-1)^k p^2(k)$$ (4.21)

where

$$\left[\left(b^5\right)^{-1}\right](k) = Z^{-1}\left\{120/\left(z^2 + 26z + 66 + 26z^{-1} + z^{-2}\right)\right\}$$ (4.22)

and

$$p^2(k) = \frac{1}{2^2}Z^{-1}\left[1 + 3z^{-1} + 3z^{-2} + z^{-3}\right]$$ (4.23)

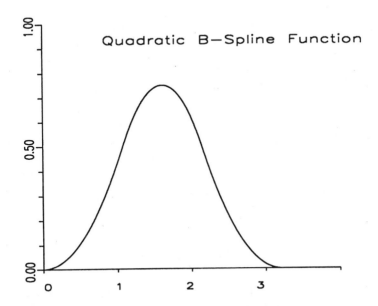

FIGURE 4.3   Quadratic B-spline functions.

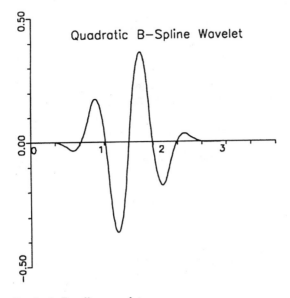

FIGURE 4.4   Quadratic B-spline wavelet.

It may be shown that $[b^5(k)]^{-1}$ can be expressed in factorized form by

$$\left[\left(b^5\right)^{-1}\right](k) = Z^{-1}\left[\frac{\alpha_1\alpha_2}{\left(1-\alpha_1 z^{-1}\right)\left(1-\alpha_1 z\right)\left(1-\alpha_2 z^{-1}\right)\left(1-\alpha_2 z\right)}\right] \quad (4.24)$$

where $\alpha_1 = -0.04309$ and $\alpha_2 = -0.43057$.

Now $[b^5(k)]^{-1}$ can be determined as

$$\left[\left(b^5\right)^{-1}\right](k) = \alpha_1\alpha_2 /\left(1-\alpha_1^2\right)\left(1-\alpha_2^2\right)\left(1-\alpha_1\alpha_2\right)\left(\alpha_1 - \alpha_2\right).\left(\alpha_1\left(1-\alpha_2^2\right)\alpha_1^{|k|} - \alpha_2\left(1-\alpha_1^2\right)\alpha_2^{|k|}\right)$$

$$(4.25)$$

The coefficient values for these filter kernels h(n) and g(n) are given in Table 4.1 for the quadratic spline wavelets.

TABLE 4.1

Coefficients of the Truncated Decomposition Filters h, g (IIR) and Reconstruction Filters p^2, q^2 (FIR) for Quadratic Spline Filters.

k	h(k)	g(k)	$p^2(k)$	$q^2(k)$
-10	+0.00157	-0.00388		
-9	+0.01909	-0.03416		
-8	-0.00503	+0.00901		
-7	-0.04440	+0.07933		
-6	+0.01165	-0.02096		
-5	+0.10328	-0.18408		
-4	-0.02593	+0.04977		+ 1/480
-3	-0.24373	+0.42390		-29/480
-2	+0.03398	-0.14034	+1/4	+147/480
-1	+0.65523	-0.90044	+3/4	-303/480
0	+0.65523	+0.90044	+3/4	+303/480
1	+0.03398	+0.14034	+1/4	-147/480
2	-0.24373	-0.42390		+29/480
3	-0.02593	-0.04977		- 1/480
4	+0.10328	+0.18408		
5	+0.01165	+0.02096		
6	-0.04440	-0.07933		
7	-0.00503	-0.00901		
8	+0.01909	+0.03416		
9	+0.00157	+0.00388		

4.2.6 THE FAST ALGORITHM

The initial step is to find the B-Spline coefficients $c(k)$ at the resolution level 0. This is efficiently implemented using

$$c^+(k)f(k) + b_1 c^+(k-1), (k = 2, \ldots, K) \tag{4.26a}$$

$$c^-(k)f(k) + b_1 c^-(k+1), (k = K-1, \ldots, 1) \tag{4.26b}$$

$$c(k) = b_0\big(c^+(k) - f(k)\big) \tag{4.26c}$$

where $b_0 = -8a/(1-a^2)$ $b_1 = a = \sqrt{8} - 3$ with $c^+(1) = c^-(K) = 0$. The wavelet coefficients are then computed iteratively for I octaves by filtering and decimating by a factor of 2

$$c_{i+1}(k) = [h * c_i] \downarrow_2 (k) \tag{4.27a}$$

$$d_{(i+1)}(k) = [g * c_i] \downarrow_2 (k), \quad i = 0, 1, 2, \ldots, I-1 \tag{4.27b}$$

where $\downarrow_2$ indicates down-sampling by 2 and where h and g are the low-pass and the high-pass filter kernels for decomposition respectively and computed as

$$h(k) = 1/2 [b^{2n+1}]^{-1} \uparrow_2 * b^{2n+1} * p^n(k) \tag{4.28a}$$

$$g(k) = 1/2 [b^{2n+1}]^{-1} \uparrow_2 * (-1)^k p^n(k) \tag{4.28b}$$

where $\uparrow_2$ indicates up-sampling by 2. The basic computational block diagram for the I octave wavelet decomposition is given in Figure 4.5. The input to the block diagram is the signal at resolution level (0). The $H(z)$ and $G(z)$ denote the z-transform of the low-pass and high-pass filter kernels h and g, respectively. The discrete wavelet transform by the spline wavelets is performed by these two filters, as in the form of multiresolution decomposition. In each resolution level, the output of the high-pass filtering yields the wavelet coefficients for that resolution (i), and the output of the low-pass filtering yields the coarser input to the next resolution level ($i+1$). For the purpose of scaling, the discrete sequences decimation by two is performed after each time filtering is applied. Various fast algorithms for the implementation of the recursive filtering and decimation schemes for wavelet decomposition are proposed in Reference 17.

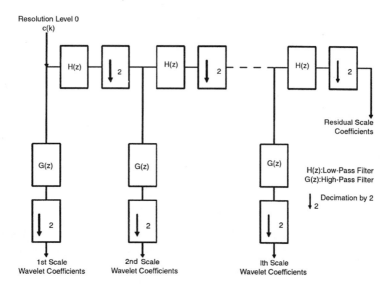

FIGURE 4.5   The basic computational block diagram of one octave wavelet decomposition algorithm.

The wavelet transform is a sequential band-pass filtering operation using filters with logarithmically ordered band-pass characteristics in frequency domain. In the time domain, this operation corresponds to projecting a signal onto a subspace spanned by the dilated and translated versions of prototype functions called wavelet and scaling functions. Dilation yields logarithmically ordered filters, and translation yields the information about where in time any frequency band activity in the signal occurs.

In the multiresolution scheme, the number of wavelet coefficients halves from one scale to the next, which implies that longer time windows for lower frequencies and shorter time windows for higher frequencies are employed in the analysis. The wavelet coefficients can be used to investigate the amplitude and phase characteristics of oscillations in various frequency bands forming the waveform or the image under consideration.

For the exploration of different oscillatory components in a waveform using multiresolution analysis, the waveforms are decomposed into six or more scales by the Quadratic Spline Wavelets. The decomposition filters are designed as described in Section 4.2 for the corresponding multiresolution approximation. As an example, if the sampling rate is 1KHz, each scale covers the frequency band:

first Scale	250–500	Hz
second Scale	125–250	Hz
third Scale	62.5–125	Hz
fourth Scale	31.3–62.5	Hz
fifth Scale	15.6–31.3	Hz
sixth Scale	7.8–15.6	Hz
Residual Scale	0.0–7.8	Hz.

4.3   INVARIANT MOMENTS

Moments have long been used in statistical theory and classical mechanics.[28] Statisticians view moments as means, variances, skewness, and kurtosis of distributions, while classical mechanics students use moments to find centers of mass and moments of inertia.[24] In imaging, moments have been used as feature vectors for classification[11] as well as for image texture attributes and shape descriptors of objects.[16,20] In the early 1960s, Hu developed seven invariant moments from algebraic moment theory.[6] These seven moments are invariant under translation, rotation, and scaling. Perhaps the most important contribution of this work was the application of these seven invariant moments to the two-dimensional pattern recognition problem of character recognition. This was a crucial development, since many key problems in imaging and image recognition focus on recognizing an image even though it has been translated or rotated, or perhaps magnified by some means. Since that time other pattern recognition applications have included hand-printed characters, aircraft identification, chest X-rays and ship recognition, and more recently face recognition.[11]

Similar to the definition of moments in classical mechanics, the two-dimensional moments of order $(p + q)$ with an image intensity distribution of $f(x,y)$ are defined as

$$m_{pq} = \int\int x^p y^q f(x,y)\, dx\, dy \qquad (4.29)$$

where $p,q = 0,1,2,\ldots$. These moments in general are not invariant to any distortions,[28] and, therefore, the central moments are defined as

$$\mu_{pq} = \int\int (x-x')^p (y-y')^q f(x,y)\, d(x-x')\, d(y-y') \qquad (4.30)$$

where

$$x' = \frac{m_{10}}{m_{00}} \text{ and } y' = \frac{m_{01}}{m_{00}}.$$

The central moments are known to be invariant under translation, and by working through Equation 4.30, it can be shown that the first four orders of central moments can be expressed in terms of the ordinary moments defined in Equation 4.29 as

$$\mu_{00} = m_{00}$$
$$\mu_{10} = \mu_{01} = 0$$
$$\mu_{11} = m_{11} - \frac{m_{10}m_{01}}{m_{00}}$$
$$\mu_{20} = m_{20} - \frac{m_{10}^2}{m_{00}}$$
$$\mu_{02} = m_{02} - \frac{m_{01}^2}{m_{00}}$$ (4.31)
$$\mu_{12} = m_{12} - m_{02}x' - 2m_{11}y' + 2m_{10}y'^2$$
$$\mu_{21} = m_{21} - m_{20}y' - 2m_{11}x' + 2m_{01}x'^2$$
$$\mu_{03} = m_{03} - 3m_{02}y' + 2m_{01}y'^2$$
$$\mu_{30} = m_{30} - 3m_{20}x' + 2m_{10}x'^2$$

Often it is desirable to normalize the moments with respect to size. This may be accomplished by using the area, μ_{00}. The normalized central moments are

$$\eta_{pq} = \frac{\mu_{pq}}{\mu_{00}^r}$$ (4.32)

where $r = 1 + \frac{(p+q)}{2}$ for $(p + q) = 2, 3, \dots$

With these normalized moments the seven Hu invariants are found by:

$$\varphi_1 = \eta_{20} + \eta_{02}$$

$$\varphi_2 = (\eta_{20} - \eta_{02})^2 + 4\eta_{11}^2$$

$$\varphi_3 = (\eta_{30} - 3\eta_{12})^2 + (3\eta_{21} - \eta_{03})^2$$

$$\varphi_4 = (\eta_{30} + \eta_{12})^2 + (\eta_{21} + \eta_{03})^2$$

$$\varphi_5 = (\eta_{30} - 3\eta_{12})(\eta_{30} + \eta_{12})\left[(\eta_{30} + \eta_{12})^2 - 3(\eta_{21} + \eta_{03})^2\right]$$

$$+ (3\eta_{21} - \eta_{03})(\eta_{21} + \eta_{03})\left[3(\eta_{30} + \eta_{12})^2 - (\eta_{21} + \eta_{03})^2\right] \qquad (4.33)$$

$$\varphi_6 = (\eta_{20} - \eta_{02})\left[(\eta_{30} + \eta_{12})^2 - (\eta_{21} + \eta_{03})^2\right]$$

$$+ 4\eta_{11}(\eta_{30} + \eta_{12})(\eta_{21} + \eta_{03})$$

$$\varphi_7 = (3\eta_{21} - \eta_{03})(\eta_{30} + \eta_{12})\left[(\eta_{30} + \eta_{12})^2 - 3(\eta_{21} + \eta_{03})^2\right]$$

$$- (\eta_{30} - 3\eta_{12})(\eta_{21} + \eta_{03})\left[3(\eta_{30} + \eta_{12})^2 - (\eta_{21} + \eta_{03})^2\right]$$

While these seven invariants are given in terms of the normalized moments, they may be calculated from the central moments as well. In that case the formulas are the same with μ substituted for η in Equation 4.33.

One item to note is that these normalized moments assume that the image is represented by pixels, the values of which are all > 0. There is no problem in calculating the original and central moments even if some of the pixels are < 0. However, the normalized moments pose a different problem. If a substantial number of pixels have values < 0, then μ_{00} becomes negative. This causes a problem during the normalization, since μ_{00} raised to a nonintegral power becomes an imaginary number. Because this system calculates the moments of the wavelet coefficients there are situations where a substantial number of coefficients are negative. Because of this, when the normalized moments are calculated μ_{00} is treated somewhat like a vector. It is considered to have magnitude $|\mu_{00}|$ and a direction which is (+) if $\mu_{00} > 0$ and (−) if $\mu_{00} < 0$. Therefore, the normalization of μ_{pq} is then:

$$\eta_{pq} = \frac{\mu_{pq}}{\mu_{00}^r} \text{ if } \mu_{00} > 0 \qquad \eta_{pq} = -\frac{\mu_{pq}}{|\mu_{00}|^r} \text{ if } \mu_{00} < 0 \qquad (4.34)$$

Therefore, this technique actually calculates pseudo-moments to use as features.

4.4   ENTROPY

Entropy is a quantity that is widely used in information theory and is based on probability theory.[4,18] Consider first an event E that can occur when an experiment is performed. How surprised would one be to see that E actually does occur? The answer to that question depends upon the probability of E. Suppose, for instance, cards are being drawn with replacement one at a time from a full deck of playing cards. If E were defined as "the card being a heart," it would not be too surprising if E occurred as P(E) = 0.25. (There are 52 cards in a full deck broken into four equal suits, hearts, spades, diamonds, and clubs.) However, if E were defined to be "the card being the ace of hearts," then we might be rather surprised to see that occuring, as now P(E) = 0.0192 or $\frac{1}{52}$, but it is possible to quantify this concept of surprise.[18]

If this concept is extended to a random variable $\mathbf{X}$ which can be one of the values $x_1, x_2, ..., x_n$ with probability $p_1, p_2, ..., p_n$, then the expected amount of surprise upon learning the value of $\mathbf{X}$ is

$$H(\mathbf{X}) = -\sum_{i=1}^{n} p_i \log_2 p_i \qquad (4.35)$$

This quantity is the entropy of the random variable $\mathbf{X}$. Note that if $p_i = 0$, then $0 \log_2 0$ is defined to be 0.

Thus $H(\mathbf{X})$ represents the average amount of surprise associated with the value of $\mathbf{X}$. It also can be interpreted as representing the amount of uncertainty that exists in the value of $\mathbf{X}$. In information theory, $H(\mathbf{X})$ is considered to be the average amount of information received when the value of $\mathbf{X}$ is observed.[4]

It is from the information theory point of view that the entropy would be a valid data point for images. It is conjectured that perhaps normal outcomes would contain more information than abnormal, or vice-versa, and therefore, the entropy values would be of use for classification purposes. Interestingly enough, when viewed by a human observer, the distribution of entropy values of abnormal mammograms, for example, seemed no different than the distribution of the normal mammogram entropy values. However, when used as input to a neural network, they indeed added some discrimination.

4.5   CEPSTRUM ANALYSIS

Naturally, a recorded signal, such as EEG, is mixed with noise. And the relationship between the signal and noise is simply considered as an addition, i.e.,

$$x_i = s_i + n_i \qquad (4.36)$$

where x_i is the recorded signal, s_i the pure signal, n_i the noise and i the time index.

Additionally, Fourier spectrum analysis and signal filtering can be applied directly. However, if the relationship between the signal and noise is convolution, which often occurs, i.e.,

$$x_i = s_i * n_i \tag{4.37}$$

where $*$ means convolution, the system is not linear, and the Fourier analysis and filtering cannot be used directly and deconvolution is needed. Basically the relationship between the signal and noise is a mixture of addition and convolution. After the discrete Fourier transform (DFT) is obtained

$$\hat{x}_i = \hat{s}_i \cdot \hat{n}_i \tag{4.38}$$

where $\hat{x}_i$ is the DFT of x_i, s_i, is the DFT of s_i, $\hat{n}_i$ the DFT of n_i and $\cdot$ means multiplication. Then after the Log is applied, the signals become additive:

$$\tilde{x}_i = \tilde{s}_i + \tilde{n}_i \tag{4.39}$$

where $\tilde{x}_i$ is the Log of $\hat{x}_i$, $\tilde{s}$, the Log of $\hat{s}_i$, and $\tilde{n}_i$ the Log of $\hat{n}_i$.

To return to the time domain, an inverse DFT (linear transform, which maintains the addition relationship) is performed, and the output is the cepstrum c_i. Just the low part of the cepstrum, where the signal is assumed to be concentrated, is selected by cepstrum filtering through the cepstrum window. The output cepstrum will be the input to the ANN. If the real Log of the absolute values of the DFT spectrum is used, the output is the real cepstrum. If the complex Log of the complex values of the DFT spectrum is used, the output is the complex cepstrum.

4.6 FRACTAL DIMENSION

Additional features extracted from images and signals focus on texture features. The method used here was based on fractals. While fractal geometry has been around for over a century, it is thanks to Mandelbrot, who coined the term 'fractal' and popularized this class of mathematical functions.[14] Examples of fractals best explain what a fractal really is. Purely mathematical fractals include the Mandelbrot set and the von Koch snowflake. Fractals occurring in nature include clouds, trees, the coast of England, mountains, blood vasculature, cauliflower, and much more. What these all have in common is some degree of self-similarity; naturally occurring fractals are *statistically* self-similar. In other words, a whole cauliflower looks like half a cauliflower looks like one fourth a cauliflower, and so on. This example using cauliflower can be found in Peitgen et al.[12] or try it for yourself!

The measure used to describe the 'fractal-ness' of an object or image is the fractal dimension. The fractal dimension is the measure of self-similarity of an image; the basic idea is that an image with fractal properties will look the same at all scales. Many have used the fractal dimension to analyze and segment textures.

In a sense, the fractal dimension measures the roughness of an image. For example, computer generated mountains with low fractal dimension are smooth and rolling while those generated with high fractal dimension are rough and jagged.[11,14]

Fractals have been used in general purpose texture analysis, synthesis, and segmentation, particularly of natural scenes. Pentland[11] segmented images containing aerial views, mountains, and a desert scene using the fractal dimension. Keller et al.[7] used two characteristics related to the fractal dimension to distinguish silhouettes of trees from silhouettes of mountains. Keller et al.[2] and Dubuisson and Dubes[2] discuss the use of lacunarity (another fractal-based feature) in conjunction with fractal dimension to improve the segmentation of natural textures. They argue that the fractal dimension alone cannot discriminate all natural textures as well as a human observer but that other fractal-based features such as lacunarity can improve this segmentation.

In general, the dimension of a set can be found by the equation

$$D = \log(N) / \log(1 / r) \tag{4.40}$$

where D is the dimension, N is the number of parts comprising the set, each scaled down by a ratio r from the whole.[12] For a two-dimensional square, N parts scaled by the ratio $r = 1/N^{1/2}$ results in $Nr^2 = 1$ or $D = 2$. A set is considered fractal if D is a noninteger value. For example, von Koch's snowflake has four parts, each one-third the length of its parent, so $D = \log(4) / \log(3) = 1.26$.

There are many ways to calculate the fractal dimension of naturally occurring images. Most common in the literature are methods that estimate the fractal dimension based on statistical differences in pixel intensity. Voss[27] and Keller et al.[7] used a popular box-counting method similar to Sarkar and Chaudhuri[19] described in more detail next. Peleg et al.[13] used a multiresolution method to measure the fractal dimension by observing the change in surface area of a covering blanket over the topological map of an image at different scales. Super and Bovik[23] used the outputs from multiple Gabor filters and fit them to a fractal power-law curve to obtain the fractal dimension. This has the added capability of spatial localization. Pentland[14] gathered second-order statistics at varying distances, resulting in a Gaussian-shaped distribution. A fractal dimension estimation was gathered by fitting the standard deviations over scale.

The method chosen here is that proposed by Sarkar and Chaudhuri.[19] Starting with the basic equation $D = \log(N) / \log(1/r)$, an image sized $M \times M$, and G number of gray levels, the algorithm is as follows:

1. Divide up the image into size $s \times s$ where $M/2 > s > 1$ such that $r = s / M$.

2. Imagine the two-dimensional image is a topological map in three dimensions. On each grid sized $s \times s$ can be built a column of boxes sized $s \times s \times s'$ where $\lfloor G/s' \rfloor = \lfloor M/s \rfloor$ with indices starting with 1 for the bottom box.

3. Find the lowest and highest boxes intersected by the image in the current column of boxes and name them k and l, respectively.

4. Add up the differences $(l - k + 1)$ for all areas $s \times s$ for the current scale r and call it $N(r)$.

5. Do this for all scales, and the result will be a vector $N(r)$ where $1/r = 2$, 4, 8, ... $M/2$.

6. Plot $\log(N[r])$ vs. $\log(1/r)$ and calculate the slope using a least square linear fit. This is the fractal dimension.

FIGURE 4.6   Test images used to test fractal dimension algorithm, FD = fractal dimension.

Sarkar and Chaudhuri[19] used random images with increasing standard deviation to test their algorithm. These results were recreated by the authors by first generating 8-bit test images of Gaussian noise with a mean of 128 and standard deviation ranging from 8 to 128 in increments of 8. Some examples are shown in Figure 4.6. The size of all test images was 128×128 pixels. The results are shown in Figure 4.7. As demonstrated, the fractal dimension increases with increasing noise standard deviation similar to the results Sarkar and Chaudhuri show in their paper.[19]

Although it probably would have been more appropriate to generate actual fractal images with known dimension, such as fractal Brownian images, the calculation of the fractal dimension is only an estimation and can be thought of as a measure of the roughness of an image. Furthermore, when using the fractal dimension in texture analysis applications, the more salient property is that it increases monotonically with increasing true fractal dimension or roughness. Thus, by using these random

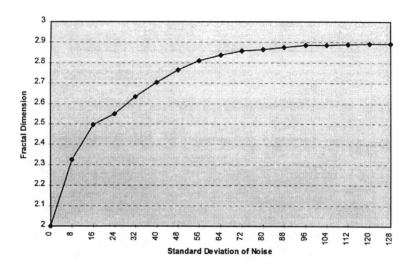

FIGURE 4.7   Fractal dimension as a function of standard deviation of test images.

images, the authors demonstrated an increase in fractal dimension with increased noise or roughness; compare their algorithm with that of others, namely Keller et al.,[7] Peleg et al.,[13] and Pentland.[14] In comparison, Pentland[14] and Peleg et al.[13] are accurate and cover the full dynamic range of fractal dimensions but are computationally expensive. On the other hand, the methods of Gagnepain and Roques-Carmes (1986) and Keller et al.[7] are computationally efficient but do not cover the full dynamic range. The method of Sarkar and Chaudhuri demonstrated both qualities and was therefore chosen for implementation.

4.7   SGLD TEXTURE FEATURES

Described in detail in Haralick et al.,[3] the gray-tone spatial-dependence matrix, as they call it, reveals the second order statistics of an image. Often applied to the segmentation and/or classification of textures, the SGLD matrix in Haralick et al.[3] was used to classify aerial images. The matrix itself contains the statistics of pairs of pixel intensities located a distance d away from each other at an angle Θ. Note that d is not the Euclidean distance but represents the number of units away in pixels; a pixel's eight nearest neighbors are one unit away. Thus, element (i, j) in the SGLD matrix $\Phi_{d,\Theta}$ is the probability that pixel valued i is a distance d away from a pixel valued j at an angle Θ. The matrix $\Phi_{d,\Theta}$ for a two bit per pixel image (four gray levels) is explained in Table 4.2.

The SGLD matrix is best explained by an example from Haralick et al.[3] Take an image I with four possible gray levels {0, 1, 2, 3},

TABLE 4.2
Calculation of Matrix $\Phi_{d,\Theta}$ for A Two Bit Per Pixel Image

i \ j	0	1	2	3
0	# pairs (0,0) w/ (d, Θ) separation	# pairs (0,1) w/ (d, Θ) separation	# pairs (0,2) w/ (d, Θ) separation	# pairs (0,3) w/ (d, Θ) separation
1	# pairs (1,0) w/ (d, Θ) separation	# pairs (1,1) w/ (d, Θ) separation	# pairs (1,2) w/ (d, Θ) separation	# pairs (1,3) w/ (d, Θ) separation
2	# pairs (2,0) w/ (d, Θ) separation	# pairs (2,1) w/ (d, Θ) separation	# pairs (2,2) w/ (d, Θ) separation	# pairs (2,3) w/ (d, Θ) separation
3	# pairs (3,0) w/ (d, Θ) separation	# pairs (3,1) w/ (d, Θ) separation	# pairs (3,2) w/ (d, Θ) separation	# pairs (3,3) w/ (d, Θ) separation

$$I = \begin{pmatrix} 0 & 0 & 1 & 1 \\ 0 & 0 & 1 & 1 \\ 0 & 2 & 2 & 2 \\ 2 & 2 & 3 & 3 \end{pmatrix}$$

The four most typical SGLD matrices, Φ_H (H = Horizontal), Φ_{RD} (RD = Right Diagonal), Φ_{LD} (LD = Left Diagonal), and Φ_V (V = Vertical) are symmetric matrices with a default d = 1 and Θ = 0°/180°, 45°/225°, 135°/315°, and 90°/270°, respectively. They are calculated by first computing the SGLD for only one angle, then taking the transpose (representing the SGLD matrix of the angle added to 180°), and adding the transpose to the original matrix as demonstrated below for Θ = 0°/180°. This makes the SGLD matrix invariant to 180° rotations.

$$\Phi_{1,0°} = \begin{pmatrix} 2 & 0 & 0 & 0 \\ 2 & 2 & 0 & 0 \\ 1 & 0 & 3 & 0 \\ 0 & 0 & 1 & 1 \end{pmatrix}, \Phi_{1,180°} = \Phi^T_{1,0°} = \begin{pmatrix} 2 & 2 & 1 & 0 \\ 0 & 2 & 0 & 0 \\ 0 & 0 & 3 & 1 \\ 0 & 0 & 0 & 1 \end{pmatrix} \text{ and}$$

$$\Phi_H = \Phi_{1,0°} + \Phi_{1,180°} = \begin{pmatrix} 4 & 2 & 1 & 0 \\ 2 & 4 & 0 & 0 \\ 1 & 0 & 6 & 1 \\ 0 & 0 & 1 & 2 \end{pmatrix}$$

The other three matrices, Φ_{RD}, Φ_{LD}, and Φ_V are

$$\Phi_{RD} = \begin{pmatrix} 4 & 1 & 0 & 0 \\ 1 & 2 & 2 & 0 \\ 0 & 2 & 4 & 1 \\ 0 & 0 & 1 & 0 \end{pmatrix}, \Phi_{LD} = \begin{pmatrix} 6 & 0 & 2 & 0 \\ 0 & 4 & 2 & 0 \\ 2 & 2 & 2 & 2 \\ 0 & 0 & 2 & 0 \end{pmatrix}, \Phi_V = \begin{pmatrix} 2 & 1 & 3 & 0 \\ 1 & 2 & 1 & 0 \\ 3 & 1 & 0 & 2 \\ 0 & 0 & 2 & 0 \end{pmatrix}$$

The above example is an oversimplified case. Typical gray-scale images have 256 gray levels; therefore, the resulting SGLD matrix would be 256×256, much larger than 4×4.

In order to convert each matrix into a probability density function, each element of the above matrices must be normalized. This is done by dividing each element by the number of pixel pairs included in the calculation of the matrix. This value depends on d and Θ. This is best described in Euclidean terms. Let D be the Euclidean distance between pixel pairs. For example, if $d = 1$ and $\Theta = 45$, then $D = \sqrt{2}$. Then, let m be the horizontal projection of (D, Θ) where $m = D\cos(\Theta)$, and let n be the vertical projection of (D, Θ) where $n = D\sin(\Theta)$. Therefore, for an image sized N_x × N_y, each element must be divided by $2(N_x - m)(N_y - n)$. The factor of 2 is only necessary for symmetric matrices invariant to 180° rotations.

Of course, if it is desired to reflect the exact relationship of pixel pairs to each other or, in other words, to detect 180° rotations, the symmetric matrix would not be used. For example, if only the relationship $d = 1$, $\Theta = 0°$ is desired to be detected, only the matrix $\Phi_{1, 0°}$ would be used instead of Φ_H.

Statistical features are then computed from the above matrices (after normalization). The most common features used in the literature include the energy, entropy, correlation, local homogeneity, inertia, cluster shade, and cluster prominence. Often, these features are averaged over the four directions (or one averaged SGLD matrix would yield the same averaged features). Variance of the features from the four directions may also be an important feature, and in this case the four SGLD matrices will need to be preserved. The equations for these features are

$$energy = \sum_{i=0}^{G-1} \sum_{j=0}^{G-1} f(i,j)^2 \tag{4.41}$$

$$entropy = -\sum_{i=0}^{G-1} \sum_{j=0}^{G-1} f(i,j) \log[f(i,j)] \tag{4.42}$$

$$correlation = \sum_{i=0}^{G-1} \sum_{j=0}^{G-1} \frac{(i-\bar{x})(j-\bar{y})f(i,j)}{\sigma^2} \tag{4.43}$$

$$local\ homogeneity = \sum_{i=0}^{G-1} \sum_{j=0}^{G-1} \frac{1}{1+(i-j)^2} f(i,j) \tag{4.44}$$

$$inertia = \sum_{i=0}^{G-1} \sum_{j=0}^{G-1} (i-j)^2 f(i,j) \tag{4.45}$$

$$cluster\ shade = \sum_{i=0}^{G-1}\sum_{j=0}^{G-1}\left((i-\bar{x})+(j-\bar{y})\right)^3 f(i,j) \tag{4.46}$$

$$cluster\ prominence = \sum_{i=0}^{G-1}\sum_{j=0}^{G-1}\left((i-\bar{x})+(j-\bar{y})\right)^4 f(i,j) \tag{4.47}$$

where

$$\bar{x} = \sum_{i=0}^{G-1}\sum_{j=0}^{G-1} f(i,j) = \bar{y} = \sum_{i=0}^{G-1}\sum_{j=0}^{G-1} f(i,j) \tag{4.48}$$

for symmetric matrices,

and

$$\sigma^2 = \sum_{i=0}^{G-1}(i-\bar{x})^2 \sum_{j=0}^{G-1} f(i,j) = \sum_{i=0}^{G-1}(j-\bar{y})^2 \sum_{j=0}^{G-1} f(i,j) \tag{4.49}$$

for symmetric matrices. Other less common features mentioned in Haralick et al.[3] are the sum average, sum variance, sum entropy, difference variance, difference entropy, and information measures of correlation.

$$sum\ average = \overline{(x+y)} = \sum_{i=0}^{2(G-1)} i f_{x+y}(i) \tag{4.50}$$

$$sum\ variance = \sum_{i=0}^{2(G-1)}\left(i-\overline{(x+y)}\right)^2 f_{x+y}(i) \tag{4.51}$$

$$sum\ entropy = \sum_{i=0}^{2(G-1)} f_{x+y}(i)\log\left[f_{x+y}(i)\right] \tag{4.52}$$

$$difference\ variance = \sum_{i=0}^{G-1}\left(i-\overline{(x-y)}\right)^2 f_{x-y}(i) \tag{4.53}$$

where

$$\overline{(x-y)} = \sum_{i=0}^{G-1} i f_{x-y}(i) \tag{4.54}$$

$$difference\ entropy = -\sum_{i=0}^{G-1} f_{x-y}(i) \log\left[f_{x-y}(i)\right] \tag{4.55}$$

information measures of correlation:

$$info12 = \frac{HXY - XXY1}{\max\{HX, HY\}} \tag{4.56}$$

$$info13 = \left(1 - \exp\left[-2(HXY2 - HXY)\right]\right)^{1/2} \tag{4.57}$$

where

$$HXY = entropy,\ HX = entropy\ of\ f_x(i) = HY = entropy\ of\ f_y(j),$$

$$f_x(i) = \sum_{j=0}^{G-1} f(i,j) = f_y(j) = \sum_{i=0}^{G-1} f(i,j) \tag{4.58}$$

due to symmetry

$$and\ HXY1 = -\sum_{i=0}^{G-1}\sum_{j=0}^{G-1} f(i,j)\log\left[f_x(i)f_y(j)\right] =$$

$$HXY2 = -\sum_{i=0}^{G-1}\sum_{j=0}^{G-1} f_x(i)f_y(j)\log\left[f_x(i)f_y(j)\right] \tag{4.59}$$

also due to symmetry.

Features can be averaged over the four directions (horizontal, vertical, right diagonal, and left diagonal) to calculate an average value. The variance between the features of the four directions is also calculated. Both these average values and variances can be used as input features in the classification of tissue ROIs.

REFERENCES

1. Daubechies, I., Orthonormal bases of compactly supported wavelets, *Comm. in Pure and Appl. Math.*, 41(7), 1988.

2. Dubuisson, M-P. and Dubes, R. C., Efficacy of fractal features in segmenting images of natural textures, *Pattern Recognition Lett.*, 15, 419, 1994.
3. Haralick, R. M., Shanmugam, K., and Dinstein I., Textural features for image classification, *IEEE Trans. Syst., Man Cybern.*, SMC-3(6), 610, November 1973.
4. Held, G., *Data Compression*, John Wiley & Sons, New York, 1987.
5. Gabor, D., Theory of communication, *J. IEE (London)*, 93, 429, 1946.
6. Hu, M-K., Visual pattern recognition by moment invariants, *IRE Trans. Inf. Theor.*, IT-8, 179, February 1962.
7. Keller, J. M., Crownover, R. M., and Chen, R. Y., Characteristics of natural scenes related to the fractal dimension, *IEEE Trans. Pattern Anal. Mach. Intell.*, PAMI-9(5), 621, September 1987.
8. Laine, A., Schuler, S., Fan, J., and Huda, W., Mammographic feature enhancement by multiscale analysis, *IEEE Trans. Med. Imaging*, 13(4), 725, December 1994.
9. Laine, A., Fan, J., and Yang, W., Wavelets for contrast enhancement of digital mammography, *IEEE Eng. Med. Biol.*, 536, September/October 1995.
10. Mallat, S., A theory for multiresolution signal decomposition: the wavelet representation, *IEEE Trans. Pattern Anal. Mach. Intell.*, 11(7), 674, 1989.
11. Micheli-Tzanakou E., Uyeda E., Ray R., Sharma A., Ramanujan R., and Doug J., Comparison of neural network algorithms for face recognition, *Simulation*, 64(1), 15, July 1995.
12. Peitgen, H-O., Jurgens, H., and Saupe, D., *Chaos and Fractals — New Frontiers of Science*, Springer-Verlag, New York, 1992.
13. Peleg, S., Naor, J., Hartley, R., and Avnir, D., Multiple resolution texture analysis and classification, *IEEE Trans. Patt. Anal. Mach. Intell.*, PAMI-6(4), 518, July 1984.
14. Pentland, A. P., Fractal-based description of natural scenes, *IEEE Trans. Pattern Anal. Mach. Intell.*, PAMI-6(6), 661, November 1984.
15. Press, W. H., Flannery, B. P., Teukolsky, S. A., and Vetterling, W. T., *Numerical Recipes in C*, Cambridge University Press, Cambridge, 1995.
16. Reeves, A., Prokop, R., Andrews, S., and Kuhl, F., Three-dimensional shape analysis using moments and Fourier descriptors, *IEEE Trans. Pattern Anal. Mach. Intell.*, 10(6), 937, Nov 1988.
17. Rioul, O., Duhamel, P., Fast algorithms for discrete and continuous wavelet transforms, *IEEE Trans. Info. Theor.*, 38(2), 569, 1992.
18. Ross, S., *A First Course in Probability*, Macmillan Publishing Co. Inc., New York, 1976.
19. Sarkar, N. and Chaudhuri, B. B., An efficient approach to estimate fractal dimension of textural images, *Pattern Recognition.*, 25(9), 1035, 1992.
20. Shen, L., Rangayyan, R., and Desautels, J., Application of shape analysis to mammographic calcifications, *IEEE Trans. Med. Imaging*, 13(2), 263, June 1994.
21. Solka, J. L., Priebe, C. E., and Rogers, G. W., An initial assessment of discriminant surface complexity for power law features, *Simulation*, 58(5), 311, May 1992.
22. Spreckelsen, M. V. and Bromm, B., Estimation of single-evoked cerebral potentials by means of parametric modeling and Kalman filtering, *IEEE Trans. Biomed. Eng.*, 35(9), 691, 1988.
23. Super, B. J. and Bovik, A. C., Localized measurement of image fractal dimension using Gabor filters, *J. Visual Commun. Image Represent.*, 2(2), 114, June 1991.
24. Thomas, G. and Finney, R., *Calculus and Analytic Geometry*, Addison-Wesley, Reading, MA, 1988.
25. Unser, M., Aldroubi, A., and Eden, M., On the asymptotic convergence of B-spline wavelets to Gabor functions, *IEEE Trans. Inf. Theor.*, 38(2), 864, 1992.

26. Unser, M., Aldroubi, A., and Eden, M., Fast B-spline transforms for continuous image representation and interpolation, *IEEE Trans. Pattern Anal. Mach. Intell.*, 13(3), 277, 1991.

27. Voss, R. F., Random fractals: characterization and measurement, *Scaling Phenomena in Disordered Systems*, Pynn, R. and Skjeltorp, A., Eds., Plenum Press, New York, 1986, 1.

28. Yu, F. T. S. and Li, Y., Application of moment invariants to neural computing for pattern recognition, *Hybrid Image and Signal Processing II, Proc. SPIE*, 1297, 307, 1990.

29. Levine, M., *Vision in Man and Machine*, McGraw-Hill, New York, 1985.

Section II

Unsupervised Neural Networks

5 Fuzzy Neural Networks

Timothy J. Dasey and
Evangelia Micheli-Tzanakou

5.1 INTRODUCTION

This chapter is divided into four components. In the first section, the concepts and background relevant to pattern recognition, some typical optimization techniques and ALOPEX, and a tutorial on the ideas and early works in neural networks are dealt with. The danger in this presentation is that these fields might be construed as disjoint problems. The truth is that a large amount of overlap exists between these conceptual divisions. Pattern recognition has benefited from the application of neural networks and optimization. Neural networks commonly use optimization routines to guide their training and have achieved many of their greatest successes in pattern recognition applications. These relationships should be kept in mind during the reading. The last section of this chapter includes a philosophical discussion explaining the rationale for the work.

5.2 PATTERN RECOGNITION

5.2.1 THEORY AND APPLICATIONS

To most individuals, a pattern recognition task involves an ability of the brain to assign labels to objects, sounds, feelings or ideas and discriminate one from another. Most of us are extremely adept at this processing task, while being unaware of the precise mechanism that provides us with this power. In fact, it is through the scientific field of pattern recognition, which relegates this task to machines, that the methods of our brain can be fully realized. Yet, there has never been a machine designed that has our capability to be a general-purpose pattern recognition machine.

Regardless of the limitations, machines perform quite well in the grouping and labeling of patterns from certain problem sets. Machines excel when the recognition task is confined to a specific application. An extensive body of literature describes the recent attempts at relegating many pattern recognition tasks to machines as the explosive growth of information overworks the human classifiers.[20,48,67] The use of automated pattern recognition machines has now touched nearly every field in an enormous variety of workplaces.

The special nature of each pattern recognition task requires selection of the best approach.[30] Heuristic approaches, which rely on the designer's intuition and familiarity with the problem, are often sufficient to provide excellent solutions to many problems. Linguistic (syntactic) approaches are often useful when numerical measurements are not sufficient to describe the problem. Many pattern recognition problems can be solved through several mathematically substantiated techniques, using statistical variability between patterns or certain pattern similarity measures.[18]

When confronting a typical pattern recognition task, three particular problems must be addressed by the designer. The first is the representation of the input data that the system will use in its classifications. When determined, these comprise the pattern vector **x** as

$$x = \left(x_1, x_2, x_3, \ldots, x_n\right) \tag{5.1}$$

where n is the total number of parameters needed for analysis. In many mathematical pattern recognition problems, it is often convenient to envision each parameter x_i as describing an axis in n-dimensional space (n-space), where each pattern then comprised a point in that space, as in the two-parameter space depicted in Figure 5.1.

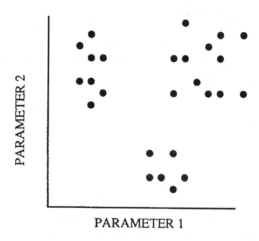

FIGURE 5.1   A two-parameter space. Each point in the space corresponds to an input pattern vector.

The second problem concerns the extraction of certain characteristic attributes from the pattern vectors and a reduction in the dimensionality (from n to m) of those vectors. This is usually termed the preprocessing and feature extraction problem. The attributes of features to be selected vary with the application but involve the selection of pattern attributes that can best be used for discrimination among the patterns. The feature extraction process can be thought of as an intermediate formulation of the more prominent goal of pattern recognition: the compression of large numbers of attributes to a small number of class determinants.[64]

The third problem involves the determination of optimum decision procedures, which are used for the identification and classification process. Many such procedures involve the separation of n-space (or m-space) into clusters, much as Figure 5.1 includes pattern points which are grouped into three similar categories.

Although mathematical formulations of pattern recognition methods have been available for several decades, many prominent problems still must be solved before great theoretical improvements can be made.[36,38] One issue, that of properly estimating the classification performance of a machine, has been largely agreed upon.[58] It is generally thought that two pattern populations are needed to ensure that a machine pattern recognizer can generalize. One group is reserved for the training of the machine (determination of the decisions involved in making a classification), and the other is used for post-training testing. This helps to ensure that the decisions formulated for the training group also apply to the similar but distinct non-training group.

5.2.2 FEATURE EXTRACTION

A feature of a given parameter set refers to an attribute described by one or more elements of the original pattern vector. In an application to imaging, the elements are each pixel, and a feature may be selected as a subset of the pixel intensity values. More commonly, a feature describes some combinations of the original pixels, as in a Fourier expansion or a spatial filtering operation. The precise meanings of a preprocessing operation and a feature extraction process overlap, but in general a feature extraction operation involves the reduction in dimensionality of the pattern vector. The primary reason for such a transformation is to provide a set of measurements with more discriminatory information and less redundancy to the classifier.

The precise choice of features is perhaps the most difficult task in pattern processing. In order to know the most successful set of features for a particular problem, the accuracy of the classifier must be known. Yet the classifier depends on the information from a feature extraction device, and thus cannot normally provide that information without a completely designed feature extractor. This enigma remains the primary reason for the difficulty in evaluating competing feature extraction methods. It has prompted many researchers to select features subjectively from an educated guess of what will be most important to the classifier. These techniques can be effective but are increasingly more difficult to ascertain as the complexity of the patterns increases, and they are always subject to personal bias.

Many common mathematical parameters are used as feature measurements.[39] A set of n-space Euclidean distance measurements is a very common example. In other situations where the identity of the patterns are known, transformation matrices to minimize intraset pattern entropy[65] or intraset pattern dispersion and functional approximation methods are commonly used. Several orthogonal expansions are also used, including the Fourier expansion and the Karhunen-Loève (K-L) expansion.[21] The K-L expansion offers certain optimal properties and will be reviewed in more detail in Section 5.3.1. Other common measurements, such as moment invariants,[55] are used because of their constancy under many common pattern transformations. The number of different schemes for feature extraction, even in similar applications (such as the processing of handwritten characters), is typically enormous.

5.2.3 CLUSTERING

"Clustering operation" may have different connotations for different people, even in the pattern recognition community. This chapter will use a terminology commonly accepted by many scientists. A clustering operation involves the grouping of like patterns with one another without any knowledge of pattern identity beforehand (an unsupervised classification operation). Classification problems generally have this information. In the clustering problem, patterns must be separated solely on their specific attributes, whereas the classification problems have access to error signals which can be generated to guide the decision making of the machine.

Continuing with the geometrical analogy outlined in Section 5.1.1, let us envision each pattern as a point in n-space, much as we see each star as a point in the sky. If we are asked to group the stars in the sky, what measurement do we use for this determination? The exact formulation of this answer often depends on the application.[50,53] In some instances, the distances between patterns can be used to separate patterns. In other cases, pattern density in regions of space are used to indicate locations where patterns likely are drawn from the same class, the techniques known as histogram approaches.[37]

It should be clear already that no clustering operation can ever be guaranteed to operate without error. The successful operation of the clustering method relies on the separability of the data from the attributes used as the pattern vectors. If two or more classes overlap in n-space, they will never be perfectly separated. All clustering problems rely on the fidelity of these input data and generally are based on the separation from highly dense regions of patterns from one another. Each of these "modes" of the distribution is assigned a particular class label at a later time.

A large class of problems relies on a hierarchical grouping of pattern data.[13] The procedures used usually have the disadvantage of a phenomenon called "chaining", where small errors in grouping at the extremes of the tree accentuate at later levels. Patterns in this scheme can be given a classification arbitrarily by choosing to "cut" at a particular level of the tree, but recent thinking is that there is significant information in leaving the class identity of patterns "fuzzy". Fuzzy clustering refers to assigning grades of membership to patterns and is currently a widely touted method.[16,23]

5.3 OPTIMIZATION

5.3.1 THEORY AND OBJECTIVES

In many situations it is desirable to find the values of a set of parameters that best define the solution to a particular problem. As a rule, it is always possible to perform an exhaustive search over the entire parameter space, choosing the parameter values that are closest to the desired operation of the system. In most cases, a measure can be formalized to assess the degree of fit of the proposed solutions to the ideal. This measure is usually termed a cost, energy, Hamiltonian, or objective function. Although an exhaustive search through all allowable combinations of system parameters is always theoretically possible, it is generally not feasible for even a moderately high number of system parameters. In fact, the number of possible choices (N) explodes exponentially with the number of parameters (q) involved in the space as

$$N = n_1 n_2 n_3 \ldots n_2 \tag{5.2}$$

where n_j is the number of samples of parameter j, or as

$$N = n^q \tag{5.3}$$

when $n_1 = n_2 \ldots = n_q$. As an example, if the system is composed of three parameters, and the search is conducted by sampling the parameters every 0.1 in the interval [0,1], then there are 11^3 or 1331 search items. With the same sampling and ten parameters, the search list includes 11^{10} or 2.6×10^{10} items.

It is obvious that this scheme is untenable for all but simple problems and is certainly impossible for the implementation of a dynamic system. It is for this reason that optimization procedures have received attention for a long period. The goal is to find the optimal (or at least close to optimal) solution with a shorter search time than the exhaustive search method. One of the conceptual means of achieving this uses a hyper-dimensional geometrical visualization of the cost function as it varies with each of the (presumably uncouples)* system parameters. This parameter space is usually widely variant, and the search over that space involves the extraction of a global minimum (or maximum, depending on the cost function used) from among all of the local minimum. In truth, the global extremum is rarely consistently attainable for realistic situations in finite time, but usually a very close approximation is both achievable and sufficient.

Two means for adjusting the exhaustive search technique readily come to mind. The first involves sampling the entire parameter space at a low resolution and finding the lowest (in the case of a minimization) region. Then that subregion can be sampled at a higher resolution ad infinitum. This procedure has occasionally been adapted but makes the major assumption that the global extremum is contained in a larger depression about it (and that the boundary of the global minimum is at least approximately funneling into that extremum). This is a gross oversimplification, and application of these methods can result in a solution far from the best choice. In many other schemes, referred to as gradient descent techniques, the effect of a parameter change on the cost function is calculated, and the parameter is adjusted so that it is moving downhill toward a better solution. This results in a rapidly converging iterative procedure, but the technique is fortuitous if the solution at which it arrives is a global, not a local, minimum. All good optimization routines work on the concept that a short-term deleterious move, moving uphill as well as downhill, is necessary to ensure the possible escape from local minima and arrive near or most preferably at the global extremum.

5.3.2   BACKGROUND

Much of the literature on optimization deals with the analogy between optimization and statistical mechanics. Perhaps the first to draw this comparison was the technique that has become known as the Metropolis algorithm.[43] This system was originally written as a means for investigating such macromolecular properties as the states of substances at the level of a set of N individual molecules. At any particular time, the potential energy of the system can be found as

$$E = \frac{1}{2} \sum_{i=1}^{N} \sum_{j=1}^{N} V\left(d_{ij}\right) \ \forall i \neq j \quad (5.4)$$

where V is the potential between molecules and d_{ij} the minimum distance between molecules i and j. The problem consists of optimizing the positions of the particles in space (in this case 2-D space) to arrive at the lowest potential energy of the system. Starting with random positions of the particles, each particle is moved a random amount in a random direction. The new energy of the configuration is checked. If the energy is decreased, the move is allowed. However, if the move results in a

$$P(\Delta E) = e^{\frac{-\Delta E}{kT}}, \quad (5.5)$$

higher potential energy, the move is allowed with a probability $P(\Delta E)$, which is the Boltzmann distribution. Notice that this is no longer a gradient descent technique, but rather there is always a finite probability that the system can move uphill, out of a local minimum. In this way, the equilibrium states of the set of molecules could be analyzed, and it was seen that the system settled in configurations which also conformed to a Boltzmann distribution. It turns out that this method is a simple modification of a Monte Carlo scheme, where instead of choosing configurations randomly and weighting those configurations with a Boltzmann factor, the configurations are chosen with a Boltzmann distribution (evident through the simulations) and weighted evenly.

The analogy between this statistical mechanics problem and optimization was explored even further with the introduction of the "simulated annealing" procedure.[34] If we examine the Boltzmann update from the Metropolis algorithm, it is clear that the higher the temperature, the more likely that an uphill move will be accepted. Conversely, at zero temperature all uphill moves will be denied, and the system will fall to an energy minimum. To ensure a ground state configuration (without crystal imperfections) in a material, the system must be carefully annealed, a process where the substance is first melted and then slowly cooled, with extra time spent near the vicinity of the phase transition. With the analogy to the optimization problem, a "ground state" (global minimum) of the system may be found by starting off at a high value of temperature. This corresponds to melting the system so that uphill moves are nearly equiprobable to downhill moves, and the system randomly wanders in parameter space. By slowly lowering the temperature (and thus reducing the probability of uphill moves), the system can slowly settle in to a minimum. It has been shown that the simulated annealing procedure can find the global minimum under certain conditions with probability 1.0, but that finding may take an inordinate amount of time. An analogous calculation to the specific heat of the system can be sued to signal phase transitions in the optimization. The simulated annealing procedure has been applied to a wide variety of pattern recognition tasks.[66] The primary emphasis of experimentation with the algorithm has been the adjustment of the cooling schedule, the process of lowering the temperature.[51] The simulated annealing procedure has received great attention over the last decade but is burdened by the application-dependent optimum cooling method and the necessity of a large number of iterations for convergence.

Another has been dubbed Mean Field Annealing.[49,54] In this scheme, the cost function $H(x)$, which may have many local minima and in other ways be "ugly", is replaced by another function $H(x,m)$ which "resembles" $H(x)$ but has components that are much easier to minimize (they could be convex functions with only one minimum). In order to make the two functions resemble each other, the set of parameters m_i must be estimated. To perform this, another technique is borrowed from statistical mechanics, the mean field approximation. The details are too intensive to consider in this synopsis, but by estimating each of the parameters m_i, the problem is reduced to a series of gradient descents at each value of temperature (note that this theory also utilizes a Boltzmann probability distribution).

Several researchers have noted that each of the above methods involves a local search about the current point in parameter space. That is, even with the capability to move uphill, all operations are still local. The odds of crossing a wide gap to a region of "better" minima is low, and thus more global search methods have been proposed. One commonly used technique is to run multiple trials on a given data set, saving the best result. Given a high number of random starts, the hope is that the global optimum will be among the optima identified. Galar[22] proposed a similar optimization routine to that of Eigen's theory of macromolecular evolution[19] which was more capable of crossing wide gaps. This method has many striking similarities, at least in concept if not in the method of application, to the ALOPEX process discussed in detail in the next section. Galar describes a two-term parameter update, one of which is a modified Markov chain and the other a random walk component. He claims that the resulting "biased random walk" is more capable of crossing wide gaps between local extrema than procedures like simulated annealing.

Another recent approach has been named the dynamic tunneling algorithm.[40] This routine uses gradient descent to go to a local minimum. At this time, the system "tunnels" through the surrounding hill (using an appropriately defined tunneling function) for the purpose of finding a point, other than the last minimum, which, when gradient descent is continued, will arrive at a point lower than the last minimum. The calculations are quite intensive, but the algorithm converges relatively often to the global minimum and may be more effective for problems with a high density of local minima.

5.3.3 Modified ALOPEX Algorithm

The optimization routine ALOPEX (ALgorithms Of Pattern EXtraction) presents an alternative to the previously reviewed algorithms. It was originally applied to the measurement of the visual receptive fields of cells in the adult frog tectum.[25,61] In the original application of the method, the cost function was referred to as the response function R.

The method normally updates the model parameters (the pixel intensity values in the original application) as

$$P_i(n+1) = P_i(n) + B_i(n) + r_i(n+1) \tag{5.6}$$

where $B_i(n)$ represents the influence of a term due to historical bias, and $r_i(n)$ is a random noise component. The bias term is calculated as

$$B_i(n) = B_i(n-1) + \gamma \Delta P_i(n) \Delta R(n) \tag{5.7}$$

where $DP_i(n)$ represents the previous change in the ith parameter value $P_i(n)$ as

$$\Delta P_i(n) = P_i(n) - P_i(n-1) \tag{5.8}$$

and $DR(n)$ indicates the similar change in the response function as

$$\Delta R(n) = R(n) - R(n-1) \tag{5.9}$$

The two terms in the modification of Equation 5.3 provide different influences on the optimization. The first term is a bias term which tends to move the parameter in the direction that has been successful in the past. It is actually an aggregation of the biases to that point in the simulation, where the direction of the latest addition to the bias is determined by the change in the response function due to the last move. The second term is a random number, generated for each parameter at each iteration, which provides the opportunity for the parameter to move against the direction of recent success. As mentioned earlier, this capability to move "uphill" is what provides a good optimizer with the ability to escape local extrema.* The term $r_i(n)$ in the ALOPEX update equation is typically implemented as a Gaussian random number with zero mean and standard deviation σ.

The accumulation of the biased terms in Equation 5.7 must be controlled in order to prove helpful. Without this regulation, the magnitude of B_i due to past iterations may overpower the relatively smaller change from the current iteration. In this scenario, the system has effectively gained "mass", so that the "momentum" of the movement in one direction will not allow the system to stop quickly enough at the sites of the extrema. In all simulations in this chapter, the magnitude of B_i is constrained to the limits $[-a,a]$. The first two iterations of the simulation supply random numbers for the forthcoming update statements. The responses are found for each, and the update Equations 5.6 through 5.9 are applied to all of the parameters. This process repeats itself until the simulation is finished. Note that

---

* The form of Equation 5.6 is correct for the maximization of the response function R. For minimization, the sign of the bias term should be changed.

the ALOPEX process provides for the simultaneous update of all parameters at once, which the simulated annealing algorithm does not. This generally makes the ALOPEX process more time conservative. In addition, the magnitude of the random component does not depend on the amount by which that component raises of lowers the response (there is a dependence via the Boltzmann distribution in simulated annealing). This makes it easier for parameters to traverse wide gaps between the extrema.

Even with the differences previously indicated, there are some analogies between the parameters of simulated annealing and those of ALOPEX. If the magnitude of the random component is much higher than that of the biasing component, then the parameters will be overwhelmingly driven by randomness, a situation analogous to the "melting" process in simulated annealing. Conversely, with no noise, the ALOPEX process simplifies to a gradient descent.* This indicates that the choices of γ and σ are critical for controlling the speed and accuracy of the convergence.

The suspicion that the ALOPEX process could be run under similar conditions of "annealing" was confirmed by earlier[60] and later work,[14] which showed that slowly shrinking the magnitudes of both the noise and bias components in the update of the parameters could result in a great improvement in both the speed and accuracy of the optimization. In this work, the values of γ and σ were initially high and were lowered during the course of the simulation by the schedule

$$\gamma(n) = \left(\gamma_0 - \gamma_{infinity}\right)e^{-\frac{n}{\tau}} + \gamma_{infinity} \qquad (5.10)$$

and

$$\sigma(n) = \left(\sigma_0 - \sigma_{infinity}\right)e^{-\frac{n}{\tau}} + \sigma_{infinity} \qquad (5.11)$$

where τ was used to control the rate of the "cooling", and the initial and final parameter values are user entered.

Many other improvements have also been suggested, including a parallel implementation of the algorithm,[42] averaging between multiple ALOPEX processes,[60] and an interleaved formulation of the algorithm to work on multiple response functions.[10] Recent work has used distributed ALOPEX processes working on overlapping "fields" of an image to enhance convergence speed.[41] In a situation where the algorithm is used for noise removal or the correction of pattern imperfections and there exist a set of templates to guide the optimization, it has been shown[42] that multiple response functions from each of the m templates $R_j(n)$ can be used to get a single response function as

---

*Note that there is not complete freedom of movement of the parameters with no noise. This is due to the fact that only one response function is used.

$$R' = \sum_{j=1}^{m} \left(\frac{R_j^2(n)}{\sum_{k=1}^{m} R_k(n)} \right). \qquad (5.12)$$

A similar function, with the inversion of both the numerator and denominator of Equation 5.12, was used for minimizing a particular response function.[15]

The ALOPEX process has been applied successfully to many application areas since its introduction, in large part because of its general and flexible form. The ALOPEX process is interesting in that the pattern recognizer can be converted to a pattern extractor.[44,17] Other applications include curve fitting to waveforms such as Visual Evoked Potentials,[45,62] crystal growth,[26] the traveling salesman problem,[22] and pattern recognition applications.[15] Using ALOPEX in perceptual tasks has also been addressed.[28] Recent applications include the use of ALOPEX in reconstructing compressed images,[60] reducing motion artifacts,[11] and use of the VEP as a generator through ALOPEX of patterns of stimulation.[31]

5.4   SYSTEM DESIGN

A pattern recognition system comprises four essential components, as labeled in Figure 5.2. The preprocessing module is an application dependent stage. The feature extracting and clustering modules are the trainable commodities in this scheme and comprise the bulk of the discussion of this chapter. As indicated by Figure 5.2, these modules are under the training control of the ALOPEX process, although the depiction of ALOPEX as a control external to the individual module is merely used as a convenience. A more accurate depiction would place an independent ALOPEX process within each stage. The final module is a labeling stage, in which the clusters formed in the previous stage are assigned an identity.

5.4.1   FEATURE EXTRACTION

A feature of an input pattern refers to any measurement from a set of pattern measurements that characterizes some attribute of that pattern. It was previously mentioned that a good feature extraction routine will compress the input space to a lower dimensionality while still maintaining a large portion of the information contained in the original pattern space. Although this is the most often-cited advantage of feature extraction, it is also true that an appropriate choice of features can help eliminate redundant and irrelevant information from the data set, thereby reducing the overhead for the classifier.[56] Most feature extraction routines can significantly aid in the performance of a classifier.

In unsupervised situations, the only information available to the feature extraction module is the statistical distribution of the patterns. In such a scenario, it is impossible to quantitatively analyze the effectiveness of a feature extraction routine in improving pattern classification. However, there are operators designed to main-

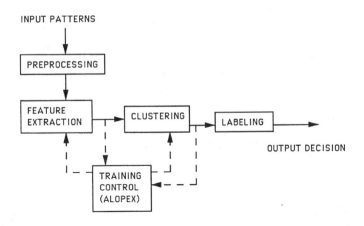

FIGURE 5.2   A component diagram of the pattern recognition system used in this research. The dotted lines indicate control signal input to the modules, whereas the solid lines denote transfer of data.

tain high information content in the features (as compared to the original measurement pattern space) with a minimum number of dimensions.

5.4.1.1   The Karhunen-Loève Expansion

Perhaps the most widely used feature extraction routine with some of these information-conserving properties is the Karhunen-Loève (K-L).[9,32,63] If it is desired to represent the kth N-dimensional input pattern x with an M-dimensional feature pattern y, then an $M \times N$ matrix Φ can be chosen so that

$$\vec{y}^k = \Phi \vec{x}^k \tag{5.13}$$

The matrix Φ is actually a set of M orthonormal vectors (meaning they lie perpendicular to one another in N-dimensional space and have unit magnitude) constructed by taking the M eigenvectors corresponding to the M largest eigenvalues of the covariance matrix C constructed by the input patterns. This matrix C is formed by

$$C = \sum_{k=1}^{P} \sum_{j=1}^{N} \left(\vec{x}_j^k - \vec{\mu} \right)^T \left(\vec{x}_j^k - \vec{\mu} \right) \tag{5.14}$$

where P is the number of patterns in the set and the vector μ is the mean vector of the patterns set as

$$\vec{\mu} = E(\vec{x}) \tag{5.15}$$

It has been shown that the eigenvectors of this covariance matrix exhibit certain optimal properties as a feature extractor when they are ordered with their correspondence to the eigenvalues of the matrix from highest to lowest.[9] One of these properties is that the mean square representation error is the minimum for any choice of M orthogonal vectors, meaning that the approximation error (ε) of reconstructing the original pattern space with only M features

$$\varepsilon = \sum_{k=1}^{P} \sum_{i=1}^{N} \left(x_i^k - \sum_{j=1}^{M} \Phi_{ij} y_j^k \right) \tag{5.16}$$

is the smallest for any choice of M vectors in Φ. This means the expansion answers a key requirement in the information compression problem of feature extraction. Its other optimum property is that this choice of vectors associates with the coefficients of the expansion a minimum measure of entropy or dispersion. The borrowed concept of entropy is often used in the pattern recognition field as a clustering measure,[57,65] and so this minimum entropy property characterizes the K-L expansion as likely to contain clustering transformational properties. The crux of the theory is that the features contain the most information (without knowing the pattern identities) for the price and probably retain the existing pattern groups in the population.

An implementation of the K-L expansion as a feature extractor[21,35,33] generally proceeds in the opposite direction to the above analysis as the following steps:

1. Calculate the covariance matrix in Equation 5.14 using all available patterns and the mean vector from Equation 5.15.
2. Find the eigenvalues and eigenvectors of that covariance matrix.
3. Select the eigenvectors corresponding to the M largest eigenvalues and store them in the matrix Φ.
4. Find the feature values for each of the patterns via Equation 5.13.

The primary advantage in using the K-L expansion for selection of features is that it requires no previous knowledge of pattern labels and thus is perfectly suited to unsupervised tasks. Many people confuse the aforementioned optimum properties of the expansion with an assumption that the features generated for the expansion provide optimum performance of the classifier. This is certainly not the case. A feature extractor can never provide optimum classifier information without information from the classifier about its historical performance. Nevertheless, the K-L expansion is useful in situations where that information is not reliable or available. The K-L expansion is also a linear operation, and considerable evidence suggests that other nonlinear features can often provide more useful information to the classifier.[12]

There is an additional point which should be mentioned here. It turns out that the primary eigenvector points in such a direction that the variance of the patterns in that feature space is maximum for all vectors. Subsequent eigenvectors find other locally maximum variance features in orthogonal directions. Furthermore, the eigenvalue representing each eigenvector is exactly equivalent to the variance of projected patterns onto the corresponding eigenvector.[18] It is this realization that provides the impetus for the enactment of the K-L expansion onto a neural network, as described in the next section.

5.4.1.2 Application by a Neural Network

The linear projection of a pattern vector onto one of the K-L expansion vectors is a simple inner product operation, as denoted by Equation 5.13. Conveniently, this is the same operation that is most commonly given to units of an artificial neural system, as seen by the comparison with Equation 5.13 for all $Q_k = 0$. In concept then, an artificial neuron can use its connection weightings to act as one of the eigenvectors contained in the matrix Φ in the last section. It only remains to consider the method of training a cell to retain that specific pattern of connections. After training, the output of the neuron is a real number corresponding to the feature value from the input pattern.

A hint has already been given about the means for training a cell to retain the "maximum" eigenvector. It was mentioned that the primary K-L expansion vector had the property that the output features generated by it contained the maximum variance of any features generated by any other choice of vectors. In direct relation to the neural network, if the variance of the output of the cell for all training patterns is maximized during training, the connection vector retained after training corresponds to the primary K-L vector. The scheme for training any one cell follows the steps outlined below.

1. Set the connections to the cell to random values.
2. Calculate the output value of the cell for every training pattern.
3. Find the variance of the output over all patterns used in step 2.
4. Update the connection weights to the cell via an ALOPEX update equation.
5. Normalize the connection weights to the cell to a vector magnitude of 1.0.
6. Go to step 2 until a convergence criterion has been met.

In particular, the output y of the jth cell from the input of the kth N-dimensional pattern x is found as

$$O_j^k \sum_{i=1}^{N} C_{ij} x_i^k \tag{5.17}$$

and the variance of the jth cell output (V_j) is calculated over all P patterns by

$$V_j = \frac{\sum_{k=1}^{P} \left(O_j^k - \mu_j \right)^2}{P}, \quad \mu_j = E\left[O_j \right] \tag{5.18}$$

The ALOPEX update equation uses the changes in the variance (ΔV_j) and connections (ΔC_{ij}) from the current iteration (n) to the previous iteration ($n-1$) to change the connections by

$$C_{ij}(n+1) = C_{ij}(n) + \Delta B_{ij}(n) + r_{ij}(n) \tag{5.19}$$

where

$$\Delta B_{ij}(n) = \gamma \Delta V_j(n) \Delta C_{ij}(n) \tag{5.20}$$

[see Equations 5.17 through 5.20]. The term $r_{ij}(n)$ is a Gaussian random number with zero mean and standard deviation. The factors γ and σ are adjusted as in Equations 5.21 and 5.22.

To illustrate the performance of the algorithm, a simple two-dimensional pattern space was constructed, and one neuron was trained on 60 patterns in this space using ALOPEX output variance maximization. The performance of the algorithm can be tested easily, since the "ideal" maximum is known (it can be found by the analysis of Section 5.4.1). The resultant vectors of the ALOPEX method and the K-L expansion are shown in Table 5.1. Also included in that table are two other commonly used methods for unsupervised neural network training. The first is a simple normalized Hebbian scheme, where the connections are changed during each iteration by

$$\Delta C_{ij} = \eta O_j x_i \tag{5.21}$$

where η is a gain factor. The second is a widely touted method employed by Oja,[47] which is a variant of the Hebbian proposal and includes output feedback as

$$\Delta C_{ij} = \eta O_j \left(x_i - O_j C_{ij} \right) \tag{5.22}$$

Table 5.1 shows some clear results, the first of which is that the variance maximization scheme with ALOPEX was able to mimic very accurately the optimal K-L vector. Second, it clearly points out a weakness in the Hebbian proposal. As further experiments will show, the Hebbian training[29] (and the Oja training, which is nearly identical in most real world situations) cannot optimize to the "best" vector because the input patterns are not zero-mean centered (they do not share a center of mass at the origin of the coordinate system). Thus, the Hebbian mechanism is essentially unable to compensate for DC offsets, a situation that must be tolerated in nearly all real-world applications. Obviously this weakness can be compensated for by centering the data before it enters a Hebbian-trained module, but this requires additional post-training computation and is impractical in a highly connected system. Oja claims that this scheme will result in the cell retaining the principal component of the input pattern distribution.*

---

* The principal component is identical to the first K-L expansion vector when the center of mass of the patterns in data space is at the origin.

TABLE 5.1
Converged Connection Vectors

Principal K-L Vector	ALOPEX Variance Maximization	Normalized Hebbian	Oja Scheme
0.2011	0.1920	0.6305	0.6299
0.9796	0.9810	0.7760	0.7770

Note: For the two-dimensional sample data space shown in Figure 5.3. The first row in the table refers to the first connection value in the neuron, while the second row denotes the second connection value.

Figure 5.3 shows the pattern space that was used for the training results in Table 5.1. Clearly the space is composed of three quite distinct clusters. However, not all choices of vectors allow all of those clusters to be clearly evident at the output of the feature cell. The illustrations of Figure 5.4 make it obvious why the K-L and ALOPEX-trained connections are superior choices to the Hebbian scheme. The diagrams in Figure 5.4 are histograms of the output values of the cell for each of the different vectors in Table 5.1. It is easy to see why the K-L expansion and ALOPEX-trained vectors are preferable to the Hebbian and Oja schemes: the increased range of the neuron output levels increases the information content of the output by allowing all three clusters to be evident on the output line. So the concept of variance maximization indirectly promotes the retention of cluster forming information. Note that this is not a proof of superiority of the methods, but it is a valid observation with an example training set.

With the one-cell implementation described thus far there is still the question of how a network of cells is to optimize to other vectors in the K-L expansion. Since each cell is searching for the "optimal" vector, all cells in a network will arrive at or near the same vector when using the same input pattern set. There are several possible means for forcing other neurons to optimize to other K-L vectors when more than one feature output is necessary. The training routine can be altered so that the optimization is not a global search. This will force each cell to arrive at a local optimum that differs depending on the initial conditions, but this does not guarantee that redundant cells will not be formed, and it imposes a strong likelihood that some cells will arrive at local maxima which are not K-L expansion vectors. A second possibility is for an additional term to be added to the ALOPEX cost function which uses feedback connections between other feature extracting cells in the network to impose a constraint on the uniqueness of each cell. There are several problems with this choice. The insertion of inter-network feedback requires that additional care be taken to prevent instability, and even when stability of the output of the cells is guaranteed, computation time is drastically increased because of the

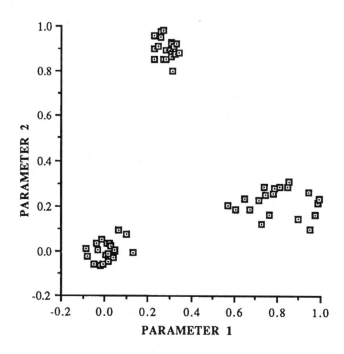

FIGURE 5.3   The input pattern space used for the training of the vectors in Table 5.1. There are 60 patterns in the space, with 20 grouped in each cluster.

need to wait until the outputs of the cells "settle". Additionally, the weighting between the terms is most likely problem-dependent and is certainly optimally different for each cell. This possibility was examined in detail, though, and the interim results indicate the possibility of training cells to recognize features that were highly nonlinear and yet often retained some of the optimal aspects required by the problem. The work on this possibility was suspended when the training times became exorbitant. Instead, a more viable solution was obtained. The actual method used imposes the constraints on the ALOPEX optimization in a more implicit way. Instead of adding extra qualifying terms to the ALOPEX update equation, the input pattern space itself was altered to exclude the information that was already retained by other cells. In the resulting architecture (Figure 5.5), the cells are "chained" to one another in series by a weighted feedback of the output of the previous cell to the input of the current cell. The first cell receives the original pattern space unaltered and will perform exactly as the single cell simulations shown previously. Subsequent cells receive the original pattern space (x), minus the component of that space extracted by the previous cells as

$$\hat{x}_i = x_i - C_{i,j-1}O_{j-1} \tag{5.23}$$

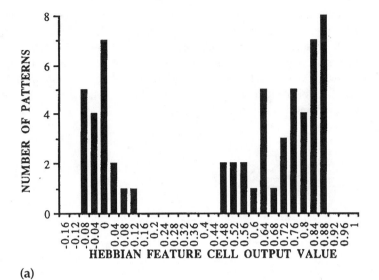

FIGURE 5.4 The feature cell output value histograms for the space of Figure 5.3 for (a) the normalized Hebbian training and (b) the ALOPEX variance maximization scheme.

for the ith input to the jth cell in the chain. The output of the jth cell is then found in the same way as before, only now using the modified input space as

$$O_j = \sum_{i=1}^{N} C_{ij}\hat{x}_i. \qquad (5.24)$$

Using this method, only the first cell, which receives no interference from other cells in the network, can optimize to the principal vector, while all others are relegated to finding secondary yet potentially important vectors. When a network such as this is trained, it can be seen that the first cell optimizes to the first Karhunen-Loève eigenvector, and the other cells locate in order the remaining eigenvectors. Sanger recently published an architecture similar to this scheme,[52] but his training relies on the gradient descent method of Oja.[47]

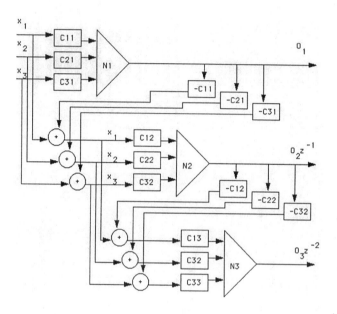

FIGURE 5.5 The feature extraction architecture for three feature cells. After the first cell, all inputs are modified by a subtracted weighted input from the outputs of the previous cells, as in Equation 5.1.11. The triangle symbol represents the neuron integrator, while the symbol C12 indicates the connection weight from the first input to the second cell.

To illustrate the performance of this architecture, let us use the pattern space employed in the one neuron simulation in Figure 5.3. Since the first cell in the chain architecture above does not receive any influence from other cells, it will optimize in the same way as in Table 5.1. If we now find the new pattern space to provide to the second cell according to Equation 5.24, we get a different pattern space. All information in the direction of the first cell vector of Table 5.1 has been removed from the pattern space given to the second cell. It is then a trivial matter for the second cell to optimize on this pattern space. Note that we were, in this example, extracting two features from what was originally a two-dimensional pattern space. This is not a compression scenario. In a more general case where the number of features is less than the number of pattern dimensions, the input space would never compress to a line, as in Figure 5.6. Rather, with each cell in the chain, there is essentially a virtual removal of a dimension from the pattern space in the direction of the previously optimized vectors (not a physical removal, since the feature extraction is still performed over the same number of dimensions). The application of this feature extraction method to a more complicated pattern space is shown in subsequent chapters, where it can be clearly seen that this architecture and training allow for the network cells to converge on the K-L expansion vectors in order.

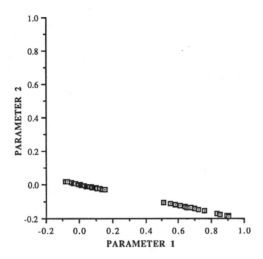

FIGURE 5.6 The revised pattern space of Figure 5.3 as seen by the second cell in a feature extraction network chain.

5.5 CLUSTERING

5.5.1 THE FUZZY c-MEANS (FCM) CLUSTERING ALGORITHM

The concept of fuzzy logic was first introduced by Zadeh, whose classic paper has become the philosophical bible in the field.[68] The concept is simple: set membership, and indeed reasoning of any sort, carries more information when there is a continuum of grades of membership. The reasoning is based on Zadeh's Principle of Incompatibility, which maintains that high precision is incompatible with high complexity. The suggestion is that the complexity of a system and the precision with which it can be analyzed bear a roughly inverse relation to one another. He asserted that since real world ideas appear to be fuzzy in nature, there is reasonable cause for adapting this approach to machines. Since that time, the number of applications to decision making and pattern clustering in particular have been numerous.[16]

One of the first to apply fuzzy reasoning to pattern recognition was James Bezdek.[2] The method that he and his colleagues have introduced, the Fuzzy c-Means clustering (FCM) algorithm,[3-5] has seen great popularity as a flexible and easily implemented method. The method itself is actually a spinoff of the venerable ISODATA algorithm.[1] The ISODATA clustering method is one of a set of techniques that assumes that the optimal cluster partitioning is described as the minimum (or maximum) of an objective function. For the ISODATA algorithm and others like it, which use a set of c prototype "centers" of clusters around which patterns are grouped by their resemblance to these centers, the most common choice of objective function J is of the form

$$J = \sum_{k=1}^{P} \sum_{i=1}^{c} u_{ik} d_{ik} \tag{5.25}$$

where u_{ik} is the membership strength of pattern k in cluster i and d_{ik} is the squared distance from pattern k to cluster center i in m-dimensional feature space. The ISODATA algorithm is normally used to generate *hard* partitions of the data. A hard partition is one in which each pattern is allocated entirely to one cluster or another, so that the membership strengths take on the values of zero or one.

In most scenarios, the assignment of patterns to any one cluster prototype in exclusion of all others is a gross simplification of the complexity of the pattern space. Bezdek used the concept of fuzzy logic, where decisions are made through analog weightings, and applied it to this objective function J. In doing so, J was defined as

$$J = \sum_{k=1}^{P} \sum_{i=1}^{c} \left(u_{ik}\right)^{q} d_{ik} \tag{5.26}$$

It is easiest to think of this objective function as representing the sum of the errors (the distances) in representing the patterns by a set of c cluster centers, weighted by the membership of the patterns to those clusters. The exponent q controls the sharpness of the decision boundaries, so that when $q = 1$, hard clusters are constructed, and when $q = \infty$, all patterns share the same membership to each cluster.

Most importantly for the mathematical analysis of this function, the use of continuous memberships means that the decision space is now continuous for all $q > 1$. It now becomes possible to examine the conditions for minimization of this function. It was demonstrated that J could be locally optimal for any one q only if

$$\bar{v}_i = \frac{\sum_{k=1}^{P} \left(u_{ik}\right)^{q} \bar{x}_k}{\sum_{k=1}^{P} \left(u_{ik}\right)^{q}}; \, 1 \le i \le c; \tag{5.27}$$

and

$$u_{ik} = \sum_{j=1}^{c} \left(\frac{d_{ik}}{d_{jk}}\right)^{\frac{-2}{q-1}}; \, 1 \le k \le P; 1 \le i \le c \tag{5.28}$$

where V_i is the ith cluster center and x_k the kth pattern. By iterating through these conditions, Bezdek[3] claimed that a local minimum of the function J would be achieved. It was later seen that this iteration could only guarantee stationary points and not necessarily local minima;[69] nevertheless, the FCM method was found widely useful in practically achieving rapid (usually < 25 iterations) and "good" clusterings of data from many application areas.[6,7,8,24]

There are two factors in the use of the FCM procedure which still require discussion. One of these points is the optimal selection of the parameter q. To this point, there is no automated way of selecting the best value of q for any one pattern set, but most applications seem to find reasonable values as lying somewhere between 1.2 and 4.0.[3]

The second factor is the way in which the distance d_{ik} is calculated. In general, the distance d can be calculated through the quadratic form

$$d_{ik} = \left\| \vec{x}_k - \vec{v}_i \right\|_A^2 = \left(\vec{x}_k - \vec{v}_i \right)^T A \left(\vec{x}_k - \vec{v}_i \right) \tag{5.29}$$

which is termed the A-norm distance. If the matrix A is chosen to be the identity matrix, then the distance is the squared Euclidean distance from pattern x_k to center v_i. This causes the FCM algorithm to form roughly hyperspherical clusters. This makes convergence simpler (since each cluster shape is identical) but is not optimal for most data sets with clusters of unequal shape. If the matrix A is chosen as C^{-1}, where C is found as the fuzzy covariance matrix

$$\vec{C}_i = \frac{\displaystyle\sum_{k=1}^{P} \left(u_{ik} \right)^q \left(\vec{x}_k - \vec{v}_i \right) \left(\vec{x}_k - \vec{v}_i \right)^T}{\displaystyle\sum_{k=1}^{n} \left(u_{ik} \right)^q} \tag{5.30}$$

then the axes of the cluster are effectively scaled according to the distribution of the data points within those clusters. This is a modification of the Mahalanobis distance measure for "hard" data sets.[56] Other forms of the matrix A are also popular, including a diagonal matrix of the eigenvalues of the matrix C which Equation 5.30 calculates.[5]

Using $A = C^{-1}$, and with the same calculation of membership strengths as in Equation 5.28, clusters of essentially hyperellipsoidal shape can now be found. Furthermore, the cluster shapes can be variant from one cluster to another, since each cluster has its own covariance matrix C. The incidence of local optima in the use of a variant A matrix such as this has been shown to rise drastically, affecting almost every problem, even with small data sets.[3] To compensate for this, elaborate means of choosing initial conditions have been used, with unproven ability to guarantee global success.[23]

Finally, every clustering algorithm must develop a proven means for determining the optimal number of clusters in the data set and whether a converged set of clusters is a "good" clustering. This is usually termed the *cluster validity* problem, and there are at least as many opinions as to what is the best set of parameters to provide this information as there are clustering routines.

The originators of the FCM routine usually use an entropy measure to characterize the effectiveness of the clustering operations, which is given by

$$H_c = -\left(\frac{1}{P}\right)\sum_{i=1}^{c}\sum_{k=1}^{P} u_{ik} \log_a(u_{ik})$$

(5.31)

$$a \in (1, \infty)$$

where $H_c = 0$ for hard partitions and $H_c = \log_a(c)$ for an entirely "blurred" (or indecisive) clustering. Another related parameter, termed the partition coefficient (F), is found by

$$F = \left(\frac{1}{P}\right)\sum_{i=1}^{c}\sum_{k=1}^{P}(u_{ik})^2$$

(5.32)

Both of these parameters rely on one of the major paradoxes of fuzzy logic. That is, although the pretext of fuzzy clustering is to incorporate more information via using analog decision criteria, heuristically the "best" clustering is one in which the resultant clusters are hard (have binary membership strengths). This idea is the basis behind the use of the entropy (H) and partition coefficient (F) measures and assumes that the optimal number of clusters is the choice for which H is minimized and F is maximized.

The FCM algorithms already fit many of the requirements for ALOPEX training. There is an explicit cost function which determines the "optimal" choice of clustering, a requirement for an optimization routine such as ALOPEX. Further, there is really only one set of independent parameters which must be varied in order to minimize the cost function of Equation 5.26: the set of cluster centers (c times m parameters in total). All other information for the determination of membership strengths results from the specification of cluster centers. The distances (Equation 5.29) are found in reference to the cluster centers, and the membership strengths are based entirely on distance information (Equation 5.28).

It remains only to justify the use of ALOPEX to this application. ALOPEX will reduce the likelihood of arriving at a locally optimum solution at the price of increased computation time. However, most situations demand accuracy, even when having to sacrifice increased computation time. This is especially true when you consider that the decision of these clustering partitions is usually needed only once, after which those partitions are used to make rapid decisions about new data.

The danger of locally optimal solutions becomes especially apparent when clusters of nonhyperspherical shape are assumed. The distance measurements often become very local in certain directions from the cluster center. The result is a much more localized cost function, which is therefore much more volatile. The resulting distances generated can often exceed the real number range of most software languages, and special care must be taken to ensure the stability of the algorithm as it iterates.

As mentioned before, all of the FCM family of algorithms share the danger of locally optimal solutions. Even with a Euclidean distance measurement, it is easily apparent that multiple runs of the FCM algorithm arrive at different solution points for pattern spaces of reasonable complexity. A more complete example of this behavior is shown in the context of the classification of handwritten characters.

In order to incorporate a global optimizer into the fuzzy c-means family, ALOPEX is used to adjust the cluster centers iteratively in the steps outlined below.

1. Randomly choose initial cluster centers.
2. Find squared distances between patterns and centers.
3. Calculate membership strengths via Equation 5.28.
4. Find the current cost function J from Equation 5.26.
5. Use ALOPEX to update the centers based on recent change in the cost function and the centers.
6. Go to step 2 until a convergence criterion is met.

The performance of this routine, and that of the feature extractor, is illustrated in the context of two application domains. These are described in detail in Section 5.6.

Most pattern recognition schemes need to consider the assignment of labels or pattern identities to decision codes generated by the pattern recognition system. Most commonly, this consideration is important for supervised schemes, in which the pattern identities are known without ambiguity. In unsupervised methods, the notion of pattern labeling is somewhat self-defeating. That is, if the identities of the patterns used in the training were known before training, then a supervised method would have been more productive. If, however, the pattern labels are suppositions or decisions with an amount of uncertainty, it would be more useful to assign labeling based only on cluster membership strengths.

The classification of handwritten digits by this unsupervised system is an unusual task. As mentioned before, it is motivated by a desire to improve the algorithm without biasing the answer toward concurrence with the medical diagnoses. Paradoxically, the labels of each of the characters used in the study (which were never subjected to a segmentation process) are known unambiguously before training, but the knowledge is only used in the assignment of pattern labels after training is completed. This allows for a quantitative description of the performance of the algorithm.

The labeling of such a scenario is performed in this research by analyzing the constituent memberships of each cluster into an array Ω.

$$\Omega_{ij} = \frac{\sum_{k=1}^{P} \left\{ u_{ik\cdot} : \forall k \in PATTERN\ TYPE\ j \right\}}{\sum_{l=1}^{c} \sum_{k=1}^{P} u_{lk}}, \quad 1 \le i \le c, 1 \le j \le R \qquad (5.33)$$

where Ω_{lj} is the percentage of cluster l membership from pattern type j (for R pattern types in the simulation, i.e., 10 digits in the character recognition problem), and there were c clusters formed from the P training patterns. Then the degree (ψ_{kj}) to which pattern k belongs to pattern type j is calculated as

$$\psi_{kj} = \frac{\sum_{i=1}^{c} u_{ik}\Omega_{ij}}{\sum_{l=1}^{R}\sum_{i=1}^{c} u_{ik}\Omega_{il}}, 1 \le k \le P, 1 \le j \le R \tag{5.34}$$

The label of pattern k (L_k) is then the maximum of the degrees of memberships to the pattern types as

$$L_k = \max\{\psi_{kj}\}. \tag{5.35}$$

In this way the labeling is performed not only on the membership strengths of the patterns given by the clustering module but also on the specificity to a single pattern type demonstrated by the clusters. That is, the labeling of a pattern with a high membership strength to a cluster with a high population of more than one pattern type will downplay that cluster membership strength in favor of other clusters with more "pure" pattern types.

Most neural network decisions formulate their decisions in a highly intertwined and complicated way. Even if the network is purely feed-forward (as is the multilayer perceptron used in the backpropagation algorithm), there is usually only a limited idea of the criteria that the network used in making its decision.

The neural network just presented is an intriguing exception to this category of systems. The primary finding of the clustering module is the set of "centers" around which the cluster boundaries are formed. Since the coordinates of this center reside in the same space as the feature vectors of the input patterns, the cluster center coordinates can be thought of as the feature values that would have been extracted if there were a corresponding input pattern.

The primary question is whether, knowing these feature values, we can find out what the input pattern would have looked like. The answer is a resounding yes! The feature extracting neural network implements the K-L expansion, as we have already mentioned. Since the K-L expansion is really a linear expansion of the input pattern and since the K-L expansion is used both as a feature extractor and as a data compression method, the feature vector can be reconstructed to find the input pattern with the knowledge of the K-L vectors that derived the features. This is essentially the same concept that was used to <u>remove</u> information from subsequent cells in the feature extraction network applied to a different task.

Given an m-dimensional feature vector y and a desired representation of the input pattern x, we can reconstruct an approximation to the input pattern (x) as

$$\hat{x}_i = \sum_{j=1}^{m} C_{ij} y_j \qquad (5.36)$$

where x_i is the ith input of the vector x and C_{ij} is the network connection strength from input l to feature extracting cell j. The vector x is an approximation to m terms of the K-L expansion the original input pattern x.

The realization that the cluster solutions (the centers) can be reconstructed into the corresponding input pattern (with hopefully a small error) allows the system to be used in an entirely new light. Not only can the system provide unsupervised classifications of a set of patterns, but through the reconstruction of the input pattern, a glimpse of the reasoning of the decisions can be made. This is made more apparent when the reconstructions of specific applications are displayed, as is done in Chapter 6 with Figure 6.8.

REFERENCES

1. Ball, G. H. and Hall, D. J., A clustering technique for summarizing multivariate data, *Behav. Sci.*, 12, 153, 1967.
2. Bezdek, J. C., Fuzzy Mathematics in Pattern Classification, PhD Diss., Cornell University, Ithaca, NY, August 1973.
3. Bezdek, J. C., *Pattern Recognition with Fuzzy Objective Function Algorithms*, Plenum Press, New York, 1981, 65.
4. Bezdek, J. C. and Dunn, J. C., Optimal fuzzy partitions: a heuristic for estimating the parameters in a mixture of normal distributions, *IEEE Trans. Comput.*, August 1975.
5. Bezdek, J. C., Ehrlich, R., and Full, W., FCM: the fuzzy c-means clustering algorithm, *Comput. Geosci.*, 10(2), 191, 1984.
6. Bezdek, J. C. and Fordon, W. A., Analysis of hypertensive patients by the use of the fuzzy ISODATA algorithm, *Proc. JACC*, 3, 349, 1978.
7. Bezdek, J. C., Trevedi, M., Ehrlich, R., and Full, W., Fuzzy clustering: a new approach for geostatistical analysis, *Int. J. Sys., Meas. Decisions*, 1982.
8. Cannon, R. L., Dave, J. V., Bezdek, J. C., and Trivedi, M. M., Segmentation of thematic mapper image data using fuzzy c-means clustering, *Proc. 1985 IEEE Workshop Lang. Automation IEEE Comput. Soc.*, 93, 1985.
9. Chen, Y. T. and Fu, K. S., On the generalized Karhunen-Loève Expansion, *IEEE Trans. Inf. Theor.*, 15, 518, 1967.
10. Chon, T. and Micheli-Tzanakou, E., A Probabilistic Approach to the ALOPEX Process using Moment Invariants of Images, Proc. Int. Joint Conf. on Neural Networks, Vol. II, 1989, 611.
11. Ciaccio, E. J. and Micheli-Tzanakou, E., The ALOPEX Process: Application to Real-Time Reduction of Motion Artifact, Proc. 12th Annu. Int. Conf. IEEE/EMBS, Vol. 12, 1990, 1417.
12. Cover, T. M., The best two independent measures are not the two best, *IEEE Trans.*, SSC-6, 33, 1974.

13. Dante, H. M. and Sharma, V. V. S., Optimum Decision Tree Classifiers for Classification in Large Populations, Proc. IEEE Intl. Conf. on Cybernet. and Soc., 1985, 559.

14. Dasey, T. and Micheli-Tzanakou, E., A Pattern Recognition Application of the ALOPEX Process on Hexagonal Images, Proc. Int. Joint Conf. on Neural Networks, Vol.II, 1989[a], 119.

15. Dasey, T. and Micheli-Tzanakou, E., Efficiency Exploration of ALOPEX Based Recognition and Hexagonalized Images, Proc. of the Fifteenth Annu. Northeast Biomed. Eng. Conf., 1989[b], 177.

16. Davis, J. C. and Economou, C. E., A review of fuzzy clustering methods, *Adv. Eng. Software*, 6(4), 1984.

17. Deutsch, S. and Micheli-Tzanakou, E., *Neuroelectric Systems*, NYU Press, New York, 1987.

18. Devijver, P. A. and Kittler, J., *Pattern Recognition: A Statistical Approach*, Prentice-Hall, Englewood Cliffs, NJ, 1982.

19. Eigen, M., Self-organization of matter and the evolution of biological macromolecules, *Naturwissenschaften*, 58, 465, 1971.

20. Fu, K. S., *Application of Pattern Recognition*, CRC Press, Boca Raton, FL, 1982.

21. Fukunaga, K. and Koontz, W. L. G., Application of the Karhunen-Loève expansion to feature selection and ordering, *IEEE Trans. Comput.*, C-19(4), 311, 1970.

22. Galar, R., Evolutionary search with soft selection, *Biol. Cybern.*, 60, 357, 1989.

23. Gath, I. and Geva, A. B., Unsupervised optimal fuzzy clustering, *IEEE Trans. PAMI*, 11(7), 773, 1989.

24. Granath, G., Application of fuzzy clustering and fuzzy classification to evaluate provenance of glacial till, *Math. Geol.*, 16, 283, 1984.

25. Harth, E. and Tzanakou, E., A stochastic method for determining visual receptive fields, *Vis. Res.*, 12, 1475, 1974.

26. Harth, E., Kalogeropoulos, T., and Pandya, A. S., ALOPEX: A Universal Optimization Network, Proc. Spec. Symp. on Maturing Technol. and Emerging Horizons in Biomed. Eng., 1988, 97.

27. Harth, E. and Pandya, A. S., Dynamics of the ALOPEX process: applications to optimization problems, in *Biomathematics and Related Computational Problems*, Ricciardi, L., Ed., Kluwar Acad. Publ., 1988, 459.

28. Harth, E., Pandya, A. S., and Unnikrishnan, K. P., Perception as an optimization process, in *Proc. of IEEE Computer Society Conference on Computer Visual and Pattern Recognition*, IEEE Computer Society Press, Washington, DC, 1986, 662.

29. Hebb, D., *The Organization of Behavior: A Neurophysiological Theory*, John Wiley & Sons, New York, 1949.

30. Highleyman, W. H., The design and analysis of pattern recognition experiments, *Bell Syst. Tech. J.*, 41, 723, 1962.

31. Iezzi, R., Jr., Micheli-Tzanakou, E., and Cottaris, N., Effects of Pattern Convergence and Orthogonality on Visual Evoked Potentials, *Proc. 12th Annu. Int. Conf. IEEE/EMBS*, Vol. 12, 1990, 897.

32. Karhunen, K., Uber lineare Methoden in der Wahrscheinlichkeitsrechnung, *Ann. Acad. Sci. Fennicae*, Ser. A137 (trans. by I. Selin in *On Linear Methods in Probability Theory*, T-131, The RAND Corp., Santa Monica, CA, 1960), 1947.

33. Kirby, M. and Sirovich, L., Application of the Karhunen-Loève procedure for the characterization of human faces, *IEEE Trans. PAMI*, 12(1), 103, 1990.

34. Kirkpatrick, S., Gelatt, C. D., and Vecchi, M. P., Optimization by simulated annealing, *Sci.*, 220, 671, 1983.

35. Kittler, J. and Young, P. C., A new approach to feature selection based on Karhunen-Loève expansion, *Patt. Recog.*, 5, 335, 1973.
36. Kohonen, T., Problems in practical pattern recognition, *Neural Netw.*, 1(suppl.), 29, 1988.
37. Leboucher, G. and Lowitz, G. E., What a histogram can really tell the classifier, *Patt. Recog.*, 10, 351, 1978.
38. Lerner, A., A crisis: in the theory of pattern recognition, in *Frontiers of Pattern Recognition*, Watanabe, S., Ed., Academic Press, New York, 1972, 367.
39. Levine, M. D., Feature extraction: a survey, *Proc. IEEE*, 57(8), 1391, 1969.
40. Levy, A. C. and Montalvo, A., The tunneling algorithm for the global minimization of functions, *SIAM J. Sci. Stat. Comput.*, 6(1), 15, 1985.
41. Marsic, I. and Micheli-Tzanakou, E., Distributed Optimization with the ALOPEX Algorithms, Proc. of the 12th Int. Conf. IEEE/EMBS, Vol. 12, 1990, 1415.
42. Mellissaratos, L. and Micheli-Tzanakou, E., The Parallel Character of the Alopex Process, *J. Med. Syst.* 13(5), 243, 1990.
43. Metropolis, N., Rosenbluth, A. W., Rosenbluth, M. N., and Teller, A. H., Equation of state calculations by fast computing machines, *J. Chem. Phys.*, 21(6), 1087, 1953.
44. Micheli-Tzanakou, E., Non-linear characteristics in the frog's visual system, *Biol. Cybern.*, 51, 53, 1984.
45. Micheli-Tzanakou, E. and O'Malley, K. G., Harmonic Content of Patterns and their Correlation to VEP waveforms, *IEEE 17th Annu. Conf. of EMBS*, 1985, 426.
46. Micheli-Tzanakou, E. and Dasey, T. J., Pattern Recognition with Neural Networks on Compressed Images, 6th IASTED Int. Conf. on Expert Systems and Neural Networks, 1990, 9.
47. Oja, E., A simplified neuron model as a principal component analyzer, *J. Math. Biol.*, 1982.
48. Oja, E., *Subspace Methods of Pattern Recognition*, Research Studies Press, Ltd., Letchworth, 1983.
49. Peterson, C. and Hartman, E., Explorations of the mean field theory learning algorithm, *Neural Net.*, 2, 475, 1989.
50. Romesburg, H. C., *Cluster Analysis for Researchers*, Lifetime Learning Publications, London, 1984.
51. Rutenbar, R. A., Simulated annealing algorithm: an overview, *IEEE Circ. Devices Mag.*, 5(1), 19, January 1989.
52. Sanger, T. D., Optimal unsupervised learning in a single-layer linear feedforward neural network, *Neural Net.*, 2, 459, 1989.
53. Scoltock, J., A survey of the literature of cluster analysis, *Comput. J.*, 25(1), 130, 1982.
54. Snyder, W., Bilbro, G., and Van den Bout, D., New Techniques in Optimization: A Tutorial, Technical Report NETR-89-12, Center for Communication and Signal Processing, North Carolina State University, Raleigh, NC, November, 1989.
55. Teh, C.-H. and Chin, R. T., On digital approximation of moment invariants, *Comp. Vis. Graphics Image Proc.*, 33, 318, 1986.
56. Tou, J. T. and Gonzalez, R. C., Automatic recognition of handwritten characters via feature extraction and multilevel decision, *Int. J. Comput. Inf. Sci.*, 1, 43, 1972.
57. Tou, J. T. and Heydorn, R. P., Some approaches to optimum feature extraction, in *Computer and Information Sciences - II*, Tou, J. T., Ed., Academic Press, New York, 1967.
58. Toussaint, G. T., Bibliography on estimation of misclassification, *IEEE Trans. Inf. Theory*, IT-20(4), 472-479, 1974.

59. Tucker, W. T., Counterexamples to the convergence theorem for the fuzzy c-means clustering algorithms, in *Analysis of Fuzzy Information, Vol. III - Applications in Engineering and Science*, CRC Press, Boca Raton, FL, 1987, 109.

60. Tzanakou, E., Principles and Design of the ALOPEX Device: A Novel Method of Mapping Visual Receptive Fields, Ph.D. Diss., 1977, International Publication No. 77-30, 771.38/8, 1978.

61. Tzanakou, E., Michalak, R., and Harth, E., The Alopex process: visual receptive fields by response feedback, *Biol. Cybern.*, 35, 161, 1979.

62. Wang, J.-Z. and Micheli-Tzanakou, E., The use of the ALOPEX process in extracting normal and abnormal visual evoked potentials, *IEEE-EMBS Mag. Spec. Issue DSP*, V9(1), 44, 1990.

63. Watanabe, S., Karhunen-Loève Expansion and Factor Analysis — Theoretical Remarks and Applications, Proc. of the 4th Conf. on Inf. Theory, Prague, 1965.

64. Watanabe, S., Pattern recognition as information compression, in *Frontiers of Pattern Recognition*, Watanabe, S., Ed., Academic Press, New York, 1972, 561.

65. Watanabe, S. and Kaminuma, T., Recent Developments of the Minimum Entropy Algorithm, *9th Int. Conf. Pattern Recognition*, 1988, 536.

66. Xu, L., Some Application of Simulated Annealing to Pattern Recognition, *Proc. Int. Conf. Pattern Recognition*, Rome, 1988, 1040.

67. Young, T. T. and Calvert, T. W., *Classification, Estimation, and Pattern Recognition*, American Elsevier, New York, 1974.

68. Zadeh, L., Fuzzy sets, *Inf. Control*, 8, 338, 1965.

69. Tucker, N. D. and Evans, F. C., A Two-Step Strategy for Character Recognition Using Geometrical Moments, *Proc. 2nd Int. Joint Conf. Pattern Recognition*, 1974, 223.

6 Application to Handwritten Digits

Timothy J. Dasey and Evangelia Micheli-Tzanakou

6.1 INTRODUCTION TO CHARACTER RECOGNITION

Among the widely varied applications of pattern recognition techniques, perhaps none has been studied more intensively than the machine recognition of character data.[1] The number of potentially profitable uses for such systems is nearly limitless, since so much of the information resident in today's industrial society is textual. This is one reason that an application of the methods developed in this book is devoted to the character recognition arena. Also, the nature of the task permits the experimenter a concrete success formulation since the correct classes can be determined unambiguously. This is a much more desirable environment for the development of a new method than the less clearly formulated class memberships of medical data is treated in Chapter 7.

The industrial applications of character recognition (CR) systems fall into several broad categories. One area is certainly that of data entry of handwritten information into conventional computer systems. Such arenas are typically constrained to data sets with limited character sets and constrained paper format (i.e., banking). This overlaps with the text entry area, which is more concerned with the input of typewritten characters into a word processing or publishing environment. These systems can only recognize characters of certain fonts, but with very high success (> 99.9%). Other character recognition systems use the deciphered information to control a process, as would happen in a post office branch with a CR system that sorts mail. A final application area deals with providing an interface with the visually impaired, which often involves both a recognition procedure and a translator into speech.

Any comparison of character recognition tasks must be approached cautiously, since the difficulty of the task is determined largely by the constraints imposed on the data and the information available to the machine. It is certainly much easier for a machine to recognize typewritten characters than handwritten characters, since the typewritten characters would usually follow more standard guidelines and be less variable. Similarly, a signature verification system would likely be more successful if the machine had access to the pen pressure, velocity, and acceleration information at the time of the writing, as well as the shape characteristics of the signature.

The recognition of handwritten characters is a subset of the much more extensive optical character recognition (OCR) problem. It deals with the recognition of single hand-drawn characters of an alphabet that is unconnected. It must be differentiated from script recognition, which is concerned with the recognition of handwritten characters that may be connected and cursive. In this sense the developers of handwritten character recognition schemes do not need to concern themselves with the extremely challenging task of segmenting the characters.[2] Still, handwritten character recognition is not as simple a task as it may appear, since some claim[3] that even human beings can make up to 4% of mistakes when reading certain characters in the absence of context. Errors in reading handprinted characters, in addition to deriving from the algorithm and scanning methods, can also arise because of variations in shape due to the habits, style, mood, health, and other conditions of the writer.[4]

The recognition of handwritten characters must consider at least two problems: the means of scanning the image and the method for its recognition. The choice of a scanning device is not considered in this discussion, but the methodology of the recognition has been categorized as[5,6]

(i)   Point-by-point global comparison with stored images;[7]
(ii)   Global transformations such as Karhunen-Loève,[8,9] Fourier,[10,11] Walsh,[12] moments of inertia,[13] and others;[14,15]
(iii)   Extraction of the local properties, such as endpoints, line crossings, and angles;[16,17,18]
(iv)   Use of curvature and stroke information for analysis;[19,20,21] and
(v)   Structural methods, including decomposition of the character into graphs or other constituent elements.[22,23]

Many techniques contain portions that overlap between these categories. Each technique must be assessed by its ability to "ignore" deformation of the image caused by noise, translation, rotations, style variations, and other distortions, as well as practical considerations of the implementation, such as speed and complexity.

The work of Grimsdale et al. represents one of the earliest attempts at character recognition.[24] In this scheme, each digitized pattern is analyzed for shape by a computer, which extracts heuristic features and compares them to feature values stored on the computer. A few years later the notion of "analysis-by-synthesis" was presented by Eden.[25,26] He initially proposed that all Latin characters could be formulated by only 18 strokes, which in turn could be generated by a subset of four strokes, called segments. More generally, the concept was that handwritten characters are formed by a small, finite number of schematic features, which, when known, can be used for recognition of character data.

Perhaps more than Latin character recognition, the study of the Chinese alphabet is a stringent test for any algorithm.[27] One of the first attempts at this problem was made by Casey and Nagy at IBM.[28] A step-by-step approach was used for this large character set, in which the first stage grouped similar characters, and then "group masks" and "individual masks" were employed to further specify the character. This type of method, in which a hierarchical decision process is employed, is characteristic of several OCR schemes.[29]

Other researchers implemented a more mathematically formulated process for their systems. Tou and Gonzalez used a two-stage system, the first stage performing a series of measurements for subgroup separation and the second extracting a set of specialized features.[30] Pavlidis and Ali used a "split-and-merge" algorithm to produce polygonal approximations of the characters which could provide enough information for decision making,[31] while others used clustering procedures on the task.[32]

The review paper of Suen et al.[33] discusses the efforts in the recognition of handprinted numerals. The best classifications of over 30 studies ranged from 85 to 99.79%, but direct comparison of methods is rarely feasible. This is due to the large discrepancy in the experimental setups. Not surprisingly, the 85% success rate used a realistic data set collected from the U.S. Postal Service[34] while the 99.79% accuracy was derived after training writers to write numerals in specified shapes and sizes.[35] In addition to this widely varying data quality, the number of training patterns was different in each case, and some studies never reserved any patterns for testing of the system after training.

The arrival of neural network concepts into the pattern recognition field has spurned some wonderful successes and great disappointments in the character recognition field. Most studies use the backpropagation algorithm for the training and network architecture.[36] The recent study at AT&T Bell Laboratories is one of the more notable projects, in which postal zip code numerals were trained with a modified backpropagation algorithm.[37] Using this challenging data set of 7291 training patterns, they were able to show 0.14% error on the training set and 5% on the 2007 pattern test set. Fukushima's Neocognitron network[38] also has demonstrated an invariance in recognizing characters of different rotations, sizes, and translations. Other works, such as the ART topologies, have very limited success in detecting any discrepancies among patterns.[39]

The testing of newly established algorithms often relies on a realistic data base for the development of the method. Several popular data sets are widely available for this purpose, the most popular of which are those created by Highleyman,[40] Munson,[41] and Suen.[33] The Munson data set seems to be the most popular, chiefly because of its difficulty. A more recent large data set was created in which the optimal writing style for recognition was also examined.[42] Still, many researchers choose to construct their own data sets, and this is the strategy used in this work.

6.2 DATA COLLECTION

Digits were collected from 13 subjects, who were instructed to write several of each of the digits on a clean sheet of paper. No limitations were imposed as to the style, size, clarity, thickness, or slant of the digits. The numbers were then digitized using a Hewlett Packard Scanjet II digital scanner using the Scanning Gallery software package on an IBM PC. Each of the digits was segmented by hand, scanned at 300 DPI with 16 gray levels, and saved in separate files. The Scanning Gallery program saved the files in the TIF binary file format, and a C program was necessary to convert these files to an ASCII file format, which could be read by subsequent programs. A total of 1500 digits were collected in this manner, approximately 150 of each of the 10 digits types.

There are only a few assumptions that were imposed on the data. Among these are

1. The digits were to be clearly segmentable from one another. That is, a rectangular box could be drawn around each digit so that the entire content of one digit resided inside that box, and no portion of any other digits were contained in that region.
2. The background was relatively noiseless so that there exists a clear threshold between background and digit intensities.
3. No character was rotated more than 45° from what is normally considered its upright position (the character was not upside down).

Figure 6.1 displays many of the digits collected with this process. It is clear that they were written without regard to neatness, and in fact some of the digits appear ambiguous to human classifiers. This variety was encouraged to provide a realistic environment for the training process.

6.2.1 PREPROCESSING

The networks used for feature extraction and classification are highly dependent on spatial overlap of digits of the same class for their success. The original digits were written without regard to this constraint, and so it was necessary to process the digits to alleviate differences in size, thickness, rotation, location, and intensity. This was not expected to destroy the recognition capabilities of either humans or machine, since the information content of the digits is largely contained in the form of the digits. In addition, the resolution of the digits was reduced to prevent prohibitive training time for subsequent modules.

The preprocessing was conducted in the following steps: intensity thresholding to remove noise, center of mass adjustment, line thinning, simultaneous rotation to standard axis and translation to standard center of mass, size determination and fixation, reduction in resolution, and smoothing of digits as a form of anti-aliasing. This sequence is depicted in Figure 6.2 and the methods for these steps are described in the following paragraphs. The inputs to the preprocessing stages come from the digitized characters from the digital scanner, while the outputs of the preprocessing feed into the inputs of the feature extraction network.

6.2.2 NOISE THRESHOLDING

A threshold was applied to each pixel of the original digits, creating a binary image for further processing. The threshold served a dual purpose. It eliminated weak and extraneous information from the digit, thereby aiding a separation from the background. Second, it eliminated intensity variability from within the contour of the digit. Each pixel of the digit was checked against the threshold value. If it was lower, it was set to zero. If the pixel value was equal to or higher than the threshold, it was set to a maximal value (assigned to be 2). An effective threshold was found to be

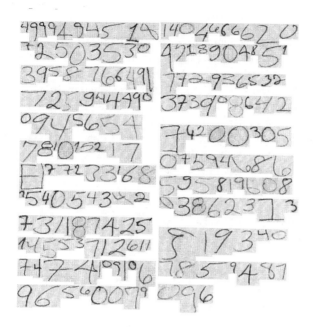

FIGURE 6.1   Random samples of the original unprocessed characters used in this study.

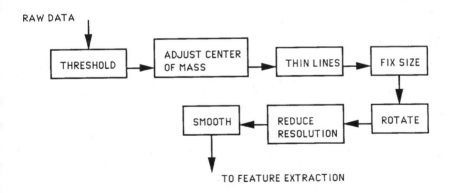

FIGURE 6.2   The sequence of steps in the preprocessing of the handwritten digits.

at a gray level of 4, and this value was used for the processing of all digits. A more flexible approach would have been to use an adaptive threshold, whereby the deciding value is based on the content of each digit by analyzing an intensity histogram. In part because of the controlled lighting conditions of the digital scanner and also due to the assumption of a clear separation of the digit from its background, it was felt that this additional computation was not necessary. This analysis was confirmed by the high quality of the digits after thresholding was applied.

6.2.3 CENTER OF MASS ADJUSTMENT

Particular problems were encountered when some digit types (nines, eights, sixes) had small loops or regions of high density of pen marks. In such instances, the center of mass of the digit was highly skewed toward that region, and overlap with similar digits was often small. This also often resulted in an abnormal rotation when that routine was applied. An adjustment was applied to each digit to expand small regions as this and so move the center of mass toward the absolute center of the digit. In this method, the center of mass ($CM = [x_c \, y_c]^T$) was located and the digit split into quadrants about this point. Each quadrant was then mapped into its corresponding quadrant in absolute space (using the absolute center $AC = [x_a \, y_a]^T$) by scaling each of the regions as

$$x' = \frac{\left(x - x_c\right)\left(x_{max} - x_a\right)}{\left(x_{max} - x_c\right)} \tag{6.1}$$

where the old x coordinate is mapped to the new location x'. The y coordinate is changed in the same way.

6.2.4 LINE THINNING

A thinning routine was used on the binary level digits to reduce the effects of line thickness. The method used is familiar in the literature.[43,18] Basically, the algorithm pares away all boundary points in the digit until it is left with only skeletal pixels, which must be kept in order to preserve the integrity of the digit contour. A pixel is considered to be a skeletal pixel if it is part of the digit (has a nonzero value), one of its four neighbors is zero valued, and it passes either of two conditions as described thoroughly in a previous article.[18] Several passes of this procedure are necessary to reduce the image to one consisting of only skeletal pixels, since the above criteria will remove only boundary pixels with each pass through the digit.

6.2.5 FIXING TO SIZE

Prior to the use of the rotation routine, the image is fixed to a standard size (60×100 pixels). This is necessary to avoid errors in the calculation of the digit principal axes caused by distortions in portions of the digit. To perform this task, the corners of the digit are located and scaled to the new size. Pixels are mapped into the nearest pixel after the scaling factor has been applied. The operation is performed in the same way as Equation 6.6 (a),(b), where the x coordinate magnifier is 60.0 and the y magnifier is 100.0.

6.2.6 ROTATION

The rotation algorithm uses the coordinates of each of the nonzero valued image pixels to find a principal vector of the image. This vector specifies the angle of the principal axis of the digit in two-dimensional space, which can then be manipulated to create a transformation matrix which will rotate the image. Each of the digits is rotated and translated in this space to a standard location and primary axis. The center of mass of the digit ($M = [m_x\ m_y]^T$) is located and used to find a correlation matrix for the digit, calculated as

$$C = \left(\frac{1}{r} \sum_{i=1}^{r} \vec{P}\vec{P}^T \right) - \vec{M}\,\vec{M}^T \tag{6.2}$$

where the summation is over all r nonzero pixels in the image and the vector P is the coordinate vector of the pixel ($P = [p_x\, p_y]^T$). An eigenvector matrix E is calculated from the 2×2 matrix C, encoding the angle (f) through which the primary axis of the digit runs as

$$E = \begin{bmatrix} \cos(\phi) & -\sin(\phi) \\ \sin(\phi) & \cos(\phi) \end{bmatrix} \tag{6.3}$$

Each digit was rotated so that the primary axis lies vertically (90°). Each pixel location $P' = [p_x'\ p_y']^T$ of the rotated image is calculated from the original image as

$$\vec{P}' = E'\left(\vec{P} - \vec{M} \right)^T \tag{6.4}$$

$$E' = \begin{bmatrix} \cos(90° - \phi) & -\sin(90° - \phi) \\ \sin(90° - \phi) & \cos(90° - \phi) \end{bmatrix} \tag{6.5}$$

and the vector P contains the original pixel coordinates. The elements of P' are rounded to the nearest integer locations.

6.2.7 REDUCING RESOLUTION

The corners of the rotated image (smallest rectangle which completely encloses the digit) are found and used to scale the digit to a new resolution of 16×16. This is performed by calculating a new pixel coordinate $[x'\, y']$ by

$$x' = nint\left(\frac{x*16.}{x_{max} - x_{min}} \right) + 1 \tag{6.6a}$$

and

$$y' = nint\left(\frac{x*16.}{y_{max} - y_{min}} \right) + 1 \tag{6.6b}$$

where the *nint*( ) operation nearest integer takes the nearest integer of the resultant division. A pixel in the new 16 × 16 digit is assigned a value of 2 if *any* of the positive valued pixels of the higher resolution are mapped into that location. An alternative is to assign an additional threshold to turn on a pixel in the lower resolution image if the number of original pixels mapping into that location exceeds the threshold. The resultant 16 × 16 images were considered generally to be of good enough character (by subjective analysis) to avoid this additional complication.

6.2.8 BLURRING

As was mentioned previously, one of the assumptions fundamental to the success of the subsequent neural network processors is that of a high degree of spatial overlap of similar digits. That is, because of the hard wiring of neural inputs to image locations, the neural networks are not position invariant. The aforementioned pre-processing steps can aid in creating an invariance but is by no means invincible in this task. To assist in the overlap of the digit contours of similar digits, a simplified smoothing operation was applied to the 16 × 16 images. This operation can also be thought of as an anti-aliasing operation. Basically, if a zero valued pixel has one or more of its four primary neighbors with a nonzero value, that pixel is turned on with a value of 1 (it should be remembered that the pixels on the contour were given values of 2).

6.3 RESULTS

The feature extraction routine (ALOPEX variance maximization of network node outputs using the architecture of Figure 5.5) was applied to each of the digits in the data set. A random 1000 digits were selected for the training of this module, and 32 features were extracted from each 256 dimensional input image.* Since the feature extraction module has an architecture reminiscent of a pipeline, it was more efficient computationally to allow each neuron in the module to complete training before any subsequent nodes were altered. Training appeared most efficient with ALOPEX parameters of $\gamma_0 = a_0 = 5.0 \times 10^{-3}$, $\sigma_0 = 7.5 \times 10^{-3}$, $\gamma_\infty = a_\infty = 5.0 \times 10^{-5}$, $\sigma_\infty = 7.5 \times 10^{-5}$, and $\tau = 1000$, and typically required between 8,000 and 12,000 iterations per node for a good convergence, as seen by the response curve of Figure 6.3. The vast bulk of the processing time is due to the large number of patterns (1000) used in the training, since each pattern must be presented to the neuron during each iteration.

The number of features to retain was calculated by plotting the eigenvalues in descending order as generated from the conventional K-L expansion, as in Figure 6.4. The magnitude of the eigenvalues is identical to the optimum variances of the cell outputs in the Feature Extraction (FE) network, and it is convenient to relate the magnitude of the eigenvalue with the amount of information the corresponding K-L vector carries. The number of features to extract (32) was subjectively obtained

---

* Note that the neural networks in this study have gloval inputs and are not spatially interdependent. This means that the preprocessed 16×16 digits are viewed as a 256-dimensional vector by the networks.

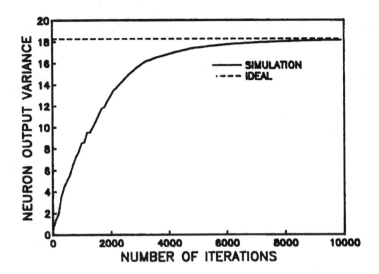

FIGURE 6.3   A response curve vs. iteration number. A convergence can be observed in about 10,000 to 12,000 iterations.

from Figure 6.4 as the point in which the information given by an additional eigenvector reduces to near zero. Another way of finding the optimum number of neurons in the feature extracting network is to set a threshold. If a neuron optimizes to an output variance below this threshold, the node is not logically added to the network, and the training simulation is stopped. In this way both the extent of the network and its connectivities can be adaptable in the training.

Figure 6.5 depicts the feature cell vectors as they would appear in image form. Each connectivity strength is given a corresponding intensity (relative to the strength of the connectivity) in the spatial position where the input to the connection arose. Very high intensities (white) indicate large positive connections, and large negative connections are shown as a low intensity value (black). Some of the feature "filters" have regions of high contrast, which remind us of features in the character data set. It is clear that the last few feature images are quite "noisy", and this is consistent with their low information content. Moreover, it is very obvious that these "optimal" vectors would be very difficult to specify heuristically.

The clustering operation needed to incorporate some understanding of the number of clusters necessary to accurately describe the data. Since the ALOPEX optimization for the clustering operation was quite time consuming, the standard FCM algorithm with Euclidean distance measurements was used to find the cluster validity measures described in Section 5.4 for as few as 2 and as many as 40 clusters. These simulations typically required no more than 30 iterations for convergence.

In Section 5.2.1 the determination of the number of clusters necessary for any given data space was discussed (the cluster validity problem), and the cluster validity measures F and H were introduced. Figure 6.6 illustrates the change in the validity measure H for the converged clusterings of the FCM algorithm from

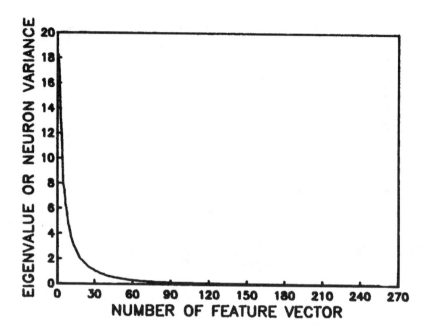

FIGURE 6.4   The eigenvalue as a function of the number of the K-L expansion vector. The eigenvalue can also be thought of as the optimum nodal output variance for the ALOPEX-trained feature cell.

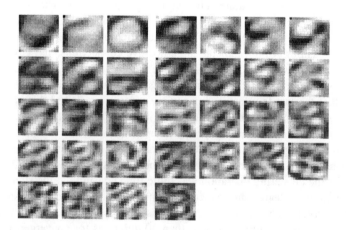

FIGURE 6.5   The 32 feature cell connection vectors displayed in image form. The vectors are shown in descending order of their output cell variance as you view from left to right and top to bottom.

2 to 40 clusters (q = 1.2). It is hoped that a distinct minimum in the entropy (H) measure and a distinct maximum in the partition coefficient (F) measure will present themselves definitively around a certain value of c. This is not the case in any of the plot of Figure 6.6. In fact, the data space created by these digits appears rather homogeneous in nature, with few well-separated regions for simple cluster identification. This may be an artifact of the high number of patterns used in the cluster formation, which may "fill in" many of the less dense regions of feature space used for simple cluster identification.

The credibility to the notion that the data space is highly uniform is enhanced by the extremely low value of q which was necessary to form clusters. At q = 1.2, the clusters are formed with very sharp decision boundaries. When a more commonly used value (q = 2) was used, the FCM algorithm converged every cluster center to the same point, so that the class memberships were entirely fuzzy, and no distinguishing information was provided.

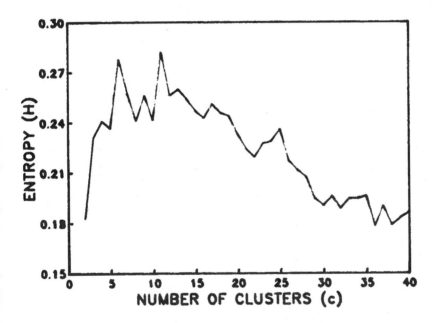

FIGURE 6.6 The variation in the entropy (H) for choices for the number of clusters (c) from 2 to 40.

There are two regions in the curve of Figure 6.6 in which it is reasonably safe to assume that there are relatively "better" clusterings than for other c values. The first is for the value of c = 2, which the curve of Figure 6.6 shows as locally optimal. For the purposes of this study, the value of c = 2 had to be rejected simply because of the understanding that there are at least 10 clusters desired. This is because of the 10 digit types (zero through nine) used in the data set.

The second region occurs for values of c > 30. In this region of the curve of Figure 6.6, there begins a "plateau" region, beyond which a mental extrapolation of the curves would anticipate little improvement for a much higher number of clusters.* The region from c = 30 to c = 40 is heuristically an acceptable region and still maintains an adequate number of average samples (25–33) per cluster. The heurisitic basis for the credibility of this range of c values resides in the belief that each of the digit types can be written in, on average, 3 to 4 different styles. For example, a one can be written as a single vertical line, or with additions of an upper diagonal line alone or with an accompanying lower horizontal line.

It is interesting to note that there is indication that the FCM algorithm was not finding the globally optimal solution to the clustering problems it was presented with. One evidence of this was that the cost function value of the converged solution was often higher than one of the intermediate solutions through which the simulation had passed. But by far the simplest determination of locally optimal solutions is to run the program several times with the same parameter set and the same patterns. When this was performed at c = 30, the FCM algorithm obtained different solutions each time, as evidenced by discrepancies in the cost function value** and cluster membership distributions (the array F from section 5.3). This lends further credence to the use of an optimizer in the FCM routines.

The ALOPEX-trained FCM algorithm was trained on 30 clusters for the same 1000 training patterns. In order to reduce the computational overhead, the simulation was started by using the center coordinates converged upon by the standard FCM algorithm. The simulation typically required between 1,000 and 2,000 iterations for a "good" convergence and seemed to perform best with ALOPEX parameters of γ_0 = a_0 = 0.2, σ_0 = 0.3, γ_∞ = a_∞ = 2.0 × 10^{-3}, σ_∞ = 3.0 × 10^{-3}, and τ = 2500. Primarily for computational reasons, the Euclidean distance measure was used in the ALOPEX-trained FCM algorithm. The use of a non-Euclidean measure would have, for this application, resulted in exorbitant execution times, since the calculation of the covariance matrices results in substantial computational overhead. Another reason for not using the fuzzy-covariance matrices in the formulation of a non-Euclidean distance metric was that later simulations showed that such a selection seemed to result in an unusually hard membership assignment, which may be disadvantageous for medical applications in particular.

Table 6.1 shows the classification results for the ALOPEX-modified FCM scheme with the labeling method described in Section 5.4. The total classification accuracy is 86.3% for the 1000 training digits, and 86.0% for the 500 post-training digits, as indicated in Table 6.2.

Figure 6.7 depicts the cluster centers as images (using the method of Section 5.4), to give us a flavor for the aspects of the characters which each cluster emphasizes. As Figure 6.7 shows, most clusters have fields that are strongly reminiscent of one of the digit types, but there are a few clusters which are blends of portions of several types.

---

* This was partially confirmed with a simulation performed at c=50, which indicated a continuation of this trend.
** When the FCM algorithm was run twice at c=30, final cost function values of J=52422.17 and J=56321.15 were obtained.

TABLE 6.1
A Comparison of the Classification Results of the ALOPEX-Trained Network with the Actual Pattern Identities of the Digits Used in the Training. Of the 1000 Training Digits, 863 were Correctly Classified (86.3%).

	Assigned Digit Class									
	0	**1**	**2**	**3**	**4**	**5**	**6**	**7**	**8**	**9**
0	96	0	0	0	0	0	3	0	1	0
1	0	85	3	0	6	1	0	2	3	0
2	0	2	88	0	1	0	0	6	2	1
3	0	0	1	86	0	2	0	3	7	1
4	0	0	0	0	84	6	1	0	0	9
5	0	0	0	0	0	96	2	0	2	0
6	1	1	0	0	2	14	81	0	1	0
7	0	0	0	0	11	0	0	89	0	0
8	0	1	0	5	1	1	1	0	90	1
9	2	0	0	5	12	3	0	5	5	68

Figure 6.8 shows the misclassified digits as they appeared in their original unprocessed form, grouped by the digit type with which they were incorrectly identified with. Some of the misclassifications can be connected directly with pre-processing problems (i.e., improper rotations, noise in the image retained), while others are probably due to the strong overlap between certain characteristics of the digit types.

The data set was also tested with the backpropagation neural network training algorithm. This technique was described thoroughly in Chapter 2. A direct comparison of the backpropagation results with the ALOPEX-trained network developed in this study is not equitable, since the backpropagation algorithm is a supervised technique. However, since backpropagation is so widely used, and since it has been used in the specific application of character recognition, the results that it provides can give a calibration of the difficulty of the data set. These results can also demonstrate the degree of additional accuracy which can be extracted by knowing the pattern identities *a priori*.

The training was conducted with a network comprising 256 input nodes, 100 hidden nodes, and 10 output nodes on 1000 input patterns (consisting of the same preprocessed character training set as was used in the ALOPEX-trained system). The desired low value of the output lines was set at 0.1 and the desired high value at 0.9. The network was trained for 300 epochs with values of h = 0.1 and a = 0.75. Upon the completion of training, the training pattern classification error was determined by assigning it the class identity of the output node with the highest activity.

TABLE 6.2
A Comparison of the Classification Results of the ALOPEX Trained Network with the Actual Pattern Identities of the Digits Not Used in the Training. Of the 500 Digits, 430 were Correctly Classified (86.0%).

	Assigned Digit Class									
	0	**1**	**2**	**3**	**4**	**5**	**6**	**7**	**8**	**9**
0	46	0	0	0	1	0	1	0	0	0
1	0	41	3	0	3	0	1	1	1	0
2	0	6	44	0	0	0	0	2	0	0
3	0	0	0	41	0	2	0	2	4	1
4	0	0	0	0	41	2	1	0	0	4
5	0	0	0	0	0	49	1	0	0	0
6	0	0	0	0	0	5	47	0	0	0
7	0	0	1	0	2	0	0	46	0	1
8	0	0	0	2	1	1	0	0	48	0
9	2	0	0	3	12	1	0	1	2	27

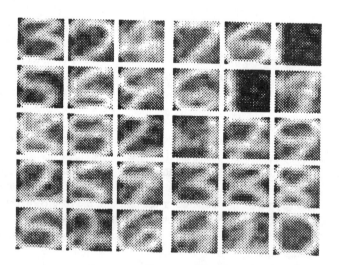

FIGURE 6.7   The 30 cluster centers displayed in image form.

For the 1000 training patterns, the backpropagation network correctly classified all but two of them, for an accuracy of 99.8%. The 500 patterns not used in the training were classified with 93% accuracy, as shown in Table 6.3 below.

6.4   DISCUSSION

Our primary interest in this application is to be able to fine-tune the training algorithm so that it is of maximum efficiency and accuracy for subsequent medical applications. In this regard, the classification of handwritten digits tests the limits of the applicability of the method. This is because the large number of clusters, features, and patterns stress the algorithm to its maximum load. The computing times for all phases of the ALOPEX-trained algorithm were significant, but the accuracy of the optimization was nearly ideal in the feature extraction training. For the clustering module, the ALOPEX simulation for c=30 provided a moderate improvement over the standard FCM algorithm. Clearly a much more substantial computational demand is caused by the use of a non-Euclidean distance metric, particularly when the calculation of a fuzzy covariance matrix is required. For a large cluster, large pattern set application such as this, the Euclidean metric becomes one of the only feasible possibilities.

TABLE 6.3
A Comparison of the Classification Results of the Backpropagation Trained Network with the Actual Pattern Identities of 500 Digits Not Used in the Training. Of the 500 Digits, 465 were Correctly Classified (93%).

	Assigned Digit Class									
	0	**1**	**2**	**3**	**4**	**5**	**6**	**7**	**8**	**9**
0	46	0	0	0	1	0	1	0	0	0
1	0	42	2	1	0	0	1	3	1	0
2	0	5	46	1	0	0	0	0	0	0
3	1	1	0	44	0	0	0	1	0	3
4	0	0	0	0	47	0	1	0	0	0
5	0	0	0	0	0	47	1	0	1	1
6	0	0	0	0	0	0	51	0	1	0
7	0	1	1	0	0	0	0	47	0	1
8	0	0	0	1	0	1	0	0	49	1
9	0	0	0	0	1	0	0	0	1	46

One of the largest problems appears to be the determination of the number of clusters necessary for an accurate depiction of the data space. Both of the cluster validity measures we used, along with about a half dozen others not included in this document, were not able to give us a definitive idea of the proper number of clusters. There is a natural tendency for all of the measures to drift toward their ideal values as the number of clusters increases, since when the number of clusters equals the number of patterns, we have a trivial but perfect set of clusters. Whether the plateau region of the curves of Figure 6.6 is an artifact from this tendency is unknown, but since the number of clusters was still substantially lower than the number of patterns (30 << 1000), the assumption is that the saturation of the validity measures after c=30 contains real information.

FIGURE 6.8   A sample of some of the characters misclassified as (a) nines, (b) eights, (c) sevens, (d) sixes, (e) fives, (f) fours, (g) threes, (h) twos, (i) ones, and (j) zeroes. Each character is shown in its unprocessed form, with the preprocessed character shown directly beneath it.

As was mentioned earlier, the low value of q and the lack of a distinct local extremum in the cluster validity measures are indicative of a data space with few if any well-separated and compact clusters. The data space appears to be naturally "fuzzy". It is curious that the FCM algorithm requires a rather hard decision to partition this fuzzy data space, while the theoretical reasoning for fuzzy logic presumes a fuzzy algorithm as ideal for this type of decision making. In any event, it is unclear whether the ALOPEX-trained FCM algorithm really arrives at a good clustering of the data space, or whether there is just a large enough number of cluster centers to "fill in" the data space.

This low value of q may account for many of the classification errors presented in Figure 6.8. Many of the misclassified digits are visually clearly a member of another class. The sharp decision boundaries created by the low value of q can push the cluster membership strengths toward their limits (zero and one). Thus, even though a character lies near the boundaries between clusters, it is given a strong membership to a cluster. This pushes a marginal pattern (i.e., a pattern with 50% similarity to each of two clusters) to become decisively incorrect! With this information misgiven in the clustering module, no labeling scheme can reclaim a correct classification.

The errors in the classification of the handwritten digits appear to arise from multiple sources. A few of the characters are of uncertain identity to the authors and several others who have viewed the data, and so it is unreasonable to expect that the computer algorithm should perform any better. Many of the erroneous classifications can be traced to the preprocessing of the characters. This is perhaps the most crucial and controllable portion of the system, and yet very difficult to design and improve. Alterations to any of the algorithms seemed to improve the operations on some characters to the detriment of several others. There are some adjustments that can and should be made to any implementable system, including the addition of thresholds to the resolution reduction and a contour tracing program to alleviate noisy elements which cannot be removed by threshold analysis and yet contribute to improper rotations.

Even with these contributions, the most significant source of error is from the decisions made by the classification system itself. The question becomes how these erroneous decisions can be reduced. The answer seems to lie in the selection of the training set and other parameters of the simulation, including the selection of the resolution of the digit representation. Subjectively, using an 8×8 image would probably result in even larger classification errors, as there is barely enough resolution to represent unambiguously most of the digits with a 16×16 image. A slightly higher resolution would probably provide some relief but was not feasible computationally with the speed and memory of the machines available.

Most importantly, the composition of the data set seems to be in question. Most other studies with digit recognition have used a much larger training set, up to 10 times the size. This larger training set helps to generalize the networks trained with supervised algorithms. The data set used in this study may have more than a "normal" share of unusual and exceptional characters, as several of the providers of the data (notably the authors) intentionally wrote the characters using various writing styles, slants, and sizes.

The use of the backpropagation algorithm allows a calibration of this method with a more widely used strategy. It appears that this data set is of comparable difficulty to the postal zip code data used by LeCun et al.,[18] since both data sets were tested with the backpropagation technique and performed similarly. Both data sets performed nearly perfectly in classifying the training set, and the untrained set classified 93% accurately in this study, as opposed to 95% accuracy for that AT&T Bell Labs group.*

Given that the unsupervised system developed in this study was not privy to the class identities during training, the 86.3% accuracy of the method is, in our opinion,

---

* The slightly higher accuracy of their study is probably attributable to the slightly variant neural architecture they used to accentuate certain inherent aspects of the digits.

outstanding. Even more striking is the ability of the unsupervised system to generalize its decision capabilities more easily than the supervised system, as indicated by the 86.0% error in classifying the untrained data. The inherent ability of unsupervised decision making to generalize more readily, and with fewer training patterns, was hypothesized as one of the motivations for the construction of the system. It is clear, at least for this application, that the decisions made by the unsupervised classifier are more general than that of backpropagation, in that there is little loss of information after training. It is also clear that the inherent accuracy, whether with regard to the training set or test set, is greater for the backpropagation system.

This understanding gives credence to a recent trend to incorporate unsupervised and supervised decision making in the same system. Each scenario is beneficial at different times. It may be particularly useful to use an unsupervised decision until the time that there are enough training patterns to construct supervised decisions. One would expect that if the training set for this study were reduced (say from 1000 patterns to 500 patterns) that the unsupervised method would have classification accuracy on the untrained data closer to the supervised decision accuracy on the same data set.

There is one other important aspect which should be considered in the comparison of these two techniques: the amount of hardware resources necessary to implement these methods in a true parallel form. For the backpropagation system, the decisions were made with 110 computing nodes and 26,600 connections, versus 93 nodes and 17,178 connections for the ALOPEX-trained unsupervised system. The savings in connections for the unsupervised method are the most critical, since this is the most challenging aspect of implementing these systems in parallel (the hardware interconnects require a great deal of space). Additionally, the unsupervised system requires only 18,015 additions and 17,388 multiplications, while backpropagation requires 26,490 additions and 26,600 multiplications.

This discussion would not be complete without some speculation about the strategy used for the character recognition process. It is our belief that the primary limitation of the classification is the assumption that the spatial form of the digit is the most important aspect of character recognition. It is more likely that an analysis of the stroke curvatures and other contour-based principals will provide more information to the classifier. As an example, notice how much confusion there was (Figure 6.8) in classifying the number two from the way many people write the number two. As difficult as this was for the software, it is remarkably easy for humans, even though the only real distinction is the smooth curvature of the number two versus the sharp vertices of the number one. The most promising systems appear to be those that can retain this contour-based information of curvature and stroke direction, or if shape information is still important, preserve the pixel neighborhoods in the analysis.

6.5 SUMMARY

An unsupervised pattern recognition system has been introduced, largely based on methods originating elsewhere, but bound by their application to a parallel architecture and their ability to be trained by a single optimization algorithm. The method has been tested in two widely varying application domains: in the classification of handwritten digits and the diagnosis of Visual Evoked Potential signals of normal and abnormal subjects. The scope of the second application is given in the next chapter.

The ALOPEX-trained system has proven itself capable of strong generalizations, as is evidenced by the application to handwritten digits, and is able to extract a significant amount of information without the advice of an omnipotent instructor. When properly tuned, the clustering module can make decisions of an analog nature, so that an understanding of the certainty of its decision can be analyzed.

To artificial neural network purists, this architecture can only loosely be referred to as a neural network. In the sense that the system is trained by example, highly parallel, and comprised of highly interconnected elements performing simple computations, the neural network label is quite fitting. If, however, the label also connotes a system trained through local information sharing and nodal units based only on inner product variations,* then a more general label should be applied. In the truest sense, the backpropagation algorithm is not a "local" training regime, since the errors propagate through the layers imaginatively and do not actually reside on any signal lines.

The reason for mentioning this small labeling problem addresses the direction of the neural network community as a whole. Only recently has the field begun to merge with other more well-established disciplines, partly because of limited utility alone and partly because of a reluctance to share in the spotlight. In the context of parallel processing systems in general, the neural network extension has a good deal to gain by attributing more computing power to the processing elements. From a hardware perspective alone, the high connectivity of "pure" neural network systems has been a technological stumbling block. Providing more power to the "neuron" means releasing the tight relationship to neurobiology upon which many researchers rely.

One of the fields that has long been tightly interwoven with the neural network field is that of combinatorial optimization. The usual hope is that the optimization schemes, such as the ALOPEX technique used in this chapter, will provide a higher probability of reaching a globally optimal solution. That ALOPEX, in this network construction, can provide this is certain. Whether it is always computationally necessary is less certain.** More overlooked but of primary importance is the utility of a proven optimization scheme as a flexible and reliable design tool. Regardless of the architecture, if the information which the user desires to retain in the network after training is expressible through the minimization or maximization of a function of the network, then ALOPEX can be used to find that information. In our laboratory,

---

* The clustering module differs from this in that some of the units perform squaring and difference operations.
** The ALOPEX simulation results for the clustering module in the application to the VEP's performed no better than the standard FCM simulation.

ALOPEX has been used as an alternative to the backpropagation training on the multilayer perceptron architecture, as well as in the training of Hopfield nets. It seems reasonable that specialized problems can be solved with this general tool. In this sense ALOPEX can perform, as it has in this study, as a conversion of the desired information onto a parallel architecture.

As hoped, the conclusion of this study has produced more questions and promising directions than answers. The character recognition arena was an interesting demonstration, but it is unlikely that this scheme can ever compete equally with the supervised methods. Still, the blending of unsupervised and supervised training methods at varying times in the learning process is intriguing and probably beneficial if a suitable application is available.

REFERENCES

1. Ullmann, J. R., Advances in character recognition, in *Applications of Pattern Recognition*, Fu, K. S., Ed., CRC Press, Boca Raton, FL, 1982, 197.
2. Taxt, T., Flynn, P. J., and Jain, A. K., Segmentation of document images, *IEEE Trans. PAMI*, 11(12), 1322, 1989.
3. Suen, C. Y., Shinghal, R., and Kwan, C. C., Dispersion Factor: A Quantitative Measurement of the Quality of Handprinted Characters, *Proc. Int. Conf. Cybernet. Soc.,* 1977, 681.
4. Suen, C. Y., Factors Affecting the Recognition of Handprinted Characters, *Proc. Int. Conf. Cybernet. Soc.*, 1973, 174.
5. Mantas, J., An overview of character recognition methodologies, *Patt. Recog.*, 19(6), 425, 1986.
6. Gaillat, G. and Berthod, M., Panorama des techniques d'extraction de traits caracteristiques en lecture optique des caracteres, *Rev. Tech. Thomson - CSF*, 11, 943, 1979.
7. Golshan, N. and Hsu, C. C., A recognition algorithm for handprinted arabic numerals, *IEEE Trans. Syst. Sci. Cybern.*, 6, 246, 1970.
8. Gudeson, A., Quantitative analysis of preprocessing techniques for the recognition of handprinted characters, *Patt. Recog.*, 8, 219, 1976.
9. Krause, P., Schwerdtman, W., and Paul, D., Two Modifications of a Recognition System with Pattern Series Expansion and Bayes Classifier, *Proc. 2nd. Int. Joint Conf. Pattern Recognition*, 1974, 215.
10. Granlund, G. H., Fourier processing for hand print character recognition, *IEEE Trans. Comput.*, 21, 195, 1972.
11. Persoon, E. and Fu, F. S., Shape discrimination using Fourier descriptors, *IEEE Trans. SMC*, 7, 170, 1977.
12. Andrews, H. C., Multi-dimensional rotation in feature selection, *IEEE Trans. SMC*, 7, 537, 1977.
13. Tucker, N. D. and Evans, F. C., A Two-Step Strategy for Character Recognition Using Geometrical Moments, *Proc. 2nd Int. Joint Conf. Patt. Recog.*, 1974, 223.
14. Niemann, H., A Comparison of Classification Results in Character Recognition by Man and by Machine, *Proc. 3rd. Int. Joint Conf. Patt. Recog.*, 1976, 144.
15. Ott, R., On Feature Selection by Means of Principal Axis Transform and Nonlinear Classification, *Proc. 2nd Int. Joint Conf. Patt. Recog.*, 1974, 220.
16. Spanjersberg, A. A., Combinations of Different Systems for the Recognition of Handwritten Digits, *Proc. 2nd Int. Joint Conf. Patt. Recog.*, 1974, 208.

17. Kwon, S. K. and D. C., Lai, D. C., Recognition Experiments with Handprinted Numerals, *Proc. Joint Workshop Patt. Recog. Artificial Intelligence*, 1976, 74.

18. Beun, M., A flexible method for automatic reading of handwritten numerals; *Phillips Tech. Rev.*, 33, 89, 130, 1973.

19. Iwata, K., Yoshida, M., and Tokunaga, Y., High Speed OCR for Handprinted Characters, *Proc. 4th Int. Joint Conf. Patt. Recog.*, 1978, 826.

20. Hosking, K. H., A contour method for the recognition of handprinted characters, in *Machine Perception of Patterns and Pictures*, The Institute of Physics, London, England, 1972, 19.

21. Toussaint, G. T. and Donaldson, R. W., Algorithms for recognizing contour-traced handprinted characters, *IEEE Trans. Comput.*, 19, 541, 1970.

22. Watt, A. H. and Beurle, R. L., Recognition of Handprinted Numerals Reduced to Graph-Representable Form, *Proc. 2nd Int. Joint Conf. Artificial Intelligence*, 1971, 322.

23. Sue, T.-J. and Chen, Z., Skeleton Chain Code Approach to Recognition of Handwritten Numerals, *Proc. Nat. Computer Symp.*, 1976, 4.15.

24. Grimsdale, R. L., Sumner, F. H., Tunis, C. J., and Kilburn, T., A system for the automatic recognition of patterns, *Proc. IEE*, 106B, 210, 1958.

25. Eden, M., On the formalization of handwriting, in *Structure of Language and its Mathematical Aspect*, American Mathematical Society, Providence, RI, 1961, 83.

26. Eden, M., Handwriting generation and recognition, in *Recognizing Patterns*, Kolers, P.A. and Eden, M., Eds., MIT Press, Cambridge, MA, 1968, 138.

27. Mori, K. and Masuda, J., Advances in Recognition of Chinese Characters, Proc. 5th Int. J. Conf. Patt. Recog, Miami, 1980, 692.

28. Casey, R. and Nagy, G., Recognition of printed Chinese characters, *IEEE Trans. Elec. Comput.*, 15, 91, 1966.

29. Parks, J. R., An Articulate Recognition Procedure Applied to Handprinted Numerals, *Proc. 2nd Int. J. Conf. Patt. Recog.*, Copenhagen, 1974, 416.

30. Tou, J. T. and Gonzalez, R. C., Automatic recognition of handwritten characters via feature extraction and multilevel decision, *Int. J. Comput. Inf. Sci.*, 1, 43, 1972.

31. Pavlidis, T. and Ali, F., Computer recognition of handwritten numerals by polygonal approximation, *IEEE Trans. SMC*, 5, 610, 1975.

32. Suen, C. Y., The Role of Multi-Directional Loci and Clustering in Reliable Recognition of Characters, *Proc. 6th Int. J. Conf. Patt. Recog.*, Munich, 1982, 1023.

33. Suen, C. Y., Berthod, M., and Mori, S., Advances in Recognition of Handprinted Characters, *Proc. 4th Int. Conf. Patt. Recog.*, Kyoto, 1978, 30.

34. Neill, J., Numeric Script Mail Sorter, *Proc. Automat. Patt. Recog.*, 1969, 49.

35. Masterson, J. L. and Hirsch, R. S., Machine recognition of constrained handwritten arabic numerals, *IRE Trans. Hum. Factors Electron.*, 3, 62, 1962.

36. Rajavelu, A., Musavi, M. T., and Shirvaikar, M. V., A neural network approach to character recognition, *Neural Net.*, 2, 387, 1989.

37. LeCun, Y., Boser, B., Denker, J. S., Henderson, D., Howard, R. E., Hubbard, W., and Jackel, L. D., Backpropagation applied to handwritten zip code recognition, *Neural Comput.*, 1, 541, 1989.

38. Fukushima, K., Neocognitron: a new algorithm for pattern recognition tolerant of deformations and shifts in position, *Patt. Recog.*, 15(6), 455, 1982.

39. Carpenter, G. A., Neural network models for pattern recognition and associative memory, *Neural Net.*, 2, 243, 1989.

40. Highleyman, W. H., An analog method for character recognition. *IRE Trans. Elec. Comput.*, 502, 1961.

41. Munson, J. H., Experiments in the recognition of handprinted text: part I - character recognition, *Proc. AFIPS*, 33, 1125, 1968.
42. Shingal, R. and Suen, C. Y., A method for selecting constrained hand-printed character shapes for machine recognition, *IEEE Trans. PAMI*, 4(1), 74, 1982.
43. Pavlidis, T., A thinning algorithm for discrete binary images, *Comput. Graph. Image Proc.*, 13, 142, 1980.

7   An Unsupervised Neural Network System for Visual Evoked Potentials

Timothy J. Dasey and Evangelia Micheli-Tzanakou

7.1   INTRODUCTION

Visual Evoked Potentials (VEPs) have been used in the clinical environment as a diagnostic tool for many decades. Stochastic analysis of experimental recordings of VEPs may yield useful information that is not well understood in its original form. Such information may provide a good diagnostic criterion in differentiating normal subjects from subjects with neurological diseases, as well as provide an index of the progress of the diseases. VEPs are assumed to reveal basic functional entities in the brain that correspond to the peaks observed in their waveforms. These peaks are the so-called N1 (60–80 msec), P1 (90–110 msec), N2 (135–170 msec), P2 (180–200 msec) and P3 (280–350 msec).[10] The ranges in the parentheses denote the time period over which each peak is to appear in the VEP waveform of a normal subject.

VEPs have been used in the detection of certain diseases which result in cortical malfunctioning, such as Multiple Sclerosis (MS), Alzheimer's disease, and others.[11,12] The detection is currently based on distortions of the waveform and mainly on peak occurrence outside its normal range and even peak disappearance.

More recently, the use of color vision tests on patients with MS has become more common. Harrison et al.[6] used a battery of color vision tests in their study, with 65% of the MS subjects failing at least one of the tests. They also used VEPs using standard black and white stimuli in their study. No correlation between the VEPs and color vision abnormalities was found. Frederiksen et al.[5] performed similar studies. Their color vision tests revealed mostly BLUE/YELLOW defects as well as some RED/GREEN. In our studies,[8,9] the circular checkerboard pattern was shown to be a more effective stimulus than the standard checkerboard, and some MS color defects were detected. A prolongation of the N1 peak and a reduction of the N2 peak were found for the circular checkerboard (CCB) pattern. Although the color patterns produced the same overall latency effect as the black and white patterns, there were distinct differences in the responses for the color VEPs. The BLUE/YELLOW combination gave results most similar to that of black and white patterns in terms of peak latencies. The other combinations, RED/GREEN and

RED/BLUE, caused rightward shifts of the main peaks, including P1. Although there is no guarantee that all MS patients will experience visual defects, the VEP has been shown to be a more sensitive indicator of MS than either the Somatosensory Evoked Potential (SEP) or the Brainstem Auditory Evoked Potential (BAER).[7]

In this study, we report on results from the training of an unsupervised neural network system in the detection of differences in VEPs between normal subjects (control) and MS patients. Since an unsupervised system does not rely on the accuracy of any external experts but rather analyzes only the statistics of the patterns and how they relate to one another, it is not sensitive to errors from a clinical diagnosis. As such, it will be able to provide a more accurate generalization with fewer training patterns.

7.2 DATA COLLECTION AND PREPROCESSING

The Visual Evoked Potentials (VEP) used in the study were collected and analyzed in previous works.[8,9] Patterns presented included the typical checkerboard but also some atypical patterns such as "windmills", "concentric circles", and "circular checkerboards". These patterns are shown in Figure 7.1. In addition to black and white presentations of these patterns, BLUE/YELLOW, RED/GREEN, and RED/BLUE color combinations were used, for a total of 16 patterns presented to each patient. The conditions for the experiment were kept as consistent as possible, with careful control given to background noise and room lighting, stimulus size and luminance, and patient positioning.

The normal subjects were volunteers from the Rutgers University community. The MS patients were volunteers from around the New Jersey area (28 to 58 years old, with a mean of 48 years), and all consented to the experimental procedure in writing. A prescreening of the patients was made in order to determine most closely the important factors in their medical and visual history. The MS patients were consulted about the progress of the disease. Most of the subjects could be grouped under that of control subjects (NL) or as afflicted with some degree of MS (as determined by their physicians). We used 13 control, 9 definite multiple sclerosis (DMS), and 3 probable multiple sclerosis (PMS) subjects, as classified by their physicians. The exceptions included one patient who was originally diagnosed with MS and later found to be suffering from neurosarcoidosis (NSD), a disease that can mimic MS. Additionally, one of the control subjects was also categorized separately from the other normals, since that patient had difficulty controlling the movements of the left eye. The patient disease states were hidden from the unsupervised classifier until all stimulations were completed.

The signals were collected with a bipolar midline electrode placement (Oz–Cz), sampled at 1000 Hz, and averaged over 64 stimulus presentations. Patterns were presented for 512 ms; then the pattern was reversed for another 512 ms during each presentation. Color VEP collection was preceded by a subjective determination of the patient's isoluminence points.

FIGURE 7.1 The stimulus patterns used for collection of the VEPs, (a) checkerboard (CB), (b) concentric circles (CC), (c) windmill (WM), and (d) circular checkerboard (CCB).

7.3 SYSTEM DESIGN

A block diagram of the components of the system was shown in Figure 5.3. This section summarizes the design of the unsupervised pattern recognition system used in this research. For more details on the system, the reader should consult Chapter 5. As seen earlier, the system first uses a feature extractor to reduce the dimensionality of the input pattern set and derive a subset of features which a classifier can later use more readily to assign patterns to a particular class. Both the feature extractor and classifier are trained by the optimization routine ALOPEX[11] (for a discussion of this method see also Chapter 5).

Figure 5.5 showed a subset of the neural network used for feature extraction. Each node in the network is trained by ALOPEX to find a set of weightings which maximize the variance of the output of each node. Successive nodes are inhibited by their predecessors, so that the feature axis, which prior nodes had chosen, is excluded from consideration. The result is that the network of nodes finds the n highest variance features of the input pattern space, where n is the number of nodes in the feature extraction network. The n highest variance nodes correspond to an n-dimensional Karhunen-Loève feature extraction. This is shown to be an optimal linear solution when other information about the pattern identities is not available.[2] In this study we chose to hide any information about the pattern classes (such as a physician's diagnosis) from the pattern recognition system to preclude the system from any biases.

The vectors of feature values output from the feature extraction network for each input pattern is subjected to a fuzzy classifier, based on an ALOPEX modification of the fuzzy C-mean clustering algorithm.[1] This module seeks to select a set of centers in the n-dimensional feature space which are representative in a geometric sense of a set of patterns which are closely related. The number of cluster centers to be found was hand selected based on a cluster entropy measure, which helps describe the degree of organization of the input patterns about the cluster centers. The result of the clustering module is a set of numbers for each pattern signifying the degree of similarity of each pattern to each cluster center. These cluster membership strengths are then analyzed against the known pattern identities and assessed for correlation.

Note that the entire pattern recognition system is a linear operation on the input patterns. This may not be entirely optimal for some pattern distributions, but it does allow the cluster centers to be reconstructed by an inverse linear operation as if they were input patterns. This is important for an unsupervised system in helping the users understand the reasoning of the system in making a decision about a pattern's class identity.

7.4 RESULTS

The number of training patterns varied between 27 and 21, depending on the stimulation pattern used to collect the VEP. When training the feature extraction module, as discussed in Chapter 5, each neuron in the module was allowed to complete training before any subsequent nodes were altered. Training was successful in 20,000 to 50,000 iterations with ALOPEX parameters of $\gamma_0 = \alpha_0 = 5.0 \times 10^{-3}$, $\sigma_0 = 7.5 \times 10^{-3}$, $\gamma_\infty = \alpha_\infty = 1.0 \times 10^{-4}$, $\sigma_\infty = 1.5 \times 10^{-4}$, and $\tau = 7500$.

Figure 7.2 shows the eigenvalue (or optimal output variance) of the ith feature node as a function of the number of features i for the patient waveforms generated through stimulation with a BLACK/WHITE checkerboard (BWCB). The number of features was chosen to be six, since this number seemed to be at a point after which little additional information could be gleaned from subsequent features, as seen in Figure 7.2. This happens to coincide with the number of K-L vectors chosen for another study with VEPs. This choice of the number of features was used for all patterns of stimulation, since the construction of plots similar to Figure 7.2 for other patterns indicated the value of six also to be near the information saturation region.

Cluster module training was successful in 700 to 1500 iterations for a two-cluster task with ALOPEX parameters of $\gamma_0 = \alpha_0 = 25.0$, $\sigma_0 = 37.5$, $\gamma_\infty = \alpha_\infty = 0.5$, $\sigma_\infty = 0.75$,

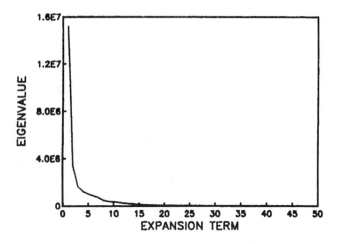

FIGURE 7.2   The change in converged output variance (eigenvalue) of the ith feature node (expansion term) as i varies for the VEPs generated from the black/white checkerboard pattern.

and $\tau = 750$ when using 27 patterns in a six-dimensional feature space. Manual examination of an entropy cluster validity measure confirmed the best cluster separation for a two-cluster stimulation for all training pattern data sets.

The two-cluster results for a sample of the stimulation patterns are shown in the form of a histogram in Figure 7.3. A one-dimensional display of the results is possible for this two-dimensional output space since the cluster membership strengths for each waveform are not independent by virtue of the fact that the sum of these cluster membership strengths must equal 1.0. Encouragingly, most of the histograms seem to indicate a good separation of most of the normals from the MS patients. In all cases, the NSD and VIS subjects are grouped near the DMS patients.

As the separation for most patterns is good, there is little subjectively to decide what are the more discriminating patterns for detecting patient abnormalities. The exception appears to be the RED/BLUE patterns, which in general result in a larger overlap between the patient populations. This is not true for the other color combinations or for the BLACK/WHITE patterns, which in general result in little overlap in cluster membership strengths for the subject populations.

In order to understand what waveform qualities the system is using for its discrimination, the cluster centers are projected back into pattern space and plotted in Figure 7.4 for a representative sample of the stimulation patterns. It should be remembered that these are *not* actual waveforms, and that the six-feature data space is missing information, although hopefully a small amount. From Figure 7.4 and other similar plots (not shown), it appears that much of the clustering analysis is based on amplitude determinations. This is in contrast to the common heuristic, that demyelinating diseases present most of the information in the VEP as prolonged latencies. A few of the reconstructed cluster centers show differing P1 latencies from one another, but most of them indicate more pronounced differences in N2 latency. This is in agreement with a previous study by Regan[10] that alluded to the additional diagnostic information in the N2 peak.

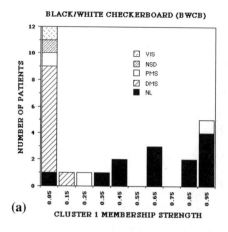

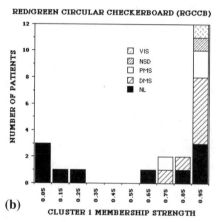

FIGURE 7.3   The cluster membership histograms for all of the subjects used in the study for stimulation with (a) a black/white checkerboard and (b) a red/green circular checkerboard.

The clustering analysis for the VEP signals is carried one step further. Since the cluster membership strength outputs of the system can by themselves be thought of as a feature of the original patient disease status as extracted by stimulation with a particular pattern, those numbers can be used for a final clustering sequence. The aim of this step is to discount unusual attributes from an individual VEP signal through the use of all 16 patterns. Thus if the VEP signal from any one pattern is abnormal due to the lack of patient attentiveness or other uncontrollable factors while signals deriving from the other patterns are "normal", the clustering process over all patterns will largely overlook the discrepancies from any one pattern.

The clustering of the pattern cluster 1 membership strengths was performed with 16 features (one cluster membership strength for each pattern used) with c = 2 and q = 4.0 (see Chapter 5). The value of c was verified as before using the cluster validity measures. The unusually high value of q = 4.0 (as opposed to the use of q = 2.0 for the single pattern clusterings) was necessary to provide at least some degree of fuzziness in the final outcome membership strengths. When clustered at $q = 2.0$, the memberships were entirely hard (took values of either zero or one), indicating a very distinct pair of clusters.

The histogram for the final clustering step is shown in Figure 7.5 for $q = 4.0$. Only 20 patients were used in this final clustering, since some patients were not tested on all 16 patterns. This depicts a clear separation between the subject populations. In a clinical setting, a "normality" threshold might be set near 0.45, if it were necessary. Significantly, if a diagnosis were to be based on this membership solely, there would be no false negatives from the DMS population, and only one PMS patient would be grouped with the normal population.

Unlike the application of this system to the classification of handwritten digits (Chapter 6), there was not a significant amount of "leftover" data that could be used to test the generalization capabilities of the clustering decisions. However, a few patients were stimulated a multiple number of times with a pattern during one experimental session. In each case (a total of 18 instances), when the test was repeated on a subject with a duplicate pattern, the cluster membership strengths arising from analysis of the waveform were within 0.05 of the training pattern cluster membership strength.

7.5 DISCUSSION

In this work, an unsupervised pattern recognition system has been introduced, largely based on methods originating elsewhere,[3-4] but bound by their application to a parallel architecture and their ability to be trained by a single optimization algorithm. The method has been tested in two widely varying application domains: in the classification of handwritten numerals (Chapter 6) and the clustering of Visual Evoked Potential signals.

This VEP application set clearly highlights the more advantageous aspects of the unsupervised routine. Through one system, the use of the VEP in the diagnosis of MS has been reaffirmed, and a means of automating the multiparametric classification of the VEP has been implemented. The system was able to perform this analysis without bias from other diagnostic sources, with relatively few training patterns, and with little computational overhead.

In the process, the system has reaffirmed some of the previous notions about what aspects of the VEP are most crucial for diagnosis, through the ability to visualize the features it is using and to depict the cluster centers. Figure 7.4 already, with this limited application, has pointed out some curious aspects of the decision making of the system. For one, it is obvious that the amplitude of the signal played a large portion in the cluster-making decisions. Some of the centers show little or no difference in P1 latencies (Figure 7.4a), while having markedly different N2 latencies. Also, some cluster centers (Figure 7.4b) appear to have a distinctive presence (or absence for the other center) of a P2 peak. The P2 peak is rarely used in the literature, since most researchers consider it to be too variable as a measure. Yet, when shown as a "trend" of the clusters derived from many of the pattern stimulations, one of the center waveforms has a noticeable P2 peak.

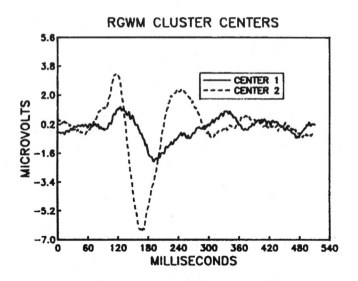

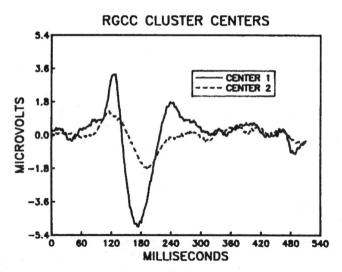

FIGURE 7.4   The cluster centers reconstructed as if they originated as pattern waveforms for the clustering results of the (a) red/green windmill and (b) red/green concentric circles. The solid curve represents the cluster with the most normals.

In a more general context, the use of a multiparametric decision in medical applications is highly desirable. Physicians often have to work with a myriad of measurements collected from several sources, decipher the relevant information from the extraneous, and make a decision on the patient status while considering all of the measurements together. In the VEP application domain alone, a conscientious clinician must consider the latency and amplitude characteristics of several peaks, the general constitution of the signal, the stimulation pattern, and other experimental

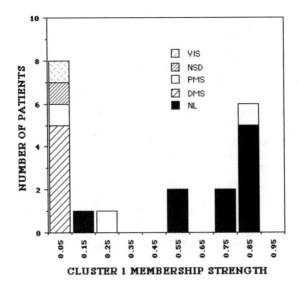

FIGURE 7.5   The cluster membership histogram for 10 NL, 5 DMS, 3 PMS, 1 NSD, and 1 VIS patient using the cluster membership strengths of each of the 16 stimulation patterns.

conditions. It seems reasonable to see that a decision of this sort is a complicated art and could be facilitated by consultation with a system like the one described.

Most computer diagnostic algorithms are of the supervised sort and typically can do no more than duplicate the past decision of physicians. As such, they are quite useful as quality control devices but will probably do little to enhance the information set of a well-trained clinician. On the other hand, an unsupervised pattern recognition scheme, such as the one in this study, can make an unbiased assessment of a set of multidimensional measurements, create a decision, and retrace the process to propose to the physician the information behind the decision.

The ALOPEX-trained system has proven itself capable of strong generalizations, as evidenced by the application to handwritten digits,[4] and is able to extract a significant amount of information without the advice of an omnipotent instructor. When properly tuned, the clustering module can make decisions of a fuzzy nature, so that an understanding of the certainty of its decision can be analyzed. As the VEP application has shown, the pattern recognition system can affirm the diagnostic usefulness of the input measurement, although the reverse is not necessarily true.

As hoped, the conclusion of this study has produced more questions and promising directions than answers. It is clear that the VEP provides a useful measure of visual defects in multiple sclerosis, but a larger patient population (including especially significant numbers of possible and probable MS subjects) would facilitate an understanding of how well this clustering scheme can indicate gradations of abnormality. Future directions include correlating the waveforms generated from the cluster centers with actual trends in the data set. Still, the blending of unsupervised and supervised training methods at varying times in the learning process is intriguing and probably beneficial in certain application domains.

REFERENCES

1. Bezdek, J. C., Ehrlich, R., and Full, W., FCM: the fuzzy c-means clustering algorithm, *Comput. Geosci.*, 10(2), 191, 1984.
2. Chien, Y. T. and Fu, K. S., On the generalized Karhunen-Loève expansion, *IEEE Trans. Inf. Theor.*, 15, 518, 1967.
3. Dasey, T. J., Unsupervised Global Optimization in the Classification of Handwritten Digits and Visual Evoked Potentials, Ph.D. Diss., Rutgers University, Department of Biomedical Engineering, New Brunswick, NJ, October 1991.
4. Dasey, T. J. and Micheli-Tzanakou, E., Neural fuzzy systems in handwritten digit recognition, *Industrial Electronics Handbook*, CRC Press, Boca Raton, FL, 1997, chap. 72, 1231.
5. Frederiksen, J., Larsson, H., Oleson, J., and Stigsby, B., Evaluation of the visual system in multiple sclerosis. II Color vision, *Acta Neurol. Scand.*, 72, 203, 1988.
6. Harrison, A., Becker, W., and Stall, W., Color vision abnormalities in multiple sclerosis, *Can. J. Neurol. Sci.*, 14, 279, 1987.
7. Jaridan, M., McLean, D., and Warren, K., Cerebral evoked potentials in multiple sclerosis, *Can. J. Neurol. Sci.*, 13, 240, 1986.
8. Mitchell-DePew, J. and Micheli-Tzanakou, E., Complex Pattern Visual Evoked Potentials in Controls and Multiple Sclerosis Patients, *Proc. 11th Annu. Int. Conf. IEEE/EMBS*, Vol. 6, 1990, 2048.
9. Mitchell-DePew, J. and Micheli-Tzanakou, E., Color Pattern Visual Evoked Potentials in Controls and Multiple Sclerosis Patients, *Proc. 12th Annu. Int. Conf. IEEE/EMBS*. Vol. 12, 1990, 874.
10. Regan, D., *Human Brain Electrophysiology: Evoked Potentials and Evoked Magnetic Fields in Science and Medicine*, Elsevier, New York, 1989.
11. Regan, D. and Spekreijse, H., Evoked potentials in vision research 1961-1986, *Vis. Res.*, 26(9), 1461, 1986.
12. Urbach, D., Gur, M., Pratt, H., and Peled, R., Time domain analysis of VEPs: detection of waveform abnormalities in multiple sclerosis, *Invest. Ophthalmol. Sci.*, 27, 1379, 1986.

Section III

---

*Advanced Neural Network
Architectures/Modular
Neural Networks*

8 Classification of Mammograms Using A Modular Neural Network

Lt. Col. Timothy Cooley and
Evangelia Micheli-Tzanakou

8.1 INTRODUCTION

Mammography has been and still remains the key screening tool for breast abnormalities. Other modalities such as thermography, ultrasound, and magnetic resonance imaging are available, but none compares with the ease and low cost of mammography.[19] However, mammography is not without its problems.[8] Mammograms miss approximately 10% of all breast cancers. Even with recent advances, mammography is as much an art as it is a science. Errors may occur at many stages, starting with the very idea itself, then in the technique of the mammographer, and then that of the radiologist who reads the mammograms. Mammograms are, simply, specialized X-rays[12] and as such an abnormality will only appear if its absorption of the X-ray waves is different from the surrounding tissue.[6] In some cases potential tumors are masked, because they appear strikingly similar to dense fibrous normal breast tissue.[8] Additionally, films from mammograms are notoriously difficult to read,[6,36] as they are low in contrast and high in noise. Because of these factors it is no wonder that a recent report showed that consistency among radiologists' interpretation of mammograms was moderate, and in several cases the readings differed substantially.[14]

In the last few years, a lot of research has been done on mammography, using different methods of processing and interpretation.[9–11] Given the difficulty in reading mammograms and the benefit of early detection of breast cancers, it is not surprising that much research has been done in recent years on computerized systems to aid in this process. This research has focused on two major areas: developing a system which enhances mammograms and makes them easier for a radiologist to read and developing systems that take regions of interest selected by a radiologist to determine whether these regions are malignant or benign. The first area falls under the category of Computer Aided Diagnosis (CAD) systems and centers on helping detect possible cancer areas. The second area focuses on reducing the number of biopsies performed each year and still requires the radiologist to perform the first step of the detection. In both cases the expert (radiologist) must be part of the system.

A different approach to this problem would be to develop a system that reads the mammogram by using the entire image and then gives a classification of that mammogram. A radiologist will classify a mammogram into one of five categories: category I is completely normal; category II shows a known abnormality that has appeared previously and is benign; category III shows an abnormality that is thought to be benign—a follow-up sooner than normal may be recommended but a biopsy should not be done; category IV shows a suspicious abnormality that needs further evaluation and probably biopsy; and category V shows a definite cancerous tumor that requires immediate attention. A computer system that could perform this type of classification could act essentially as a second opinion for a radiologist. One important benefit is that the computer system would be consistent and the inconsistencies noted by Elmore et al.[14] would no longer be a problem. However, even if the system could just delineate normal mammograms (category I) from abnormal ones (categories II, III, IV and V) this would be of help. In this case, the system could still be used as a second opinion and might detect some mammograms that are classified as normal that are really abnormal. Additionally, there is some variation between radiologists over what is a category III mammogram and what is a category IV. The difference between these two categories is somewhat subjective. There are some radiologists who choose the cautious side and recommend a biopsy on anything that is a change. Others will hold off on the biopsy but recommend a follow-up in three to six months. Because of this, any system developed to classify mammograms in all five categories would be biased toward the radiologist reading the mammograms.*

A system that could delineate between normal and abnormal would also be of use is in places where there is no radiologist. One of the most prominent places where this occurs is the mobile mammography van. These vans travel primarily to major corporations and offer women an opportunity to have a mammogram done with minimal disruption to their routine. All they need to do is schedule approximately 30 minutes for the procedure and then they are back at work. This is much more appealing and convenient than driving to a center and having to schedule perhaps a couple of hours for travel time and the appointment. However, in the mobile van, once the mammograms are taken, they are not read until the following day. This means that if there is a need for further views or perhaps for some reason the X-ray is unreadable, the woman must now schedule a follow-up appointment at a center. If the mammography technician knew at the time of the first exam whether the mammogram was normal or abnormal, the extra views could be taken right then.

This chapter discusses the design, methods, and results of a computer system that classifies mammograms as normal or abnormal. The motivation for this kind of research has been given above, but succinctly stated the system is intended to aid in the early detection of breast cancer. The goal of this research is to develop a system that can take a mammogram and classify it as normal or abnormal with no expert intervention. Additionally, the system should be mobile in the sense that it could be fielded on a mobile mammography van. For this system to function in that

---

* This system was designed when only four categories were used. Previously category I included those mammograms that were totally normal as well as those that had a known abnormality. Five categories were created when category I was split as defined above.

capacity, it must be able to produce a reading quickly enough so that the wait time for the patient is not unacceptable. Eventually digital mammography will be the standard, but until that time this system must have the capability to digitize the mammograms as well.

8.2 METHODS AND SYSTEM OVERVIEW

The system described in this paper is composed of three essential steps divided into two developed programs. The three basic elements are image acquisition, image transformation and preprocessing in order to extract features for classification, and classification itself via neural networks. Acquiring, digitizing, or scanning the image is a step which requires the use of a commercial scanning system, one that can accurately scan transparencies or X-ray films at a high resolution. The transforming and processing module and the classification process are developed software and will comprise the major part of the discussion of the next two sections. Figure 8.1 shows a block diagram of the system.

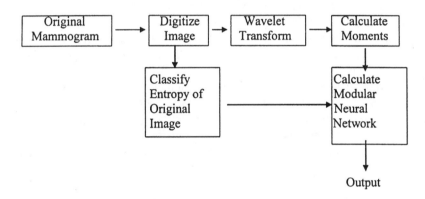

FIGURE 8.1   Overview of the system.

8.2.1 DATA ACQUISITION

As previously stated, digitizing a mammogram image can be accomplished by using one of several good scanning systems commercially available. Still, there are some issues which necessitate discussion.

A radiologist reads a mammogram looking for submillimeter microcalcifications as well as larger nodules or tumors.[32] Because of this, any system that is used to scan these images must have a high enough resolution to capture these small microcalcifications. Additionally, the scanner must be able to scan X-ray images so that they appear to have some depth and are not just flat, two-dimensional images. By using a scanner with a transparency adapter, images were obtained which more closely resembled the true mammogram film. Our scanner had a resolution of 400 dots/in., corresponding to .0625 mm/pixel (or 62.5 microns [μm]), which is adequate to capture the small microcalcifications. Additionally, the scanning system had software which allowed the contrast and brightness values to be adjusted. By using this software the images were essentially preprocessed by enhancing the contrast and obtaining the best possible image. While this does not reduce the noise in the image, it is a good first preprocessing step. This resulted in files that were 3–9 Mbytes in size. The sizes vary since mammograms vary in size, and to save space each image was windowed so that the absolute minimum area was captured. This scanning system saved the files in Microsoft Windows 3.x bitmap (BMP) format, and therefore, all processing assumes this format.

8.2.2 FEATURE EXTRACTION BY TRANSFORMATION

Since the raw digitized images are very large, some method of reducing the number of data points or extracting features must be applied prior to attempting to classify the images. Additionally, mammogram images are known to be noisy.[36] However, there are conflicting opinions as to whether it is desirable to filter the image.[2,27] On the one hand, it is difficult to distinguish noise from the microcalcifications, and any filtering may result in essential data loss. The other opinion is that the noise appears too much like the real microcalcifications and must be eliminated in order to reduce the number of false-positive readings that are obtained. The ideal situation would be to filter out the noise only and leave all the microcalcifications intact. While the perfect ideal filter may not exist, a multiresolution technique solves some of the problems.

The wavelet transformation with multiresolution analysis was chosen for the first stage of the processing. Not only does the wavelet transform tend to extract the critical areas, but the multiresolution method performs an essential task. Wavelets and multiresolution analysis together have recently been widely used. For example, they have been used to detect microcalcifications in mammograms,[22,34,27,39] to extract features in face recognition,[41] to compress image,[15,20,40] and to enhance images and simulate human perception.[26]

Mammograms are difficult images as they must be viewed at many different scales.[13] At a coarse scale, a relatively large tumor may be found which would cause a mammogram to be abnormal. These are the obvious and easier classifications. The microcalcifications appear at a fine scale, which if found in clusters will also result in an abnormal mammogram. These microcalcifications are extremely difficult to detect[32,36] and virtually impossible to find if searched for on a coarse scale. Multiresolution analysis allows the image to be analyzed at a fine scale first and then successively coarser scales at each level.[42] Additionally, wavelet functions have the ability to be translated and dilated so that in the wavelet analysis small transient signals are able to be detected.[7,31] This ability is what enables them to extract the small microcalcifications as the microcalcifications appear in an image, in much the same way a transient spike appears in a signal. Furthermore, the wavelet transform is a fast transformation. It is implemented via convolutions, which makes the complexity linear in the size of the input ($O(n)$).[4,33] While this is still a lengthy process on large images, it is the best transformation complexity that can be achieved currently. Multiresolution and wavelet transformations are discussed below.

While the wavelet transform highlights the potential problem areas, it does not reduce the number of data points. On the contrary, since the images were transformed to five octaves, the result was 33% more data points. Therefore, a feature extraction method was needed. Recent success with classifying images by using invariant moments[41] led to this direction. Mammography is not immune to some of the standard problems of imaging, i.e., translation, rotation, and scaling. It is easy to see that from one mammogram to the next, even on the same patient, it is highly unlikely that the breast will be positioned in exactly the same place on the X-ray film. Additionally, rotation is a problem since the procedure itself does not always take the mammogram from the same angle. Even without the variance in the procedure, it would be unlikely that the breast would not be rotated on the film. In order to obtain a better image of the tissue, the breast is compressed in the mammogram process. Since this is a subjective process, the amount of compression may vary slightly from one mammogram technician to another. This, of course, changes the shape or the scaling of the image. Additionally, a patient may gain or lose weight between two mammogram procedures (usually there is a year or more between mammogram appointments). This also may affect breast size. Because of all these variables, moments which are invariant to rotation, translation, and scaling appeared to be a logical choice as features for the classification process. The Hu moments[18] met these qualifications and were chosen. However, the moments were calculated not only on the original image but also on the transformed components. This is important as the wavelet multiresolution analysis has reduced the noise and highlighted the potential problem areas. In Shen et al.[30] it is noted that moments are susceptible to noise and may not perform well on noisy images. The wavelet transform has reduced the noise in most cases and increased the performance of the moments as features. Moments are discussed in more detail in Chapter 4.

Since there are seven Hu invariant moments through the third order, the number of data points was reduced at this step to 147 (i.e., five octaves of transforms with four components per octave equals 20 components. Seven moments of 20 components equals 140 moments, plus the moments of the original image equals 147). In viewing the wavelet transforms, it was observed that the fifth octave did not contain much information (due to most of the details being extracted at this point) and, therefore, was not used in the classification process. This results in only 119 data points.

In the initial design, the moments were not calculated on the original image. Because of this it was determined that some measure from the original image might aid in the classification process. One such compact and global measure is the entropy of the image. Widely used in information theory as a measure of the average number of bits required to represent each symbol in an alphabet,[43] it also can be a measure of the information content of a signal or image.[44]

With the addition of the entropy of the image there are 120 data points per image used in the classification process. This represents a huge reduction in dimensionality, which is what was needed. The average mammogram image digitizes to 4.5 Mbytes. Representing the features in 120 data points means the dimensionality has been reduced on the order of 10^5.

8.3 MODULAR NEURAL NETWORKS

Simon Haykin in his book *Neural Networks — A Comprehensive Foundation*[16] defines a modular neural network as follows:

> A neural network is said to be modular if the computation performed by the network can be decomposed into two or more modules (subsystems) that operate on distinct inputs without communicating with each other. The outputs of the modules are mediated by an integrating unit that is not permitted to feed information back to the modules. In particular, the integrating unit both (1) decides how the outputs of the modules should be combined to form the final output of the system, and (2) decides which modules should learn which training patterns.

Figure 8.2 shows an example structure of a modular neural network.

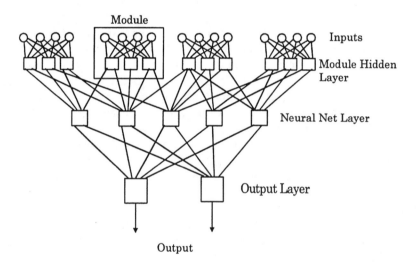

FIGURE 8.2   An example of a modular neural network. The boxed area highlights one module. The module hidden layer and the neural net layer are not fully connected just for clarity. Normally these two layers would be fully connected. Only the input layer to the module hidden layer is modularized.

The idea of modular neural networks is analogous to biological systems.[16,46] Our brain has many different subsystems that process sensory inputs and then feed these results to other central processing neurons in the brain. For instance, consider a person who meets someone they have not seen in a long time. To remember the identity of this person, multiple sensory inputs may be processed. Foremost perhaps is the sense of vision, whereby one processes what the person looks like. That may not be enough to recognize the person, as the person may have changed over the course of a number of years. However, looks coupled with the person's voice, the sensory input from the ears, may be enough to provide an identity. In addition, the olfactory system may also provide more information. In this way our biological system makes many different observations, each processed first by some module and then the results sent to be further processed at a central location. Indeed, there may be several layers of processing before a final result is achieved.

In addition to different modules processing the input, the same sensor may process the input in two different ways. While the concept of a modular neural network is based upon biological phenomena, it also makes sense from a purely practical viewpoint. Many real-world problems have a large number of data points. Using the large number of points as input to a fully connected multilayer perceptron results in a very large number of weights. Just blindly trying to train a network with this approach most often results in poor performance of the network, not to mention long training times because of slow convergence [Haykin, 1994; Rodriguez et al., 1996]. Sometimes there are feature extraction methods, such as those described in Section 2, which will reduce the number of data points. However, as was the case in this project, there are times when even then the amount of data is large. Since it is desirable to have the minimum number of weights that will yield good performance, a modular neural network may be a good solution. Each module is able to compress its data effectively and extract subfeatures which then are used as input to a fully connected neural network. Without this modularity the number of weights in the network would be far greater.

8.4   NEURAL NETWORK TRAINING

There are two types of neural network training, supervised and unsupervised.[16,23] Supervised training involves presenting examples to be learned to the network, along with the correct output (the teacher) for each example. The calculated outputs are compared to the known values, and the weights are adjusted so that the difference between the outputs and the known values is minimized. The training continues (in theory) until the output for each example is correct. Unsupervised training involves presenting examples to the neural network but without the correct outputs. The weights are adjusted based upon some well-defined algorithm in order to group or cluster like examples together in a consistent manner. Unsupervised training does not require a teacher, and because of this the network's behavior is sometimes called self-organizing. Many scientists believe that the brain operates more like unsupervised training[38] and, therefore, in order to attain learning capabilities close to those of the brain, unsupervised training must be used. In practicality, much success has been achieved with supervised training as well.

This project only considered supervised training. For that reason, the details of different unsupervised training algorithms will be left to the interested reader to research in one of the neural net references.[16,17,21,23,38] While backpropagation is perhaps the most often used supervised training algorithm, because of past performance ALOPEX was chosen for this system.

8.5   CLASSIFICATION RESULTS

As noted in Section 8.3, classification was performed by a modular neural network with nine total modules. The input to the modules were features extracted by taking the moments of the multiresolution wavelet transform of the digitized mammogram. In addition, the moments and entropy of the original mammogram were used as input as well.

Over 250 different experiments were performed with many different neural network structures and parameters. There are literally thousands of combinations that are possible. Table 8.1 lists the major variables for the classification process.

TABLE 8.1

Types of Variables in the Classification Process

Neural Net Structure (hidden nodes, output nodes)

Presentation of data (by component, by octave, entropy)

Data normalization (ln and scale, tanh, sigmoidal)

ALOPEX parameters (σ, γ, and maximum change)

Change of ALOPEX parameters (exponential, tanh, sigmoidal)

Number of files for training and testing

The kinds of things that can be changed for that type of variable or some possible values for that variable are listed in parentheses. Each one of the subvariables can take on several values of its own. Table 8.2 lists the best results achieved for certain configurations of the neural network. All of the results listed in Table 8.2 are for nine modules, eight with 29 inputs (28 for the one case where entropy was not included) and one with eight inputs (seven for the one case without entropy), with 49 training mammograms and 10 test mammograms. In all cases the ALOPEX parameters were changed by exponential decay for σ and the maximum change per epoch, and exponential growth for γ with a scaling factor of 2500 epochs. In addition the least squares function was used as the cost function in the ALOPEX algorithm.

The first column of the table gives the configuration of the eight modules which use the moments of the wavelet coefficients as their input. A 29-3-8-2 configuration means each module had 29 inputs, three hidden nodes in the module, followed by eight nodes in the neural network integration layer, and finally two output nodes. On the next line in the first column is the configuration of the module which processes the moments of the original image. An 8-2 means that there are eight inputs into the module with two hidden nodes. The neural network integration layer and output layer are the same as the other modules since all the modules feed into these layers. The third line in the first column of the table indicates whether or not the entropy values were included in each module.

The second column shows how the data were normalized. All of these use some variation of the natural logarithm (*ln*) and then scale these numbers to lie in the [−1,1] interval. Most of the experiments listed perform this process twice since that method performed much better. This is further discussed in the next section.

TABLE 8.2
Classification Results

	Neural Net Structure	Data Norm	ALOPEX Params σ, γ, Max Change	Trng %	Test %	Sensitivity % Specificity %	Epochs
1	29-3-8-2 8-2 Entropy	Double ln	0.120–0.003 0.220–25.00 0.175–0.025	93.9	100.0	100.0 100.0	36200
2	29-3-8-2 8-2 Entropy	Double ln	0.100–0.003 0.200–25.00 0.170–0.025	93.9	90.0	75.0 100.0	23400
3	29-3-8-2 8-2 Entropy	Double ln	0.150–0.003 0.275–25.00 0.200–0.025	91.8	90.0	75.0 100.0	39400
4	29-3-8-2 8-2 Entropy	Double ln	0.150–0.003 0.270–25.00 0.225–0.025	91.8	80.0	75.0 83.3	15400
5	29-3-9-2 8-2 Entropy	Double ln	0.100–0.003 0.200–25.00 0.150–0.025	91.8	80.0	75.0 83.3	28600
6	29-3-8-2 8-2 Entropy	Double ln	0.250–0.003 0.330–25.00 0.400–0.025	89.8	80.0	50.0 100.0	29200
7	29-3-7-2 8-2 Entropy	Double ln	0.150–0.003 0.250–25.00 0.225–0.025	87.8	80.0	75.0 83.3	35400
8	28-3-8-2 7-2 No Entropy	Double ln	0.120–0.003 0.220–25.00 0.175–0.025	95.9	70.0	75.0 66.7	36000
9	29-3-8-1 8-2 Entropy	Double ln	0.120–0.003 0.220–25.00 0.175–0.025	93.9	70.0	50.0 83.3	15600
10	29-3-10-2 8-2 Entropy	Double ln	0.100–0.003 0.200–25.00 0.150–0.025	89.8	70.0	75.0 66.7	24800
11	29-3-8-2 8-1 Entropy	Double ln	0.120–0.003 0.220–25.00 0.175–0.025	79.6	60.0	50.0 66.7	30800
12	29-3-8-2 8-3 Entropy	Double ln	0.120–0.003 0.220–25.00 0.175–0.025	79.6	60.0	50.0 66.7	13800
13	29-3-8-2 8-2 Entropy	Single ln	0.120–0.003 0.220–25.00 0.175–0.025	59.2	60.0	0.0 100.0	5400
14	29-4-11-2 8-2 Entropy	Double ln	0.120–0.003 0.220–25.00 0.175–0.025	98.0	50.0	25.0 66.7	8800
15	29-5-12-2 8-2 Entropy	Double ln	0.120–0.003 0.220–25.00 0.175–0.025	95.9	60.0	75.0 50.0	9200

The third column lists the initial and final ALOPEX parameters. The top line gives the σ values, the next line the γ values, and the last line the maximum change values. For the final values, these are the values that the parameters would be if the network was allowed to run long enough. The final values were reached at approximately 20,000 epochs for σ and the maximum change, and 12,000 epochs for γ.

Columns four and five give the best training and testing percentages, respectively. These percentages were taken once the network had stabilized. The graph in Figure 8.3 gives an example of the convergence of the network. As can be seen, at a low number of epochs the network oscillates and is not stable. In the graph the network starts to stabilize at approximately 6000 epochs. This point differs from configuration to configuration, but all percentages were taken after that point. Also, these percentages are the ones that were repeated and had consistently occurred over the final 600 epochs or so. This eliminates the single data point that may be an anomaly because of the random nature of the ALOPEX algorithm.

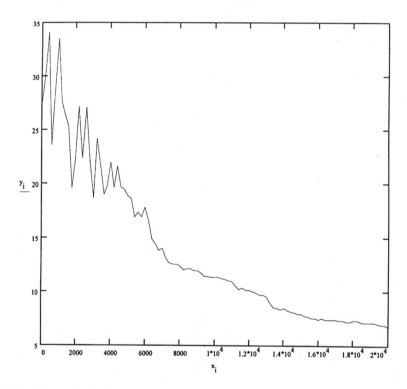

FIGURE 8.3   Graph of the convergence of one example of the neural network. The number of iterations is on the x-axis, and the total error is on the y-axis. The number of iterations only goes to 20,000, but the graph shows the early initial oscillations.

Column six of Table 8.2 lists the sensitivity and specificity of the system as measured on the test data. Sensitivity is a measure of how well the system detects abnormal mammograms, and specificity is the gauge of the system's performance on normal mammograms.[28] These quantities are calculated in the following way:

$$\text{Sensitivity} = \frac{\text{Number of True Positives}}{\text{Number of True Positives} + \text{Number of False Negatives}} \tag{8.1}$$

$$\text{Specificity} = \frac{\text{Number of True Negatives}}{\text{Number of True Negatives} + \text{Number of False Positives}} \tag{8.2}$$

where true positives are those mammograms that the system classifies as abnormal that are truly abnormal, true negatives are those mammograms that the system classifies as normal that are truly normal, false positives are those mammograms that the system classifies as abnormal that are really normal, and false negatives are those mammograms that the system classifies as normal that are really abnormal. The sensitivity and specificity values are important, but because of the small test sample, they vary widely for this system. The last column gives the number of epochs required to obtain the performance listed in columns four and five. Most of these experiments were run longer than this number of epochs. However, sometimes performance, especially the testing percentage, will decrease if the network trains beyond a certain point. Normally the experiments were run until the performance started to degrade, and then they were stopped. Therefore, the epoch numbers represent when the network peaked in its performance. Some of these issues are addressed again in the next section where reasons are given to explain many of the behaviors observed. However, many more types of experiments were run than those listed in Table 8.2. Usually these experiments showed lower performance. The next section will discuss them, since the path that led to the results is often more important than the results themselves.

8.6   THE PROCESS OF OBTAINING RESULTS

As in all research projects, the data have to be examined very carefully. First, the range of the moment values is from -1×10^{41} to 1.1×10^{41}. This is a huge span, but 85% of the raw data lie in the [−1,1] interval! In addition, the range of values in one octave is fairly large, but the range in one component is not nearly as large. This makes sense since the moments are based upon the pixel values. Clearly it is easy to see that the residual has the highest concentration of large pixel values, and therefore, these moments will be similar. Following this logic then, it would make sense that the component that contains the vertical edges would have moment values in the same general range as other like components. This is true for the diagonal and horizontal components as well. Also, the moments that were highest in absolute value normally came from the vertical and diagonal components. However, no distinguishing characteristics were gleaned from studying the entropy values. There appeared to be no difference between the normal and abnormal cases.

The above insights led to some changes in the neural network structure. However, perhaps more importantly, they led to changes in the way the data were normalized. At the start, the data were normalized by taking the natural logarithm *(ln)* of all moments and then globally scaling these data onto the [−1,1] interval. What is meant by global scaling is that a *max* and a *min* value are determined for the *ln* of all the data points and then these values are used to scale the data by the following equation:

$$\text{scaled value} = \frac{(\text{old value - min})}{(\text{max - min})} \times 2 - 1 \qquad (8.3)$$

But since most of the raw data are in the [−1,1] interval and the *max* and *min* values are so huge in absolute value, this procedure compacts most of the data into a very small interval. This makes classifying the data difficult since the separation between the data is so small. Therefore, the switch was made to scaling the data by component since the component data are similar in magnitude. In this process, maximum and minimum values for the natural logarithms were calculated for each component, and each component was scaled with its corresponding maximum and minimum values. This meant that the first component moments, which are normally very small in magnitude, did not get lost in scaling by huge numbers. These small numbers were then scaled with other small numbers, which meant that these numbers then had some impact upon classification since they were no longer compacted so tightly.

Other normalization methods were tried, such as the hyperbolic tangent and a sigmoidal function, but none worked consistently as well as the *ln* and scaling method. In addition, the best approach is to perform this method twice on the data. Potential reasons for this are discussed in the next section.

Given the above changes, the next approach was to structure the neural network with four modules each of which would process all the moments from one component, not by octave. In addition, the normalization approach was changed to scale the *ln* data by component (only once at this time), and the entropy values were added to each module. These experiments started to give some better results. Training percentages near and slightly above 80% were achieved. However, the testing percentages were normally between 60 and 70%.

Combining the two ideas presented up to this point increased the training percentage. The structure now consisted of eight modules, four of which processed the moments by octave and four of which processed the moments by component. In addition each module contained the entropy value, and the data were normalized using a single *ln* and scale method. It was now possible to obtain training percentages approaching and at times greater than 90%. However, the testing percentage continued to be between 60 and 70%.

At this point, analysis of the output from the trials thus far indicated that certain mammograms were consistently misclassified. From the start, the training had been performed on 52 mammograms, 32 normals and 20 abnormals, and the testing on the remaining 13 in the data base, 8 of which were normal and 5 abnormals. All the "problem" mammograms were normals that would be classified as abnormals. A review of these mammograms showed some interesting things. In one case there was a scar from a previous surgery and the system was picking this up as abnormal. In other cases, metallic markers were put on the breast to mark moles or other exterior-related abnormalities so that the radiologist would know that this spot was not something internal. These markers appear as very bright small dots on the mammogram, and the system was determining that this was abnormal. In yet another case, there were several marks on the mammogram that were made by a person who desired to draw attention to some area. Figure 8.4 shows an example of a mammogram with metallic markers. After this review, six normals were taken out of the data base and only 59 mammograms remained, 34 normals and 25 abnormals.

Further trials with the same structure but the new data base now quite often showed training percentages greater than 90%. Occasionally the testing percentage would reach 80%, but most likely would be 70%. At this time it was decided that more information from the original image might aid the classification process. Another module was added that only contained the seven moments from the original image. Experiments with this structure showed a better average performance. However, for the most part the best trials were the same, training greater than 90% and testing at 80%. In one instance training was performed to 96% and testing to 90%. The problem was that this result could not be recreated, but in the process of attempting to recreate this result a key concept was discovered.

Our system has the ability to gather the data from a number of files (moment files for each mammogram, entropy files and moment files for the wavelet coefficients) and then save these data to one compact file in order to simplify future experiments on the same data. These data can be saved at any time, even *after* it has been normalized. To save one processing step the data had been saved after the *ln* and scaling operation. However, in remembering the steps for that experiment it was discovered that the data had again been normalized by performing an additional *ln* and scaling operation. While those results with that neural network structure were never consistently recreated this discovery was important. By using this double *ln* and scaling method for normalization, the testing percentage was increased on the average. Comparing row 1 and row 13 in Table 8.2 shows an extreme difference between the two normalization methods. Fewer trials ended with a testing percentage of 60% and many more achieved 80% or higher. However, a reasonably consistent 90% result was not yet attainable.

8.7   ALOPEX PARAMETERS

Prior to this time the ALOPEX parameters had not been varied much. During the attempt to recreate the results, experiments were done with widely different ALOPEX parameters. Also, at this time experiments were done with changing the updating method for the parameters. Initially it seemed that relatively large starting values for the ALOPEX parameters gave better performance, and, as a result, most trials were done with initial parameters close to $\sigma = 0.25$, $\gamma = 0.35$, with the maximum change per epoch being 0.40. Once the additional module was added, better results were achieved with initial values much less than those listed above, such as $\sigma = 0.12$, $\gamma = 0.22$, and the maximum change per epoch equaling 0.175. A comparison of row 1 with rows 3, 4, and 6 shows the difference that the ALOPEX parameters can make and the effect of making the initial values smaller. Row 2 shows that making the values smaller yet does not necessarily increase performance.

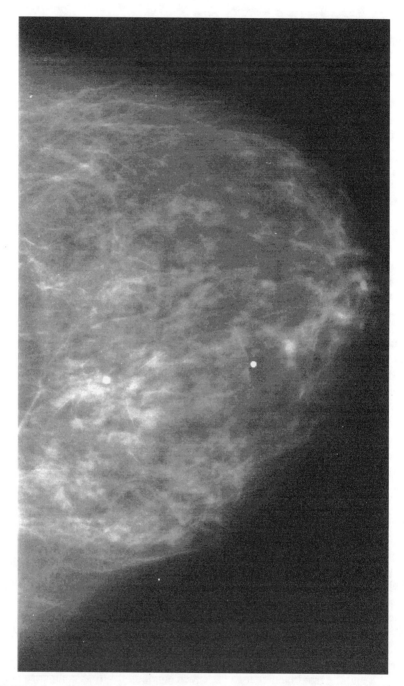

FIGURE 8.4   Example of a mammogram with metallic markers. The two metallic markers are easily seen as bright, perfectly round spots on the breast. These spots confused the classification process, and this mammogram was consistently misclassified.

While exponential decay for the maximum change and σ were always used, the rate of decay was changed. Initially, the experiments were run with a decay rate of 3500. This means that at every 3500 epochs the values are decreased by a factor of e, where e is the exponential function. The slower the decay (this translates to a larger decay rate), the more randomness exists in the system for a longer period of time. Also, since the maximum change will be greater for a longer period of time, the network has less stability in the early training stages. Initially this was thought to be beneficial. However, once the network evolved into its final structure, the longer decay rate did not perform as well. After trials with several different rates, a decay rate of 2500 was determined to provide the best results. Figure 8.5 shows a graph of different decay rates.

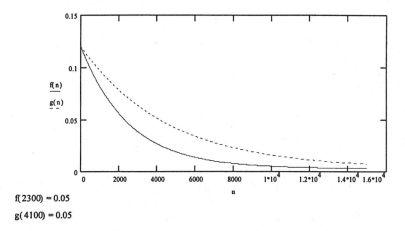

$f(2300) = 0.05$

$g(4100) = 0.05$

FIGURE 8.5   Graph of different σ decay curves. The solid line represents the decay curve with a decay rate of 2500 epochs. The dashed line represents a rate of 4500 epochs. As shown, σ is equal to 0.05 at 2300 epochs for the 2500 decay rate and at 4100 epochs for the 4500 decay rate.

The learning rate, γ, must increase as the network starts to converge. Different growth methods were tried as shown in Figure 8.6. The hyperbolic tangent, with a multiplicative factor to make γ greater than 1, would appear to be an ideal growth choice, but it initially grows too slowly. A sigmoidal function was tried, but again the performance of the network was worse with this approach as it too grows too slowly at high epoch numbers. The exponential growth method consistently performed the best, and, while not required, the growth rate was set to be the same as the decay rate for the maximum change and σ. Trials with other growth rates did not provide any improvement in convergence or performance.

Additionally, the final values for the ALOPEX parameters were varied. At the start of the experiments, these values were set to much lower numbers for σ and the maximum change, and a much higher number for γ. The low number for the maximum change led to training that would essentially stop once the number of epochs was large enough so that the final value was reached. Furthermore, the lower ending number for σ virtually eliminated the random component, thereby adding to

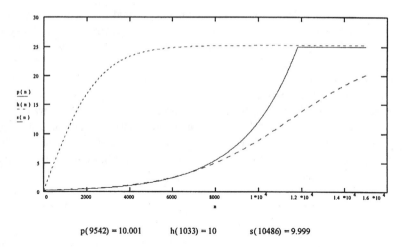

p(9542) = 10.001   h(1033) = 10   s(10486) = 9.999

FIGURE 8.6   Graph of different growth curves for γ. This graph shows the difference in possible growth curves for γ. The dotted line is the hyperbolic tangent function multiplied by the maximum value for γ, which is set at 25. The solid line is the exponential growth curve that was actually used. When γ reaches 25, it no longer increases. The dashed line is the sigmoidal-like function multiplied by 25 and translated so that when n=0, it is the initial value of γ. Note the different number of iterations needed for each curve to achieve the value of 10.

the slow convergence. Once these numbers were raised, the high number for γ caused the system to have difficulty stabilizing in the midtraining stages. Therefore the maximum γ value was lowered. The final values were not extensively varied, but the ones given in Table 8.2 seem to perform adequately.

During all the experimentation with the ALOPEX parameters, the entropy value was added as an input to the ninth module, which processes the moments of the original image. This seemed to be the final change needed and led to the best results of 93.9% training and 100% testing. While no discriminating value can be seen by examining the entropy value, that in conjunction with the moments definitely appears to provide good discrimination. Moreover, using these inputs, the neural network structure was varied. Comparing rows 1, 2, and 3 with rows 5, 7, and 10 of Table 8.2, it is evident that the structure with three hidden nodes per module (except for the ninth module with only two hidden nodes), eight integrating neural network nodes, and two output nodes gives the best performance.

Additionally, to satisfy curiosity, experiments were performed using only the moments of the original image and the entropy. While only seven different configurations were run, the best result was with five hidden nodes in the module feeding to four neural net nodes and then to two outputs. This gave 75.5% training and 50.0% testing. Experiments were also performed using a fully connected neural network. The configuration contained the same number of nodes as the best modular network, giving 240 input nodes, 26 nodes in the first hidden layer, 8 nodes in the second hidden layer, and 2 output nodes. This configuration usually did not converge. When it did converge, the outputs were all zeroes, giving 57.2% training and 60.0% testing.

The time to perform one experiment varies with the structure, the computer platform used, and the number of epochs that are required; most of these trials took between 90 and 120 min. While it may be possible to use other structures and change the parameters such that equally good results are achieved, this system appears to perform best with the configuration given above. The next section will explain why the system responds as it does to changes in configuration.

8.8 GENERALIZATION

There are three areas of the results presented in the last section that require further explanation. They are the changes in results related to the neural network structure, the changes related to the ALOPEX parameters, and the performance related to the data normalization techniques. Each of these will be discussed in detail. However, a general point that needs to be made is that ALOPEX is a stochastic process that relies upon random numbers. In addition to that, the neural network weights are initialized to random numbers. Because of this, recreating exact results is very difficult. This means that comparing the effects of parameter and neural network structure changes is also a difficult task. Most often, several trials are run with a certain configuration to get a general sense of how well that configuration performs. Therefore, the conclusions reached are based upon general trends. Because of the randomness involved, it is certainly possible that one very unique trial in one configuration may be better than a second configuration, even though in general the second configuration will give better results.

Table 8.3 shows that in 12 of the 15 cases presented, the training percentages are at least acceptable. In many cases they are very good. However, really only the first three cases show acceptable testing or validation percentages. The testing percentage tells how well the network can generalize. By generalization we mean how well the network performs on cases it has not seen but that are similar to the training data. Humans are particularly good at this. For example, once children are told what an automobile is, they can correctly identify another automobile even though it may be a different color, shape, or size. Not only that but they could be shown a picture of a boat and they will normally be able to say that it is not an automobile. The goal is to configure the neural network so it can perform as well.

TABLE 8.3
Comparison of ALOPEX Parameters

	Row 1 Value	Row 4 Value	% Change
σ	0.120	0.150	+ 25
γ	0.220	0.270	+ 23
Max Change	0.175	0.225	+ 29

If a neural network could not generalize, it would be essentially useless. A system that only recognizes data that it has already seen could be replaced by a simple look-up table.[37] Generalization depends on the size of the training set and the configuration of the network.[16] The generalization problem is closely related to the problem of polynomial interpolation.[37] Since the interpolation problem has been well studied, it will be used as an example to illustrate some of the ideas of generalization.

With enough independent variables, any number of nonconflicting* data points can be exactly fitted by an interpolating polynomial. However, often the generalization performance suffers. The same behavior is exhibited in neural networks.

As shown in Table 8.2, this system exhibits the classic generalization behavior. In a neural network the weights are the independent variables. There is a direct correlation from this application to the interpolation problem. In this problem a fixed number of data points is used for training and the rest for testing. Then a neural network structure is trained to fit those points. Neural networks with more nodes and, therefore, more weights tend to overfit the data. The ones with too few nodes and weights do not learn the data well enough. The network of the best size is the one that provides the best performance in terms of training and generalization.

In order to see the trends, experiments with similar ALOPEX parameters must be compared. The best comparison is between row 1 in Table 8.2 and rows 14 and 15. In this case we see the classic overfitting situation. The neural networks in rows 14 and 15 have more weights corresponding to more independent variables and have a better training percentage with a much lower testing percentage. Additionally, they converge in fewer epochs. This is a clear case of overfitting the data. The same situation occurs when rows 1 and 2 are compared with rows 5 and 10, but it is not as obvious from Table 8.2. Since Table 8.2 reports the best results, there are times when the training percentage is less than the best training percentage in order to report a better testing percentage. Both rows 5 and 10 trained to a higher percentage, but then the testing percentage would be around 60%. Furthermore, row 7 shows that the best performance by a network with fewer weights is not as good as equivalent networks, such as rows 1, 2, and 3, with a greater number of free parameters.

As stated above, the generalization performance is also dependent upon the size of the training set.[37] Therefore, the conclusion is that there is an optimum size of the training set to maximize generalization performance.

Indeed, much research was done on this problem in the late 1980s and early 1990s.[3,5,24] Virtually all of this research was done on standard feed-forward neural networks using backpropagation as a training mechanism. This makes the application of these results to this system difficult at best. However, if the modules of the neural network are considered to be just compressing the input features, then the classification process could be viewed as a standard feed-forward neural network with one hidden layer. This would mean that the network structure would be 26 input nodes, 8 hidden nodes and 2 output nodes, all fully connected. This gives a total of 224 weights plus a bias term at each node for a total of 260 free parameters. Wasserman[37] points out that if the neural network is viewed as a learning system,

---

* By nonconflicting, it is meant that for any given input there is only one output. For example, if the input were (1,2,3) and the output were (6,3) every time (1,2,3) was input, (6,3) would be the result.

then the Vapnik-Chervonenkis[35] or VC dimension can perhaps be used to estimate the number of training examples required for good generalization. However, for a multilayer neural network there is no explicit formula for calculating the VC dimension.[37] The VC dimension is related to the number of weights in the neural network,[3] and this number is often used as an estimate of the VC dimension. Wasserman[37] gives a lower bound on the number of examples needed by noting that for certain classes of problems the generalization error will be greater than or equal to $O(d/m)$, where $O(.)$ means "on the order of," d is the VC dimension, and m is the number of training examples. Substituting the numbers from this system if a generalization error of 10% is desired, then:

$$0.1 \geq O(260/m) \qquad (8.4)$$

or

$$m \geq 2600$$

Obviously, the system described in this paper achieved generalizations of 90% and better with far fewer training examples. This shows that these bounds are very conservative in practice.

This observed contradiction between the theoretical lower bounds and the actual behavior of the neural network in practice has been a point of contention between theorists and practitioners.[37] There may be many ways to explain the difference, one being that many times the simplifications that must be made to calculate the bounds on the number of training examples result in the bounds not reflecting reality.[24] In the case of our system, the answer may lie in the fact that the studies were done on systems much different than this with a different learning algorithm. However, another very plausible reason is that the training examples for this system may contain a high percentage of "boundary samples".[1,24] Boundary samples, or border patterns in a two-class problem, are examples that lie very close to the boundary between the two classes of data. Ahmad and Tesauro showed a marked increase in generalization performance when a large percentage of boundary samples are chosen to be in the training set. In addition, with a high enough number of boundary samples, the total number of examples needed to provide good generalization decreases. In the mammography application it is difficult to determine which examples are boundary samples (which is why most theoretical work is done on the XOR problem or the majority problem). However, given the above discussion, it would appear that the training set contains a high number of boundary samples.

As previously stated, the generalization performance is dependent on the size of the training set and the structure of the neural network. Given this, there are then two ways to attack the generalization problem. First, one could set the size of the network and then collect enough data to obtain the performance desired. Or, the data could be collected and then the network sized to provide the best results given the data. Both methods have been approached in the above discussion. However, in practicality, the second method, which is how this system evolved, is usually chosen. In many situations it would be impossible or very expensive to collect more data.

For this reason most researchers will change the network structure to accommodate the data.

Section 8.4 discussed the resulting data normalization method that this system uses. However, it was rather surprising that the *ln* and scaling approach done twice performed that much better than just a single *ln* and scale. The answer behind this lies in an analysis of the data.

As Section 8.4 listed, the moment values lie between -1×10^{41} and 1×10^{41}, with the minimum in absolute value being 1×10^{-39}. Once the *ln* is taken, the values range from approximately -90 to 94. When these values are scaled onto the $[-1,1]$ interval, the values from -90 to 0 fall into $[-1,-.02]$, which is almost half the interval. These values of natural logarithms from -90 to 0 were the original data values that were in the $[-1,1]$ interval. That is where 85% of the original moment data lie. Therefore, the majority of the data now occupies half the interval instead of $1/10^{41}$ of the interval that it did before. This results in the data becoming much more separable. To perform this operation again just further separates the data. Figure 8.7 shows this graphically. It is easy to see that the solid line, which is the *ln*(|x|), provides a good separation on the interval $[-1,1]$. However, the separation is even greater with the *ln*(|*ln*(|x|)|), as that function has a greater range. Figure 8.8 shows the same thing, only the results are now scaled to lie on the $[-1,1]$ interval to exactly match the normalization of the data. While the same result can be seen, it is clearer in Figure 8.7.

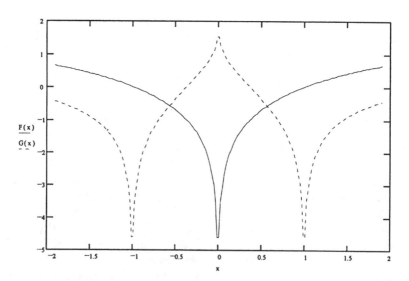

FIGURE 8.7   Graph of *ln*(|x|) and *ln*(|*ln*(|x|)|). The solid line shows the graph of *ln*(|x|) and the dotted line the graph of *ln*(|*ln*(|x|)|). The interval between -2 and 2 is shown so that the effect on the $[-1,1]$ interval can be more clearly seen.

This ability to allow the majority of the data to be more influential obviously increased the performance of the network. This says that the discriminating value

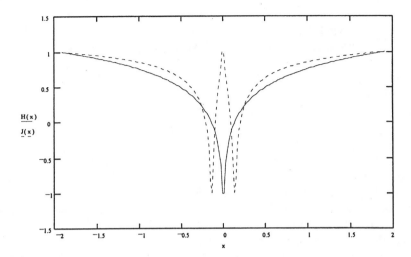

FIGURE 8.8   Graph of *ln*(|x|) and *ln*(|*ln*(|x|)|) with scaling. This graph shows the same functions as Figure 8.7, except that the result of each function is now scaled to lie on the [−1,1] interval as were the input data.

came from this part of the data and not the few data points that were large in magnitude.

Table 8.2 shows that the same neural network structure with different ALOPEX parameters will produce different results. While this is not too shocking given that the ALOPEX parameters control the amount of randomness in the system, for some cases it seems that the parameters are too sensitive to change. For example, comparing row 1 with row 2 shows that a very small change in the ALOPEX parameters led to a drop in the generalization performance. Rows 1 and 4 show an even greater performance drop for a slightly greater change in parameters. This seems to be an undesirable trait, and this situation warrants a closer look. Table 8.3 compares the initial ALOPEX parameters from row 1 and row 4.

When viewed in the perspective of Table 8.3, things do not seem so out of the ordinary. One would expect that a 25% change in parameter values would give a change in performance. Additionally, Table 8.4 compares rows 1 and 2, which had a small change in performance. Table 8.4 shows that a smaller percentage change in parameters led to this smaller change in performance. This is what would be expected and desired. The system was most sensitive to the starting point of the weights, which are randomly set. This behavior is tied to the topography of the error surface, which depends upon the cost or error function used and the number of training examples. While smaller networks have advantages in generalization and training times, as previously discussed, they tend to produce more rugged error surfaces with fewer good solutions.[29] In this case global minimization methods or methods that use at least some global information tend to perform much better than local minimization methods such as backpropagation.[29] ALOPEX uses both global and local information, and as such, it is a good choice in this instance.

TABLE 8.4
Comparison of Additional ALOPEX Parameters

	Row 1 Value	Row 2 Value	% Change
σ	0.120	0.100	−17
γ	0.220	0.200	−10
Max Change	0.175	0.170	−3

The behavior described above is exactly what was observed in this system. There were few good solutions, and the ones that existed were difficult to find. To correct this problem, additional training examples would need to be used. However, as explained earlier in this section, additional training examples may lead to a larger network required in order to learn the increased training set size. This, in turn, means longer training times. Still, it may be possible to add training examples and smooth out the error surface to alleviate some of the problem and not increase the network size. This is one of the issues for future work.

8.9 CONCLUSIONS

We have described a system that reads a digitized mammogram and classifies it as normal or abnormal. The results presented, even though they were generated on a small set of data, show great promise for a system of this type to be used in a clinical setting. With the maturation of digital mammography and the increase in processor speed of readily available and affordable personal computers, this system could easily be fielded on a mobile mammography van. Additional work on a larger database is ongoing.

ACKNOWLEDGMENTS

The authors would like to thank the United States Airforce and the Rutgers University Research Council for funding support. The mammograms were provided by the New Brunswick Radiology Group (Dr. Barry Zickerman). Our thanks are also extended to Dr. David August of the New Jersey Cancer Institute, for his insightful comments.

REFERENCES

1. Ahmad, S. and Tesauro, G., Scaling and generalization in neural networks: a case study, in *Advances in Neural Information Processing Systems I*, Touretzky, D., Ed., Morgan Kaufmann Publishers, San Mateo, CA, 1989, 160.
2. Bankman, I., Christens-Barry, W., Kim D., Weinberg, I., Gatewood, O., and Brody, W., Automated recognition of microcalcification clusters in mammograms, *Biomedical Image Processing and Biomedical Visualization*, SPIE vol. 1905, 731.

3. Baum, E. and Haussler, D., What size net gives valid generalization? *Advances in Neural Information Processing Systems I*, Touretzky, D., Ed., Morgan Kaufmann Publishers, San Mateo, CA, 1989, 81.

4. Cody, M., The fast wavelet transform, *Dr. Dobb's Journal*, 16, April 1992.

5. Cohn, D. and Tesauro, G., Can neural networks do better than the Vapnik-Chervonenkis bounds? *Advances in Neural Information Processing Systems 3*, Touretzky, D., Ed., Morgan Kaufmann Publishers, San Mateo, CA, 1989, 911.

6. Cowley, G. and Ramo J., Sharper focus on the breast, *Newsweek*, 64, May 10, 1993.

7. Daubechies, I., *Ten Lectures on Wavelets*, SIAM, Philadelphia, PA, 1992.

8. Davis, F., *How to Get the Best Mammogram*, Working Woman, 38, Oct. 1994.

9. Dhawan, A., Buelloni, G., and Gordon, R., Enhancement of mammographic features by optimal adaptive neighborhood image processing, *IEEE Trans. Med. Imaging*, M1-5 (1) 8, Mar., 1986.

10. Dhawan, A. and Le Royer, E., Mammographic feature enhancement by computerized image processing, *Comput. Meth. Prog. BioMed.*, 27, 23, 1988.

11. Dhawan, A., Chitre, Y., and Moskowitz, M., Artificial neural network based classification of mammographic microcalcifications using image structure features, *Biomedical Image Processing and Biomedical Visualization*, SPIE vol. 1905, 1993, 820.

12. Dodd, G. D., Mammography: state of the art, *Cancer*, 53, 652, 1984.

13. D'Orsi, C. and Kopans, D., Mammographic feature analysis, *Sem. Roentgenol.*, XXVIII (3), 204, July 1993.

14. Elmore, J., Wells, C., Lee, C., Howard, D., and Feinstein, A., Variability in radiologists' interpretations of mammograms, *N. Engl. J. Med.*, 331(22), 1493, 1994.

15. Goldberg, M., Pivovarov, M., Mayo-Smith, W., Bhalla, M., Blickman, J., Bramson, R., Boland, G., Llewellyn, H., and Halpren, E., Application of wavelet compression to digitized radiographs, *AJR*, 163, 463, August 1994.

16. Haykin, S., *Neural Networks — A Comprehensive Foundation*, Macmillan College Publishing Co., New York, 1994.

17. Hecht-Nielsen, R., *Neurocomputing*, Addison-Wesley, Reading, MA, 1990.

18. Hu, M., Visual pattern recognition by moment invariants, *IRE Trans. Inf. Theor.*, no. 8, 179, Feb. 1962.

19. Isard, H., Other imaging techniques, *Cancer*, 53, 658, 1984.

20. Kim, Y., Choi, I., Lee, I., Yun, T., and Park, K., Wavelet transform image compression using human visual characteristics and a tree structure with a height attribute, *Opt. Eng.*, 35(1), 204, Jan. 1996.

21. Kung, S., *Digital Neural Networks*, PTR Prentice Hall, Englewood Cliffs, NJ, 1993.

22. Laine, A., Schuler, S., Fan, J., and Huda, W., Mammographic feature enhancement by multiscale analysis, *IEEE Trans. Med. Imaging*, 13(4), 725, Dec. 1994.

23. Lippmann, R., An introduction to computing with neural networks, *IEEE ASSP Magazine*, 4-22, April 1987.

24. Mehrotra, K., Mohan, C., and Ranka, S., Bounds on the number of samples needed for neural learning, *IEEE Trans. Neural Net.*, 2(6), 548, Nov. 1991.

25. Micheli-Tzanakou, E., Neural networks in biomedical signal processing, *The Biomedical Engineering Handbook*, Bronzino, J., Ed., CRC Press, Boca Raton, FL, 1995, ch. 60, 917.

26. Myers, L., Rogers, S., Kabrisky, M., and Burns, R., Image perception and enhancement for the visually impaired, *IEEE Eng. Med. Biol.*, 594, Sept./Oct. 1995.

27. Qian, W., Clarke, L., Li, H., Clark, R., and Silbiger, M., Digital mammography: M-channel Quadrature Mirror Filters (QMFs) for microcalcification extraction, *Comp. Med. Imaging Graph.*, 18(5), 301, Sept./Oct. 1994.

28. Samiy, A., Douglas, R., Jr., and Barondess, J., *Textbook of Diagnostic Medicine*, Lea and Febiger, Philadelphia, PA, 1987.

29. Shang, Y. and Wah, B., Global optimization for neural network training, *Computer*, 29(3), 45, Mar. 1996.

30. Shen, L., Rangayyan, R., and Desautels, J., Application of shape analysis to mammographic calcifications, *IEEE Trans. Med. Imaging*, 13(2), 263, June 1994.

31. Sheng, Y., Wavelet transform, in *The Transforms and Applications Handbook*, Poularikas, A., Ed., CRC Press, Boca Raton, FL, 1996, ch. 10, 747.

32. Sickles, E., Breast calcifications: mammographic evaluation, *Radiology*, 160, 289, 1986.

33. Strang, G., Wavelets, *Am. Sci.*, 82, 250, May/June 1994.

34. Strickland, R. and Hahn, H., Wavelet transforms for detecting microcalcifications in mammograms, *IEEE Trans. Med. Imaging*, 15(2), 218, April 1996.

35. Vapnik, V. N. and Chervonenkis, A. Y., On the uniform convergence of relative frequencies of events to their probabilities, *Theoret. Probability Appl.*, 17, 264, 1971.

36. Vyborny, C., and Giger, M., Computer vision and artificial intelligence in mammography, *AJR*, vol 162, Mar. 1994, 699-708.

37. Wasserman, P., *Advanced Methods in Neural Computing*, Van Nostrand Reinhold, New York, 1993.

38. Wasserman, P., *Neural Computing: Theory and Practice*, Van Nostrand Reinhold, New York, 1989.

39. Yoshida, H., Zhang, W., Cai, W., Doi, K., Nishikawa, R., and Giger, M., Optimizing Wavelet Transform Based on Supervised Learning for Detection of Microcalcifications in Digital Mammograms, *Proc. of IEEE Int. Conf. Image Proc.*, 152, 1995.

40. Zettler, W., Huffman, J., and Linden, D., Application of compactly supported wavelets to image compression, *Image Processing Algorithms and Techniques*, SPIE vol. 1244, 1990, 150.

41. Micheli-Tzanakou, E., Uyeda, E., Ray, R., Sharma, A., Ramanujan, R., and Doug, J., Comparison of neural network algorithms for face recognition, *Simulation*, 64(1), 15, July 1995.

42. Mallat, S., A theory for multiresolution signal decomposition: the wavelet representation, *IEEE Trans. Patt. Anal. Mach. Intell.*, 11(7), 674, July 1989.

43. Held, G., *Data Compression*, John Wiley & Sons, New York, 1987.

44. Ross, S., *A First Course in Probability*, Macmillan Publishing, New York, 1976.

45. Hrycej, T., *Modular Learning in Neural Networks*, John Wiley & Sons, New York, 1992.

46. Rodriguez, C., Rementeria, S., Martin, J., Lafuente, A., Muguerza, J., and Perez, J., A modular neural network approach to fault diagnosis, *IEEE Trans. Neural Net.*, 7(2), 326, Mar. 1996.

9 Visual Ophthalmologist: An Automated System for Classification of Retinal Damage

Sergey Aleynikov and Evangelia Micheli-Tzanakou

9.1 INTRODUCTION

There is a vast variety of eye related diseases which leave visible artifacts on the retinal surface. This makes it very difficult to create any automated classification system, since the disease features vary widely. Retinal diseases are classified in two major groups: vascular diseases, caused by circulatory disturbances of the retinal vessels; and avascular diseases, where the rod and cone layers and the pigment epithelium are implicated. In this research an attempt was made to classify only diseases belonging to the first group, resulting from retinal hemorrhage.

9.2 SYSTEM OVERVIEW

The system, which was named "Visual Ophthalmologist℠" (VO) is designed for a 32-bit operating system, such as Windows NT™, or Windows 95™, running on a PC-compatible computer. The flowchart of Figure 9.1 describes the main components of the system. The system consists of five general modules:

M1. Image Acquisition Module-Image Source
M2. Database Image Management Module
M3. Image Processing Module
M4. Feature Extraction Module
M5. Neural Network Classification Module

The system can function independently just as an image data management system (IDMS), or it can also be used for image classification. Module M1 consists of a Hewlett Packard 4cx desktop scanner with a transparency adapter to scan slides. This module can be substituted with any compatible scanner capable of providing optical resolution greater than 150 dpi. Instead of using a scanner, it is also possible

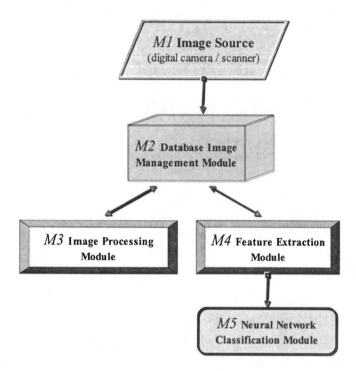

FIGURE 9.1   System components flowchart.

to incorporate a digital camera into the system's structure and use it for direct image acquisition.

The images are scanned using proprietary software into Windows Clipboard, and then pasted right into a newly created record in the Database Image Management Module (M2). The Database Image Management Module (DIMM) is a database client, which requests information from and sends information to a database server. A server processes requests from many clients simultaneously, coordinating accessing and updating of data. The advantage of this type of architecture is that the image server can be located either on a local computer or on a remote computer. The communication of the DIMM with the database server is done through an intermediate layer, called Borland Database Engine (BDE). The patients' data are organized in three independent databases connected between each other with a parent/child relationship in the following manner:

1. Patient database (this file contains general patient data: first and last name, social security number, date of birth, etc.),
2. Visit database (includes information on patient's visits, belonging to a patient in the 'patient database'. This information may include: date of a visit, description, etc.),
3. Image database (consists of several fields, which include patient's images belonging to a specific visit in the 'visit database').

9.2.1 IMAGE PROCESSING

The image processing tools included in this module are histogram equalization and stretch, image compression based on a Gaussian Pyramid,[2] image orientation, center of mass determination, and a set of convolution filters. This enables the user to acquire a more accurate and flexible classification.

9.2.2 FEATURE EXTRACTION METHODS

Module M4 in Figure 9.1 is a Feature Extraction Module. It processes selected records in the imaging database and outputs specific features extracted from the images. These features are saved in a file to be used in the image classification by module M5. The feature extraction is based on three independent methods to allow for a higher recognition rate. These methods are

1. Image Central and Invariant Moments,[5]
2. Image Power Spectrum, based on the F-CORE Decomposition,[7] and
3. Multiresolution Wavelet Decomposition Approach.[4,6]

9.2.3 IMAGE CLASSIFICATION

Module M5 (Figure 9.1) of the system consists of a Modular Neural Network, which takes the features generated and saved to a file by the Feature Extraction Module (M4), and tests them against the information on which the neural network was trained. This module is implemented in a separate program, which can be run concurrently with the Visual Ophthalmologist. The reason for not incorporating this module directly into the Visual Ophthalmologist is that since it is a versatile program by itself, it can serve for finding solutions of many independent problems of recognition/classification, similar to the ones found by the Visual Ophthalmologist project. The neural network training in this module is based on the Algorithm of Pattern Extraction (ALOPEX). ALOPEX is an optimization technique developed by Tzanakou and Harth in 1973 (for a list of references the reader is advised to look into Reference 9) to optimize receptive field mapping in the visual system of frogs. It has been applied to a broad variety of applications due to the fact that it has better convergence compared to the traditional gradient methods. Some of its recent applications include face recognition[8] and mammogram classification.[3] ALOPEX serves to minimize/maximize a system's global response R, which is a function of multiple parameters. As the parameter space of the response becomes large, it becomes more and more complicated to find an appropriate solution. A valuable characteristic feature of the ALOPEX algorithm is that at each iteration it considers both local and global effects on the response function. For the complete description of the algorithm, the reader is referred to one of the latest publications.

9.3 MODULAR NEURAL NETWORKS

The idea of building modular networks comes from the analogy with biological systems, in which a brain (as a common example) consists of a series of interconnected substructures, like the auditory and visual systems, which, in turn, are further structured on more functionally independent groups of neurons.

Each level of signal processing performs its unique and independent purpose, such that the complexity of the output of each subsystem depends on the hierarchical level of that subsystem within the whole system. Modular neural networks can be used in a broad variety of applications. Each module does its unique function, providing some output to the modules in the next level. The usage of modular neural networks is most beneficial when there are cases of missing pieces of data.[1] Since each module takes its input from several others, a missing connection between modules would not significantly alter that module's output.

With the introduction of object-oriented computer languages in the early 1990s, it has become relatively easy to implement parallelism in neural network processes, which extends the regular procedural approach to a new "biological-like" dimension.

At a level of high abstraction the network should look like the "black box" object in Figure 9.2. It receives some input from templates stored in a file, propagates it through all modules, and provides some output in a meaningful format. A module is not aware of any type of processing that is going on in the rest of the network, though it knows which particular network it belongs to in order to provide correct references, stored in the second container, to the rest of the modules. A local error of a template is the summation of a function of the absolute differences between the desired and actual values of the module's output nodes to a given template. We use different approaches for computing the local error E_i' :

$$E_i' = \left| Out_i^{desired} - Out_i^{observed} \right|$$

If $E_i' >$ threshold, then if the desired output $Out_i^{desired}$ is 1, we set E_i' equal to $\exp(2 \cdot E_i') - 1$. (This is done because we like the values on the diagonal of the output matrix to have an increased rate of convergence.) Otherwise E_i' is expressed as $\exp(E_i') - 1$.

The traditional training approach assumes that the local error is equal to $(E_i')^2$. However, we would like to make it more sensitive to a change of the argument, which is why $(E_i') - 1$ has been chosen. In fact, to get a faster convergence of the output that are set to 1, we use an even more sensitive function, namely $\exp(2 \cdot E_i') - 1$.

9.4 APPLICATION TO OPHTHALMOLOGY

This section summarizes the application of the modular neural network algorithm described in the previous section to the problem of retinal image damage classification. Once the data were acquired and stored in the database and all features were generated using approaches discussed earlier, we built a neural network as shown in Figure 9.3 to classify the obtained features.

As shown in the figure, the network consists of two levels of modules. The modules on the first level process the features generated by three feature extraction methods consecutively: a) moments, b) F-CORE, and c) wavelet histogram. The module on the second level serves as the classifier of the results generated by the

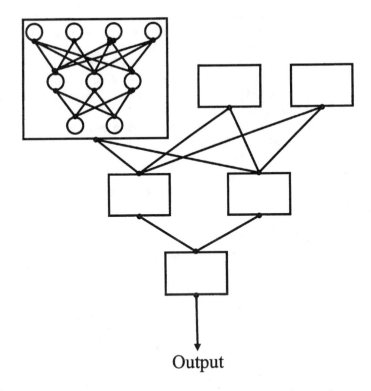

FIGURE 9.2   An example of a modular perceptron-based neural network.

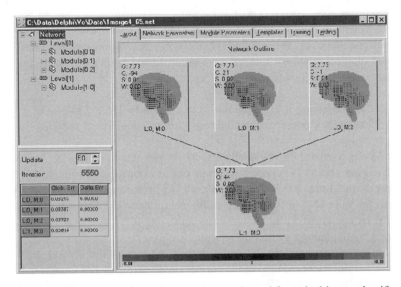

FIGURE 9.3   The configuration of a neural network used for retinal image classification.

first three modules. It combines the recognition of three different methods into a joint classification. The definition of the correct classification of the retinal hemorrhage is that the network should be able to tell whether or not any given image contains a hemorrhage. In order to test the accuracy of recognition, the original images were preclassified by a degree of hemorrhage damage to a [1, .. ,5] range, where 1 means no or very little (< 5% of retinal surface) hemorrhage, and 5 means very high degree of damage(> 80% of retinal surface). The architecture of each module is as follows: the moments processing module consists of three layers of neurons, containing respectively: 15→10→5 nodes. It contains 15*10 + 10*5 = 200 connections. The F-Core processing module consists of three layers of neurons, containing respectively: 25→15→5 nodes. It contains 25*15 + 15*5 = 450 connections. The wavelet histogram processing module consists of three layers of neurons, containing respectively: 25→15→5 nodes. It contains 25*15 + 15*5 = 450 connections. Finally, the merging module merges the classification of the previous three modules into a combined form to provide the final classification of the degree of retinal damage. It consists of 15→10→5 input/hidden/output nodes, respectively. The number of neurons in the input layer of each module is equal to the number of features in the corresponding method. The number of output neurons is equal to 5 (the number of hemorrhage classes). A hidden layer contains the number of neurons equal to the average of the neurons in the input and output layers.

9.5 RESULTS

Since the Visual Ophthalmologist is a highly integrated system, the process of obtaining results is very simple. It consists of three steps: 1) selecting the image records containing images of interest, 2) running the feature extraction algorithms on the selected data, and 3) classifying results using the modular neural network.

The system currently contains a database of 160 retinal images obtained from three different sources: a) black and white sheet slides, b) loose color slides, and c) ophthalmological atlas. All images were preprocessed before they were input into the database. Preprocessing included

a) Image re-sizing using Gaussian compression to 256 × 256 pixels
b) Histogram stretch/equalization to enhance the quality of the image

All images stored in the databases are in the Windows Device Independent Bitmap format (bmp). We chose this format due to compatibility with most Windows imaging applications. All 160 images were used for testing of the system. Twenty-five images were chosen for training of the neural network. Therefore, the network training set contained 25 templates, each consisting of 15 + 25 + 25 = 65 features provided by the feature extraction module.

The overall classification provided by the module of the second level in the network's architecture resulted in the correct classification of 127 images out of 160, which is equal to 79.38%. For each image processed by the system, the following rule was used to provide the image classification. As discussed in the previous section, each image could belong to one of five classes ordered by the degree of hemorrhage from 1 to 5. The correct recognition by the network is considered true if its most dominant output deviates by not more than a distance of two from the correct classification.

The modular network was also trained to 95%, and classified correctly 127 out of 160 images, which signifies 79.38% of recognition accuracy.

9.6 DISCUSSION

The results obtained in the process of application of the four outlined methods are mainly affected by the feature extraction criteria in each method, the normalization of data provided to a neural network, and finally the training parameters of each network. When a classical neural network is used for classification, the convergence is much slower and not as accurate. Although we chose a very small number of templates for training, we still achieved a high testing performance (80%). Once our database becomes larger, then training will be done with a larger number of templates for each class. Undoubtedly the testing performance will be improved.

REFERENCES

1. Aleynikov, S. and Micheli-Tzanakou, E., Design and implementation of modular neural networks based on the ALOPEX algorithm, *Virtual Intell., Proc. SPIE*, 2878, 81, Nov. 1996.
2. Burt, P. J., The pyramid as a structure for efficient computation, in *Multiresolution Image Processing and Analysis*, Rosenfeld, A., Ed., Springer-Verlag, New York, 1984, 6.
3. Cooley, T. and Micheli-Tzanakou, E., A Modular Neural Network for Classifying Mammograms, Proc. of the Int. Conf. On Neural Networks, 1996, 1162.
4. Daubechies, I., Orthonormal bases of compactly supported wavelets, *Comm. Pure Appl. Math.*, 41, 906, 1988.
5. Hu, M., Visual pattern recognition by moment invariants, *IRE Trans. Inf. Theor.*, 8, 179, Feb. 1962.
6. Mallat, S., A theory of multiresolution signal decomposition: the wavelet representation, *IEEE Trans. Patt. Anal. Mach. Intell.*, 11(7), 674, July 1989.
7. Micheli-Tzanakou, E. and Binge, G., F-CORE: A New Fourier based data Compression and Reconstruction Method, *MEDICON '89*, 344, 1989.
8. Micheli-Tzanakou, E., Uyeda, E., Ray, R., Sharma, A., Ramanujan, R., and Doug, J., Comparison of neural network algorithms for face recognition, *Simulation*, 64(1), 15, July 1995.
9. Zahner, D. and Micheli-Tzanakou, E., Artificial neural networks: definitions, methods, applications, *The Biomedical Engineering Handbook*, Bronzino, J., Ed., CRC Press, Boca Raton, FL, 1995, ch. 184, 2699.

10 A Three-Dimensional Neural Network Architecture

Evangelia Micheli-Tzanakou,
Timothy J. Dasey, and Jeremy Bricker

10.1 INTRODUCTION

The idea behind the presented architecture was to create a pattern recognition system using neural components. The brain was taken as a model, and although little is known about how pattern recognition is accomplished, much more is known about the cells that comprise the earliest levels of processing and analyzing the features of an environment most directly. By constructing cells with similar properties to the biological cells, we may gain an advantage in information conservation and proper utilization of neural architectures. The most important characteristic of these cells is their receptive field (RF). With this in mind, we could search for an adaptive mechanism that, by changing connective strengths, could give the desired RFs. Therefore, since we will know what information the algorithmic components are providing, when a method is found that provides the desired cell types, we may be able to trace back via the algorithm to see what information the neurons give.

10.2 THE NEURAL NETWORK ARCHITECTURE

The architecture chosen was that of a hierarchy of two-dimensional cell layers, each successive layer more removed from the environment (Figure 10.1). The first layer receives inputs from the external world and all other layers from the preceding layers. In addition, the cells may receive lateral connections from other neighboring cells within the same layer, depending on the particular choice of the architecture. The interlayer feed-forward connections are chosen so that a cell feeds its connections onto a neighborhood of cells in the lower layer. This neighborhood may have definite bounds so that all cells within it make connections, or it may have indefinite bounds in which the probability of a connection decreases as a Gaussian with distance.

The component cells themselves choose their outputs based on a weighted sum of all inputs passed through a function σ, such as

$$O_i(t) = \sigma\left[\alpha_i^* \Sigma_j C_{ij}^* O_j(t-1)\right] \qquad (10.1)$$

where $O_i(t)$ is the output of neuron i at time interval t, C_{ij} are the connection strengths, bounded from $[-\beta, \beta]$ where β is usually 1.0, and α is a constant. In the simulations, σ is usually a sigmoid of the form

$$\sigma(x) = 0.5 \cdot a \cdot \left(1 + \tanh\left(b \cdot x \cdot c\right)\right) \qquad (10.2)$$

where a,b,c are constants which fix the maximum value, steepness, and bias of the sigmoid, respectively. However, if we wish to allow the inhibitory components of the RF to be used by subsequent layers, then the sigmoid function must have a non-zero firing level for those negative inputs. This suggests the use of a spontaneous firing activity for all neurons. An additional requirement needed to keep the neurons useful and "responsive" is to keep that neuron from being pushed too far into the saturation level. If that occurs, input deviations will not be sensed well, if at all. Since each neuron receives several inputs, it is easy for this to occur. To prevent it from happening, α is usually chosen equal to the reciprocal of the number of connections to neuron i, so that the neuron simply passed a weighted average of the inputs through the sigmoid.

10.3 SIMULATIONS

A simulation usually consists of a sequence of presentations of random input patterns to the first layer and a learning rule imposed on the connections by analysis of the firings of the neurons. A random input was chosen so as to prevent the cells from being biased towards any specific environmental feature. Since neighboring inputs are uncorrelated, first layer cells that receive their influences are expected to have synapse patterns that would similarly wander aimlessly in the learning process. The first layer provides a spatial average of the overlying inputs. Since neighboring cells have the greatest overlap in their neighborhoods, they tend to have firing patterns which are most similar. This would cause cells in layer 2 to have synapses that originated from nearby cells to want to be alike. The actual training of the connections can be done in different ways.

I) Synapses can be changed based on a variation of the Hebbian rule[1] as follows:

$$C_{ij} = \delta \cdot O_i \cdot O_j \qquad (10.3)$$

where δ is a small positive constant. Due to the correlation between neighboring level 1 cells, the synapses to the cells in later layers would tend to want to be all alike without additional constraints. In order to guarantee both positive and negative synapses to every cell, an additional "resource" constraint is imposed, which takes the form of

$$\Sigma_j C_{ij} = 0 \qquad (10.4)$$

The third restriction is a bounding of the connections to the interval [−1, 1]. A synapse is allowed the freedom to switch from positive to negative and vice versa. This is not expected to alter the main results but only to prevent many of the synapses from disappearing with zero strength. Convergence usually occurs within 1000-5000 iterations, although faster convergence can be achieved with larger δ. Usually the final state of the synapses is at either the excitatory or the inhibitory limits.

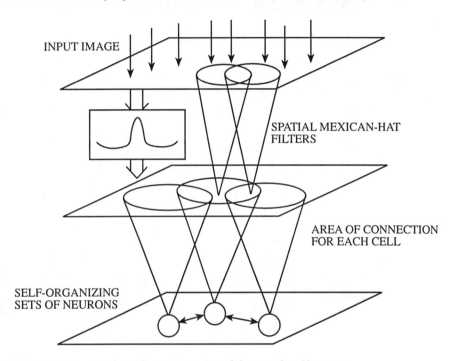

FIGURE 10.1   A schematic representation of the neural architecture.

10.3.1 VISUAL RECEPTIVE FIELDS

A network was created with three layers and 128 cells per layer. A square stimulus was assumed with 32×32 size. The maximum distance that these cells can affect is a radius of r, with minimum weight −1 and maximum weight values of +1. The network had a total of 7071 connections. In the training mode, the minimum stimulus value was assumed to be zero and the maximum equal to 10. No noise was imposed on the system. The results obtained show the emergence of cells with edge-type RFs in layer 2 (Figure 10.2a). The orientation of the edge appears to be totally arbitrary, even between neighboring cells. In layer 3, these edge cell RFs often conflict to give RFs which have oblong centers and surrounds of the opposite polarity, but many times these centers draw to the edges with further learning. Thus the final RFs often look like an elliptical center touching the outside of the field, mostly surrounded by a horseshoe-shaped region of opposite polarity (Figure 10.2b).

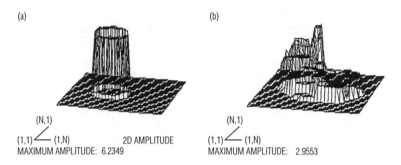

Figure 10.2   Receptive field characteristics for the neurons described in the text. (a) RF of layer 2. (b) RF of layer 3. Notice the center-surround organization of layer 2 and the elongated character of layer 3.

Figure 10.3 shows the results from a similar network, except that the minimum weight value is –0.5, i.e., less inhibitory effects. Notice that excitation spreads more and that the maximum amplitudes are much larger. Also notice that the layer 3 RF is much longer than the one in layer 2.

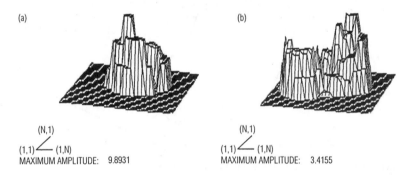

FIGURE 10.3   Receptive field organization for layer 2 (a) and layer 3 (b) when the inhibitory effects are less than in Figure 10.2. Compare the amplitudes and the spread of the RFs to those of Figure 10.2.

In the frequency domain, these RFs show more fine tuning as we move to deeper layers of the system. Figures 10.4 and 10.5 represent the power spectra of Figures 10.2 and 10.3, respectively. Also, notice that the edge effects are more obvious in the spectra of layer 3.

II) The wider the variance of the firing rate of the cells the more information the cells can carry. With such a supposition we can use an optimization routine to find the values of the synapses to a cell such that the variance in the firing rate of the cell is maximized. The optimization system is a variation of the ALOPEX process.[2] In this process two random connection patterns can be presented, and the variance (V) of the cell output is estimated with a number of random input patterns. Since we want the pattern of connection strengths to affect the variance and not the strength of the connections themselves, the variance can be modified as

$$V_i = \left[(1/N)^* \Sigma_j \left(O_i^j - O_i^{ave} \right) \right] / \left(\Sigma_j C_{ij} \right) \tag{10.5}$$

The connections are then changed based on the relation between the last change in connections and the last change in the variance, with an added noise term to prevent local minima as follows:

$$C_{ij} = \beta^* \left(C_{ij}(t) - C_{ij}(t-1) \right)^* \left(V_i(t) - V_i(t-1) \right) + \text{ noise term} \tag{10.6}$$

Amazingly, with this modification, the same edge sensitive cell RFs emerge after only about 100 iterations and remain the same until about 400 iterations. This shows that the combination of Hebb's rule and ALOPEX is something desirable. It might also mean that the way in which the architecture of the network is set up biases them toward neurons with edge detection capabilities. Work by others[3] has indicated that certain forms of the Hebb rule can be used to perform principle component analysis, a variance maximization of sorts. In addition, both feed-forward and feed-back connections are used, with feedback having a wider connective neighborhood than the feed-forward connections. All connections are variable. If the inhibitory connections are spread over a much wider area, they tend to cancel the excitatory influence, making the Hebb changes ineffective. In future work we will include feed-forward connections of cells with a Gaussian distribution, and with inhibitory connections and excitatory connections having a different spatial standard deviation. The present number of maximum synapses allowed does not give us the ability of obtaining statistical significance for initial random strength generation.

III) Both feed-forward and feedback connections can be used, with the feedback having a wider connective neighborhood than the feed-forward connections.

IV) Lateral connections on each layer are allowed and used, thus adding an extra feature of similarity to the biological system.

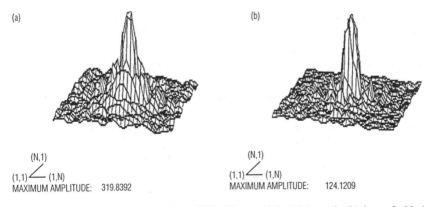

(a)   (b)

(N,1)   (N,1)
(1,1) (1,N)   (1,1) (1,N)
MAXIMUM AMPLITUDE: 319.8392   MAXIMUM AMPLITUDE: 124.1209

FIGURE 10.4   Power spectrum of the RF in Figure 10.2. (a) layer 2, (b) layer 3. Notice the fine tuning in layer 3.

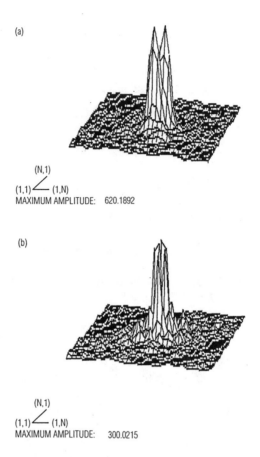

(a)

(N,1)
(1,1) ⟋ (1,N)
MAXIMUM AMPLITUDE: 620.1892

(b)

(N,1)
(1,1) ⟋ (1,N)
MAXIMUM AMPLITUDE: 300.0215

FIGURE 10.5   Power spectra of RFs in Figure 10.3. (a) layer 2, (b) layer 3. Compare with Figure 10.4. The edge effect is much more pronounced.

If each input signal value is thought of as a dimension in parameter space, any particular input will comprise a point in that space. The synapses of a neuron can then be thought of as describing a vector in the same space and the output of the neuron as the projection of the input point onto the synapse vector. If the choice of the synapses is initially random, chances are that the projections from many different inputs will lie close to one another, giving the neuron a response profile. Consider this to be the response profile of a neuron before optimization. In order to better distinguish between inputs, the synapses should be changed so that more of the neuron range can be utilized. An intriguing choice is for the neuron to perform a type of principal component analysis (PCA) (Karhunen-Loève feature extraction). Principal component analysis may be approximated by a search for the vector (described by the connection weight values), which maximizes the variance of the cell firing level. The choice of this property may serve to partition the input space into recognizable categories at the output. This analysis approximates the Karhunen-Loève search for the eigenvector of the maximum eigenvalue. For layers of neurons that have a large amount of information with near neighbors, the use of low-level lateral inhibition should prevent the system from settling on the same vector for each neuron, providing instead a graded topography to the layer.

Depending on the partitioning of the input space, this processing mode of neurons could provide many different behaviors. If the input space has clusters, the neuron may provide classification. If, on the other hand, the inputs are "randomly" distributed in space, the neuron can choose any feature vector, but could be constrained by near neighbors interactions into how it forms topographic maps.

10.3.2 MODELING OF PARKINSON'S DISEASE

We have created a network with eight neural layers in addition to a layer for stimuli. Each layer represents one physiological region of the brain or nervous system, as described by the model of DeLong et al.[4] By means of a series of excitatory and inhibitory feed-forward and feedback synapses, the brain stem (layer 7) is relatively active. Another layer was added to represent the motor neurons of the extremities (head, legs, and arms). The connections from the brain stem to the extremities are assumed to be inhibitory. Thus, in the normal state, the high activity of layer 7 subjects layer 8 to a large degree of inhibition, making this layer rather inactive. The Parkinsonian case is simulated by cutting off connections stemming from the input layer. When this happens, layer 7 is not excited to a degree as large as it is in the normal case. Sequentially, a smaller amount of activity exists with which to inhibit the extremities. This unusually high level of activation in the motor neurons of the extremities represents the tremors present in patients suffering from Parkinson's Disease.

A Pallidotomy is then simulated in the Parkinsonian scenario by destroying groups of neurons in the Globus Pallidus Internum (or GPi, layer 4). As is evident from DeLong's model (Figure 10.6), this action will reduce the total amount of activity present in the GPi, causing less inhibition to the layer following the GPi, followed by greater excitation of the cortex, greater excitation of the brain stem, and more inhibition to the motor neurons of the extremities, corresponding to a reduction in tremors. The Pallidotomy brings the degrees of activation on layers between the GPi and the extremities back to levels akin to those observed in the non-Parkinsonian scenario. The program allows the effects of different types and locations of lesions to be observed. In general, a lesion is targeted on the location in which the highest degree of activity in the GPi is recorded. The network is helpful in predicting the consequences of lesioning off-target or at a location other than the point of highest activity. Lesioning at multiple locations or on different layers may also be simulated.

The program created a network consisting of eight layers of neurons and one layer of input nodes. Each layer corresponds to one layer in DeLong's model.[4] On each layer are placed 200 neurons. This large quantity of neurons is necessary in order to visualize each layer well. The neurons are randomly scattered on each layer within the spatial bounds of $-2<x<2$ and $-2<y<2$. The input layer consists of two stimuli, each of which contains one input node. One stimulus is located at $(-1,0)$ and the other at $(1,0)$.

The input nodes connect to neurons on layer 1 within a radius of .5 units in the program's coordinate space. These connections are set at a constant value of 1. Each half of layer 1 connects to each of layers 2 or 4 with a radius of 10 units, causing each neuron on the left half of layer 1 to be fully interconnected with neurons on layers 2 and each neuron on the right half of layer 1 to be fully interconnected with neurons on layer 4 (the left of layer 1 does not connect with 4, nor the right of layer 1 with 2). The reason for this is to allow the entirety of layers 2 and 4 to play a role in the simulation while only being exposed to effects from the correct stimulus (as is seen in the figure). If the connections from layer 1 to layers 2 and 4 had a connective radius of only .5 units, then only the left-hand side of layer 2 and the right-hand side of layer 4 would be meaningful. By fully interconnecting the halves of layer 1 to these other layers, all the neurons of layers 2 and 4 are active in the model (technical note: the halves of layer 1 are connected to layers 2 and 4 by first destroying all connections between layer 1 and layers 2 and 4 and then restoring connections only between a circle on the left half of layer 1 with layer 2 and between a circle on the right half of layer 1 with layer 4). Between all other layers of the network, the connective areas have radii of .5 units. Connections are created as displayed in Figure 10.6, but the initial connection strengths (except between stimuli and layer 1) are randomly distributed between −1 and 1 (as opposed to being uniformly inhibitory or excitatory between any two layers as the figure suggests). Connections between stimuli and layer 1 are set at constant values of −1 or 1 (inhibitory or excitatory, as seen in Figure 10.6). In this figure thick arrows represent excitation and thin arrows represent inhibition. For comparison, the left side of the figure is a schematic representation of a normal subject, while the right-hand side depicts a Parkinsonian subject.

The activation function of the neurons in the network takes the form of a sigmoid with minimum value 0 and maximum value 1; when the sum of the inputs equals zero, a neuron will generate an output of 0.5. Even though this propagates a signal through the network when no stimulus exists, the conventional sigmoid (with bounds between −1 and 1, where zero input gives zero output) would not work well for the application in question. Assume, for instance, that layer 1 has a highly positive activation. Then, this would give a large degree of inhibition (a very negative input) to layer 2 (due to the negative weights of the synapses connecting these two layers). Now, if the range of neural activation reaches below zero, the neurons on layer 2 will fire with a highly negative activation. This negative activation will combine with the inhibitory (negative) synapses between layers 2 and 3 to produce an *excitatory* (positive) input to layer 3, causing the neurons on layer 3 to take a highly positive activation. Such a situation does not recreate the conditions this model attempts to emulate. Therefore, the necessity arises of using an activation function that allows the neurons only positive activation. Inhibition and excitation are thus an immediate function of the synaptic connection strengths, not of the neural activity. In this case, high activation on layer 1 causes a large negative input to layer 2, creating a small positive (near zero) output on layer 2. This small positive output on layer 2 feeds into layer 3 as a small negative input. As DeLong's model requires, layer 3 experiences an inhibitory input (though not as inhibitory an input as layer 2 experienced) and thus generates a small positive output (though larger than the output of layer 2). Evidently, restricting the activation function to a range between 0 and 1 is necessary in order to allow synapses to be inhibitory or excitatory in the fashion this model requires.

The problem now facing the network is that of configuring its connection strengths to DeLong's model (inhibitory or excitatory as seen in the figure) while producing little activation on layer 8 in the normal case and higher activation (tremors) on layer 8 in the Parkinsonian case. This configuration must come about by supervised training of the network with ALOPEX. Training is done with two templates, one representing the normal case and the other representing the Parkinsonian case. In the normal case, the input stimuli both take a positive activation (remember that the connection strengths are what create inhibition or excitation in this model; the activation of a neuron or input node is always positive) and layer 8 (which may be considered the output layer) is to show only a small amount of activity. The template for the Parkinsonian situation has input nodes, each of which has an activation of zero (which is the same as cutting off the inputs), while layer 8 is forced to a higher degree of activity than in the normal state. A special bias in the connection strengths of the network must also be implemented in order to satisfy the condition that these connection strengths be inhibitory between certain layers and excitatory between others (Figure 10.6).

If the network could be trained in such a manner, then a Pallidotomy could be simulated. A specific area (or areas) of neurons on layer 4 (the GPi) would be lesioned (killed), and the result of this lesioning would become apparent on the activation of layer 8 (reduction of tremors in the extremities).

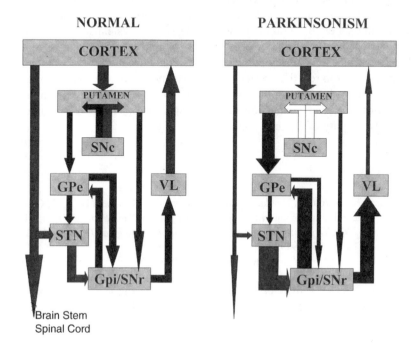

FIGURE 10.6 The DeLong model. Adapted from Reference 4.

10.4   DISCUSSION

For the neural network architecture presented, certain assumptions were made and various constraints were imposed, so that it resembled as much as possible the biological equivalents of feature detectors and edge detectors. In dealing with the above, the neural network can "learn" from stimuli alone, without a set of templates with which to compare the stimuli. To deal with this, the neural network can implement an unsupervised training with a variation of a Hebbian learning rule.[2] The connection strengths among the neurons of this network (weights) thus become the means of storing memories of the presented stimuli, where the same stimulus, if reapplied, will bring the same output to the neural network. These outputs can become the templates to a new neural network—to a different region or even the same region implementing a different function. In recollection, external stimuli must be correlated with memories already stored as templates. In the case of using another neural network for this purpose, the ALOPEX training algorithm[1] can be applied with supervision in the form of previously stored memories.

The storage/recollection process is a dynamic one, and these networks need be coordinated well in order that new "experiences" can affect both networks in a proper fashion. Damage within a network may affect storage or recognition (or both).

The results of an experiment like the simulation of Parkinson's disease as presented here would need to be compared to actual data from the operating room so that the validity of this network could be assessed. One way of comparing results from the network with those from the Operating Room (OR) is by observing (quantitatively) how a patient's motor activity is altered depending upon the location of the lesion relative to points of high and low activity in the GPi. These same observations would then be made on the network. The goal of this experiment would be to match the results of the network as closely as possible to those obtained from the OR. The network contains many parameters that would need to be "played with" in matching its results to real data. Some of these parameters are the connective areas, activation function slope, magnitude of activation function range, spacing and number of neurons on a layer, connection strength range, and other network parameters, as well as the validity of DeLong's model itself.

REFERENCES

1. Micheli-Tzanakou, E., Ueyda, E., Ray, R., Ramanujan, K. S., and Dong, J., Comparison of neural network algorithms for face recognition, *Simulation*, 64(1), 15, 1995.
2. Hebb, D., *The Organization of Behavior. A Neurophysiological Theory*, John Wiley & Sons, New York, 1949.
3. Oja, E., A simplified neuron model as a principal component analyzer, *J. Math. Biol.*, 15, 267, 1982.
4. DeLong, M. R., Activity of basal ganglia neurons during movement, *Brain Res.*, 40(1), 127, 1972.

Section IV

General Applications

11 A Feature Extraction Algorithm Using Connectivity Strengths and Moment Invariants

Tae-Soo Chon and Evangelia Micheli-Tzanakou

11.1 INTRODUCTION

The ALOPEX process has been used in the past to solve various optimization problems as in pattern recognition,[9-10] visual perception,[1-2] curve fitting, as well as face recognition.[11] One of the original applications was in the mapping of visual receptive fields (RFs) in animals.[6-7] In general when an optimization procedure is used, an optimum value of a "cost" function is sought. The function is in a well-defined domain and depends on many parameters, I_j, where $j = 1, \ldots n$. In this context, the optimum has the meaning of a maximum or a minimum, depending on whether the highest or the lowest value of the function is found. In most of the optimization procedures, a true optimum is not found: instead a local maximum or a minimum is achieved. Once the cost function is stuck at the local maximum or minimum, there is no means of getting away from it. Reaching the global optimum of the cost function thus becomes an impossibility.

Many heuristic methods have been used for different optimization problems that scan a variety of fields.[1-11] One of the most popular techniques is the simulated annealing, first introduced by Kirkpatrick et al. in 1983.[13] In this technique a "temperature effect" represents all random fluctuations involved in the problem. The temperature is gradually decreased as the system approaches an extremum. Under the assumption that the cost function representing the system is the system's energy state, the lowest energy state will be achieved at zero temperature. If the temperature is not lowered gradually and systematically but rather abruptly, a higher energy state will be reached, which does not represent the global minimum. This can be avoided by allowing the system to reach equilibrium before changing the temperature again (see also Reference 12).

ALOPEX is another such optimization technique. The cost function has been called the response of the system and represents a scalar quantity. In the application of this paper, the parameters I_j represent the intensities of a pattern. The procedure used is iterative where all I_j are changed simultaneously by small amounts and the response is computed in every iteration. The intensities depend on the changes of the response and the changes of the intensities over the previous two iterations. A stochastic element, an added noise, provides for the ability of the process to get away from a local maximum or minimum.

One of the problems that ALOPEX has encountered in the past is the large number of iterations required for convergence, especially when a χ^2 approach is used as a cost function. In pattern recognition and template matching in particular,[9] if the images used are large and complex, the speed of convergence is greatly delayed. In this study we address the issue of the convergence speed using the concept of "connectivity strengths" of the templates in conjunction with moment invariant measures.

11.2 ALOPEX ALGORITHMS

11.2.1 ORIGINAL ALGORITHM

Let us assume that in general we have an array of N^2 elements which represent the parameters. An additive noise to each of these elements is denoted by $r_j(n)$, and $R(n)$ is the cost function of the system at the nth iteration ($j = 1, 2,...,N^2$). The parameters $I_j(n)$ represent the intensities of each element in the array at iteration n and can be found as

$$I_j(n) = r_j(n) + b_j(n) \tag{11.1}$$

where $b_j(n)$ is called the bias of the jth element and $r_j(n)$ is a random variable. The biases are cumulative and represent the evolution of the pattern.

11.2.2 REINFORCEMENT RULES

The reinforcement rules can be thought of as the part of the algorithm that rewards a parameter that contributed to an increase in response. This reward should be cumulative, so that the random fluctuations from iteration to iteration will eventually be overcome by the correlation that must exist between the "global" response of the system, $R(n)$, and the value of each parameter, $I_j(n)$. The quantity to be modified is $b_j(n)$. A simple reinforcement rule that has been used most often is

$$\Delta b_j(n) = \gamma \Delta I_j(n-1) \Delta R(n-1) \tag{11.2}$$

where γ is an arbitrary constant and

$$\Delta I_j(n) = I_j(n-1) - I_j(n-2) \tag{11.3}$$

$$\Delta R(n) = R(n-1) - R(n-2)$$
$$j = 1, 2, \cdots N^2 \tag{11.4}$$

In this case Equations 11.2 and 11.3 state that the bias, and hence the expectation value of the gray level of a pixel, will be raised if, in the preceding two iterations, the gray level for that element and the response $R(n)$ are changed in the same direction. Otherwise the increment to the bias will be negative or zero. Note that the response is elicited by the whole pattern and given by the equation below:

$$R_j(n) = \sum_{j=1}^{m} \phi_{jm} \, I_j(n) \tag{11.5}$$

where ϕ_{jm} is the field strength in the RF of a template m, $m = 1, 2, \ldots k$ and $j = 1, 2, \ldots N^2$. In this case we have assumed that there exist m templates to choose from and that the process starts from a random pattern. For the first two iterations the biases are kept constant, with an initial value chosen arbitrarily.

The total response $R(n)$ of the process is computed by summing up the contributions of the responses from each template:

$$R(n) = \sum_{m=1}^{k} R_m(n) W_m(n) \tag{11.6}$$

where

$$W_m(n) = \frac{R_m(n)}{\displaystyle\sum_{m=1}^{k} R_m(m)} \tag{11.7}$$

are weighting factors for each of the templates, and $m = 1, 2, \ldots k$, indicating the individual template's contribution to the response.

11.2.3 A GENERALIZED ALOPEX ALGORITHM

For generalization of the algorithm, the intensity of the jth element of the image $I_j(n)$ is expressed as a linear combination of the template intensities I_{jm} and their connectivity strengths $P_{jm}(n)$ as well as the connectivity strengths of the noise added to the image:

$$I_j(n) = \sum_{m=1}^{k} P_{jm}(n) \cdot I'_{jm} + P_{jk'}(n) \cdot r_{jk'}(n) \tag{11.8}$$

where

$$\sum_{m=1}^{k+1} P_{jm}(n) = 1.0, \; k' = k+1,$$

and where the connectivity strengths play the role of weights. $I_j(n)$ is the intensity of the jth element of the converging image at the nth iteration, I'_{jm} is the intensity of the jth element of the mth template, $P_{jm}(n)$ is the connectivity strength (probability) that I'_{jm} could contribute to the convergence of the image at the nth iteration, and $r_{jk'}(n)$ and $P_{jk'}(n)$ are the random noise and its connectivity strengths (probabilities) at the nth iteration ($P_{jk'}(1) = 1.0$).

Regarding I'_{jm} as a variable and $P_{jm}(n)$ as its probability density function, then the image $I_j(n)$ can be represented as the first moment of I'_{jm}. (i.e., the average intensity of the mth template for pixel j). As $P_{jm}(n)$ increases while it converges to the mth template, the mth template will contribute more to the converging image. If the connectivity strengths for all elements become 1.0, then the converging image $I_j(n)$ solely contains the contributions of the responsible m^{th} template. In this case $I_j(n)$ is equal to I'_{jm} and convergence has been achieved.

Connectivity strengths lie in the range of 0.0 – 1.0, and the sum of them for each row is 1.0 (Figure 11.1b). If we imagine these connectivity strengths themselves as a separate image of a $k \times N^2$ matrix, the convergence of the actual image to a template is reflected in the changes of images from noise to a single bar in the connectivity matrix (Figure 11.1a), which is equivalent to a collection of 1.0s for the converged template, while the rest of the elements are all zeroes. Although the real images are complex, there always exists a simple way of convergence in the connectivity image, similar to the traditional ALOPEX procedure when used to imitate a simple or a complex cell receptive field (RF) in the visual system of animals.[5-7] In this particular case, a complex cell RF is reduced to a simple cell RF, using the connectivity strengths as described further.

In this study, ALOPEX is used on connectivity strengths rather than on image intensities, as if there is a receptive field for connectivities of a $k \times N^2$ matrix. The field strength for the RF represents the closeness of the converging image and the templates through an adaptive process, and can be calculated in a similar manner to the traditional model, since the image is a function of connectivity strengths (Equation 11.8).

The connectivity RF has two controlling roles in convergence: (1) maximization of differences in connectivity strengths among templates, and (2) minimization of differences in connectivity strengths within templates. Since these maximization and minimization procedures have to be satisfied concurrently during the converging process, two ALOPEX processes are applied in an interleaved manner using three arbitrary sets of "clocks", $n - t_2$, n, and $n + t_1$ where $n - 1 < (n - 1) + t_1 < n - t_2 < n < n + t_1 < (n + 1) - t_2 < n + 1$ (Figure 11.2). The overall model is described by the two processes as given by the equations below:

11.2.3.1 Process I

$$P_{jm}(n + t_1) = B_{jm}(n + t_1) + r'_{jm}(n - t_1) \qquad (11.9)$$

$$B_{jm}(n+t_1) = G_1(n+t_1)\left\{B_{jm}(n) + \Delta B_{jm}(n)\right\} \tag{11.10}$$

$$\Delta B_{jm}(n) = \gamma_1 \Delta R_1(n) \Delta P_{jm}(n) \tag{11.11}$$

$$\Delta R_1(n) = R_1(n) - R_1(n - t_2) \tag{11.12}$$

$$\Delta P_{jm}(n) = P_{jm}(n) - P_{jm}(n - t_2) \tag{11.13}$$

$$P_{jm}(n - t_2) = \frac{1}{N^2} \sum_{j=1}^{N^2} P_{jm}(n) \tag{11.14}$$

In Equation 11.14 it is assumed that an average connectivity strength is found. This part of the process is represented by arrows A and B between nodes $(n - t_2)$ and (n) of Figure 11.2.

11.2.3.2 Process II

$$P_{jm}(n) = B_{jm}(n) + r'_{jm}(n) \tag{11.15}$$

$$B_{jm}(n) = G_2(n)\left(B_{jm}\{(n-1)+t_1\} + \Delta B_{jm}\{(n-1)+t_1\}\right) \tag{11.16}$$

$$\Delta B_{jm}\{(n-1)+t_1\} = \gamma_2 \Delta R_2\{(n-1)+t_1\} \cdots \Delta P_{jm}\{(n-1)+t_1\} \tag{11.17}$$

$$\Delta R_2\{(n-1)+t_1\} = R_2\{(n-1)+t_1\} - R_2\{(n-2)+t_1\} \tag{11.18}$$

$$\Delta P_{jm}\{(n-1)+t_1\} = P_{jm}\{(n-1)+t_1\} - P_{jm}\{(n-2)+t_1\} \tag{11.19}$$

where $B_{jm}(n)$ is the bias for the connectivity strength, representing a cumulative converging factor, $r'_{jm}(n)$ is the added noise, $G_1(n)$ and $G_2(n)$ are normalization coefficients that keep the total connectivity strengths equal to 1.0, γ_1 and γ_2 are scaling constants, and $R_1(n)$ and $R_2(n)$ are the responses at the nth iteration.

Minimization of connectivity differences within a template is conducted in Process I, and the maximization of connectivity differences among templates is accomplished in Process II. If $P_{jm}(n)$ is calculated from Process II, then it generates $P_{jm}(n - t_2)$, which is the average of all $P_{jm}(n)$'s within each template (arrows A in Figure 11.2). As the response gets optimized, the reinforcement $\psi_m(n)$ gets stronger, since the variance of the connectivities $\{P_m(n) - P_{jm}(n)\}^2$ for elements in the same template is also optimized. The response is given by

$$R_1(n) = \sum_{j=1}^{N^2} \psi_m(n) \cdot P_{jm}(n) \tag{11.20}$$

$$\psi_m(n) = \sum_{j=1}^{N^2} \frac{A}{\left\{ P_m(n) - P_{jm}(n) \right\}^2} \tag{11.21}$$

where $P_m(n)$ is the average connectivity strength within a template, and A is a constant.

As the connectivities are equal for all elements within the same template, the response becomes maximum in the $(n - t_2)^{th}$ iteration. By producing $R_1(n - t_2)$ and $R_1(n)$ from $P_{jm}(n - t_2)$ and $P_{jm}(n)$ respectively, $P_{jm}(n + t_1)$ is calculated through cross-correlations in Process I (arrows B in Figure 11.2, and Equations 11.9 – 11.14). Since calculations on the biases can be conducted independently for each template, parallel processing is applied here. Parallel processing may be more effectively used if we suppose that these processes repeat many times in order to drive the connectivities closer to their averages.

In Process II, $P_{jm}\{(n - 1) + t_1\}$ and $P_{jm}(n + t_1)$ produce $R_2\{(n - 1) + t_1\}$ and $R_2(n + t_1)$ respectively and determine $P_{jm}(n + 1)$ through the ALOPEX process (arrows C in Figure 11.2 and Equations 11.15 – 11.19). This connectivity strength, $P_{jm}(n + 1)$, in turn is used in Process I to repeat the minimization of connectivities within templates. This way it is possible to keep equal time differences between consecutive iterations in each process—i.e., $n - (n - t_2) = (n + 1)-\{(n + 1) - t_2\} = t_2$ in Process I and $(n + t_1)-\{(n-1) + t_1\} = \{(n + 1) + t_1\}-(n + t_1) = 1$ in Process II. In Process II, the responses for maximizing differences in connectivities among templates are mainly obtained from the degree of closeness of moment invariants between templates (as explained below) and the converging image. As the converging image gets closer to a template, the responses become optimal.

11.3 MOMENT INVARIANTS AND ALOPEX

Moment invariants have been used successfully in pattern recognition for many years (for calculation of moments and their applications the reader is referred to References 14–21 and Chapter 4). The two-dimensional moments for discrete variables are given by

$$M_{pq} = \sum_i \sum_i i^p j^q \cdot I(i, j)$$
$$(p, q = 0, 1, 2, \ldots.) \tag{11.22}$$

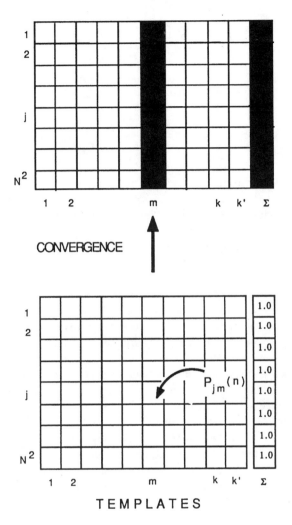

FIGURE 11.1   An image of connectivities in $k \times N^2$ field. The summation of connectivities $P_{jm}(n)$, for each row is 1.0. To emphasize the convergence to a "single bar", the column that converged is represented as black (1.0), while the others are represented as white (0.0) after convergence. Even though the images are very different their connectivity strengths are very similar.

where $I(i, j)$ is the $(i, j)^{\text{th}}$ pixel intensity, and p and q are the orders of moments.

The moment invariants are combinations of moments that have properties such as translation, rotation, and size invariance in the corresponding images. It has been found that the first seven moment invariants based on the first three moments are sufficient to help in the recognition of an image.[14]

The response for closeness of moment invariants is given by

$$R_2(n) = \sum_{m=1}^{K} W_m(n) \sum_{j=1}^{N^2} \Phi_m(n) \cdot P_{jm}(n) \tag{11.23}$$

where

$$\Phi_m(n) = \sum_{\lambda=1}^{s} \left| \frac{A'}{\theta_\lambda(n) - \theta'_{\lambda m}} \right| \tag{11.24}$$

where $\theta_\lambda(n)$ is the transformed λ^{th} moment invariant for the converging image at the nth iteration, $\theta'_{\lambda m}$ is the log-transformed λ^{th} moment invariant for the mth template, $W_m(n)$ is the weighting coefficient for the mth template at the nth iteration, s is the total number of moment invariants used in the calculation and A' is a constant. $\Phi_m(n)$ is the reinforcement factor and $W_m(n)$ is the weighting factor of the contributing template at iteration n.

Since it is difficult to use moment invariants directly due to their range differences, they are appropriately log-transformed. The ordinary moment invariants[14] are adjusted by normalizing and coding them in such a way that their absolute values lie in the range of 10^{-1} and 10^{-100} before the log transformation.

The averaging effect in Process I contributes to similar connectivities for all elements in the same template. The achievement of similarity in connectivities is reasonable, since the template requires equal chance of representation for its elements as a contribution to the converging image. Assuming that the averaging procedure in Process I is repeated enough times, the connectivity strengths for elements within a template will eventually become equal, i.e., $P_{jm}(n) = \overline{P}_m(n)$ where $\overline{P}_m(n)$ is the average connectivity for elements in the same template at the nth iteration.

Supposing we start with $\overline{P}_m(n)$ and the assumption that connectivity strengths are equal for every element within the same template, Process I may be omitted. The model is then expressed as

$$I_j(n) = \sum_{m=1}^{k} \overline{P}_m(n) \cdot I'_{jm} + P_k(n) \cdot r_{jk'}(n) \tag{11.25}$$

where

$$\sum_{k=1}^{k+1} \overline{P}_m(n) = 1.0, \ k' = k+1$$

$$\overline{P}_m(n) = B_m(n) + r'_m(n) \tag{11.26}$$

$$B_m(n) = B_m(n-1) + \Delta B_m(n-1) \tag{11.27}$$

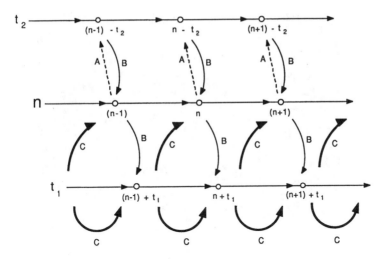

FIGURE 11.2   A computation flow for the two inter-leaved ALOPEX processes. Average connectivities within templates are generated (arrows A, dotted line) to proceed to Process I, which minimizes differences in connectivities within the same template (arrows B, thin lines carry the information). In Process II connectivities among templates are maximized (Arrows C, heavy lines carry this information).

$$\Delta B_m(n-1) = \gamma \Delta R_2(n-1)\Delta \overline{P}_m(n-1) \tag{11.28}$$

In this model, calculating the biases is simple, since only k connectivity strengths are required, while for the previous model $k \times N^2$ calculations are needed. Making the calculation independent of the number of elements makes it possible to handle large images in a much simpler way. Although the connectivity strengths are provided as a "constant (or average)" information without variability within templates, the calculated image has N^2 elements providing "local" information since $I_j(n)$ is expressed as the expected value for each element j of all templates (Equation 11.25). This causes some loss of variability in images due to the constancy of $\overline{P}_m(n)$, but the moments calculated from the constructed image are sufficiently variable.

11.4   RESULTS AND DISCUSSION

Figures 11.3 (b) and 11.4 (a) show an example of convergence with the first model (variable connectivities within a template) when $P_{jm}(n)$ is applied on five templates of 21×21 pixels each. The convergence is complete before 1000 iterations are reached, with average connectivity strength of about 80% of the maximum value possible. The speed of convergence increased roughly more than ten times compared to the speed of the traditional methods, where usually about 1×10^4 to 5×10^4 iterations are required for convergence.[9]

In order to enhance the efficiency of the system even further, another kind of response is also incorporated in Process II. Since the achievement of convergence is expressed as maximization of differences among connectivity strengths for all

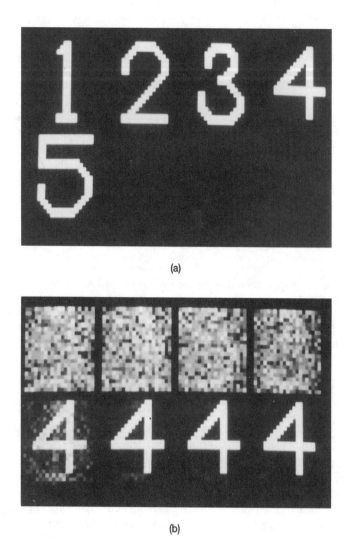

(a)

(b)

FIGURE 11.3   Examples of convergence when ALOPEX was conducted with connectivity strength $P_{jm}(n)$ on five templates of 21×21 elements (iteration numbers are 100, 150, 200, and 250 from left to right at the top row and 300, 350, 400, and 600 at the bottom row). (a) Templates. For calculating bias differences in the model responses from moment invariants and connectivity differences among templates are combined as follows: (b) Responses from moments are only used ($\lambda_1 = 1.0$ and $\lambda_2 = 0.0$ in Equation 11.29). (c) Two responses are combined "additively" ($\lambda_1 = \lambda_2 = 1.0$). (d) Both responses are activated "cooperatively" when the signs of both response ($\lambda_1 = \lambda_2 = 1.0$) differences are the same, otherwise $\lambda_1 = \lambda_2 = 0.0$. (e) Both responses contribute to the calculation "alternately" (e.g., if the iteration is an odd number, $\lambda_1 = 1.0$ and $\lambda_2 = 0.0$, otherwise $\lambda_1 = 0.0$ and $\lambda_2 = 1.0$).

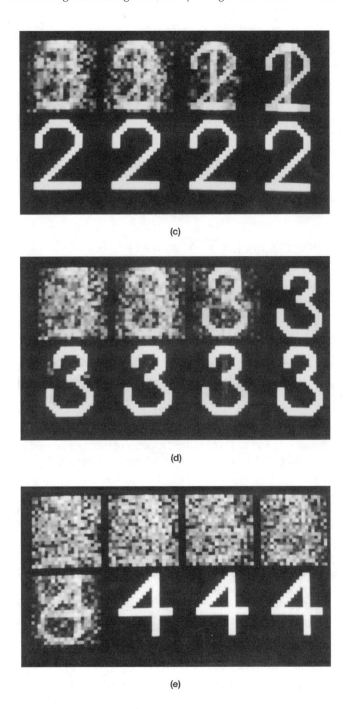

(c)

(d)

(e)

FIGURE 11.3 (Continued)

templates, the variance of connectivity can be used as reinforcement for the response function; as convergence is approached, this value increases. By adding these responses to those obtained from the closeness of moment invariants when calculating biases, Equation 11.17 is rewritten as

$$\Delta B_{jm}\{(n-1)+t_1\} = \gamma_1\Big(\lambda_2\Delta R_2\{(n-1)+t_1\}\cdot\Delta P_{jm}\{(n-1)+t_1\}\Big)$$
$$+\lambda_2\Big(\gamma_3\Delta R_3\{(n-1)+t_1\}\Delta P_{jm}\{(n-1)+t_1\}\Big)$$

(11.29)

and where

$$R_3(n+1) = \sum_{m=1}^{k}\sum_{j=1}^{N^2}\psi'_m(n+t_1)\cdot P_{jm}(n+t_1)$$

(11.30)

$$\psi'_m(n+t_1) = \sum_{m=1}^{k}\frac{\{\bar{P}(n+t_1)-\bar{P}_m(n+t_1)\}^2}{C}$$

(11.31)

where λ_1 and λ_2 determine the contributions (weighting constants) of corresponding responses, $\bar{P}_m(n+t_1)$ is the average connectivity within a template, $\bar{P}(n+t_1)$ is the overall average connectivity strength, γ_2 and γ_3 are scaling factors and C is a constant.

Since $R_3(n+t_1)$ acts as a response, it alone can contribute to the convergence without any effect from the responses from the moment invariants. By adjusting the weighting constants λ_1 and λ_2 independently, convergence is achieved in a different manner. It is assumed that $\lambda_1 + \lambda_2 = 1.0$, which imposes yet another normalization on ΔB_{jm}.

Figures 11.3 and 11.4 show some of these results. When the $R_3(n+t_1)$ is used alone (Figure 11.4e), the response curves appear smoother than those from $R_2(n+t_1)$ alone (Figure 11.4a). In this case responses from the moment invariants are not used for the calculation of biases but only for connectivity calculations ($\lambda_2 = 1$ while $\lambda_1 = 0$). In the case where responses from moment invariants are used alone, the responses from the connectivity differences among templates do not contribute to the calculations at all ($\lambda_2 = 0$ while $\lambda_1 = 1$).

The appearance of discontinuities in both the connectivity and response curves in the case of using moment invariants (Figure 11.4a) reflects the fact that the total information carried by moment invariants is variable during the whole convergence process. This may be due to the fact that appearance of small disturbances (noise) in the templates other than the converging one affects significantly the total structure of the image even though their connectivities are small. These instabilities caused by moments are most of the time advantageous since they help get away from local maxima and minima. However, in some cases it has been observed that a very abrupt divergence from the convergence point can occur. This phenomenon is under further investigation.

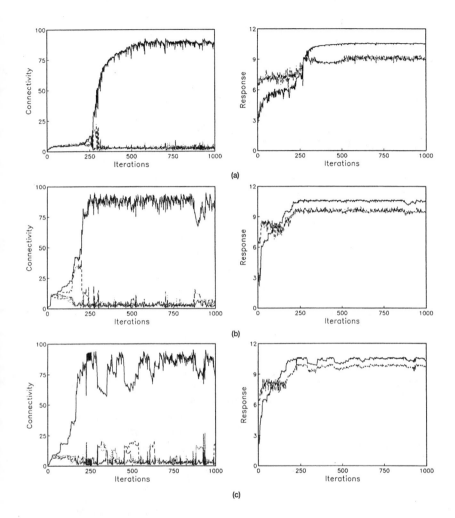

FIGURE 11.4   Average connectivity strengths and responses when ALOPEX was conducted with $P_{jm}(n)$ on the five templates of 21 × 21 elements. (a) Only moment response is used. (b) Responses from moments and connectivity differences are used "additively". (c) "Cooperatively". (d) "Alternately". (e) Only connectivity responses are activated. In (e) the connectivity occasionally drops momentarily after convergence. This is due to the fact that the signs of connectivity and response differences did not coincide; i.e., slight decrease in connectivity may produce increase in responses. Hence the product of both difference terms becomes negative, at the maximum level of connectivity. This was intensified by high connectivities and can be reduced by adjusting the amplitude of bias differences or by decreasing connectivity variabilities within templates (see Figure 11.6e).

The moment invariants have yet another advantage. Since the number of the moment invariants is much smaller than the elements of the image ($s<<N^2$), they help in time savings from excess calculations and at the same time help in responding more readily to any global changes in the image. While the responses resulting from

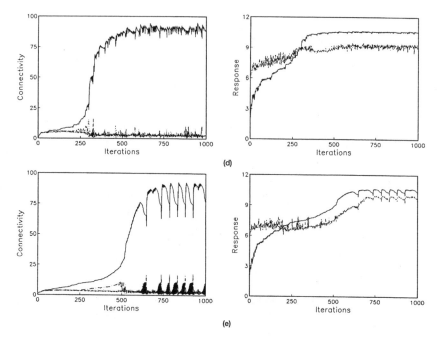

FIGURE 11.4 (Continued)

moment invariants generate some unstable situations in the ALOPEX process, the response $R_3(n)$, (Figure 11.4e) resulting from variances among template connectivities is relatively smooth, since the calculations are only based on the previous connectivities. Due to the cross-correlation properties, the process always proceeds to the direction of increased variances, guaranteeing the convergence in the connectivity of RFs to "any" one of the templates. This implies that the responses from the connectivity differences among templates are more sensitive to the detection of local extrema.

The response $R_2(n)$ (Figure 11.4a) from the moment invariants seems to perform well with respect to the overall structure of the image, being more sensitive in finding the global extremum. However, due to the frequent changes in every iteration during convergence, the information carried by the moments is in general more "ambiguous". In contrast, responses from connectivity differences among templates appear to be more responsible for slow changes in the images, and they carry more information due to the fact that the response calculations depend upon their own histories—two previous iterations—and change gradually during convergence.

As shown in Figure 11.4, the speed of convergence and the connectivity strengths increase when the two responses are combined. In this case, the amplitude of the response differences for both responses is adjusted to be roughly the same at the later stages of the process. Initially $R_3(n)$ started very low and increased rapidly as the iteration number increased. Convergence was achieved more efficiently when both responses acted "cooperatively" (Figure 11.4c) or "additively" (Figure 11.4b) than when the processes were alternating (Figure 11.4d). However, since many factors are involved in the ALOPEX process and in a complex manner, more observations may be needed under various parameter settings to generalize the combined phenomena.

Another type of experiment was also performed in order to test the feasibility of using moment invariants. In this, ALOPEX simulations are performed with five input templates of translocated and rotated images of different sizes. Figure 11.5 shows an example. Convergence is achieved in a similar manner to the previous ones since the moment invariants are set equal for the corresponding new templates. Due to the different sizes and locations of the features to be extracted, different image fields are produced, which affect the calculation of moments in each iteration producing more instabilities in the system. Better convergence is obtained when the features are larger and more toward the center of the template.

Figure 11.6 shows examples of simulation results when ALOPEX is conducted with constant connectivity strengths within the template, $P_m(n)$, using Equations 11.25–11.28. With similar values of parameters as in the previous model, the simulation generally led to a faster convergence. Convergence is achieved at about 100–200 iterations, roughly 10 times faster than the previous model. The connectivity strengths also showed an improvement.

Successful results are also obtained with combined responses from moment invariants and connectivity differences among templates in this model. To allow for the multiple responses in calculating the biases, Equation 11.28 is rewritten as follows:

$$\Delta B_m(n-1) = \gamma_1\left\{\lambda_2 \Delta R_2(n-1)\Delta P_m(n-1)\right\} + \lambda_2\left\{\gamma_3 \Delta R_3(n-1)\Delta P_m(n-1)\right\} \quad (11.32)$$

As shown in Figure 11.6, convergence is enhanced. Similar patterns are observed also in the case when variable connectivities within templates are used. Higher discontinuities are observed in the curves when responses from moment invariants are used (Figure 11.6a) and general enhancement of convergence when the responses are appropriately combined (Figure 11.6b–e).

In order to test the dependence of the model on the image size, images of 256×256 pixels are used for templates. Figure 11.7 shows some examples of convergence when one to three templates are used. Convergence is achieved at about 200–250 iterations with high connectivities similar to the previous cases using $P_m(n)$. A further enhancement of convergence is possible by using negative weights for responses from the noise pattern (Figure 11.7d). Since the converging image is initially closer to the noise pattern, it shows higher responses. The connectivity strengths decrease rapidly at the early stages of the process due to the above-mentioned negative weights of the response. Later on, the noise level drops more rapidly while convergence takes place.

The above mentioned simulations demonstrate the fact that application of the ALOPEX process on connectivity strengths improves the convergence and increases the sensitivity of finding global and local extrema. Figure 11.8 compares the flow charts for the ALOPEX on image intensities (traditional) (Figure 11.8a) with that on the connectivity strengths (Figure 11.8b). By considering the connectivities as a new image converging to the RF of a simple cell, as those found in the biological

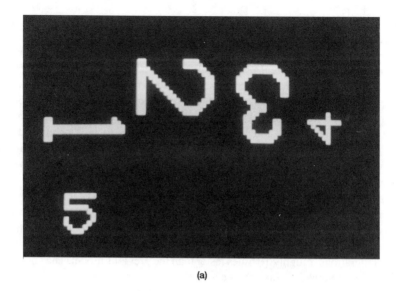

(a)

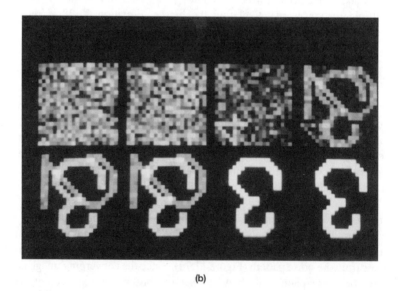

(b)

FIGURE 11.5   An example of convergence by an ALOPEX with $P_{jm}(n)$ when templates were rotated and translocated at different sizes. (a) Templates. (b) Convergence to a template. Iteration numbers are 100, 150, 200 and 250 from left to right at the top row and 300, 350, 400, and 600 at the bottom row.

visual system, the advantages of connectivity strengths and cross-correlations are effectively integrated. Cross-correlation terms in connectivities produced an intrinsic "drive for convergence" in the system. Consider the simple case where the connectivities are constant within templates and the responses are only determined from

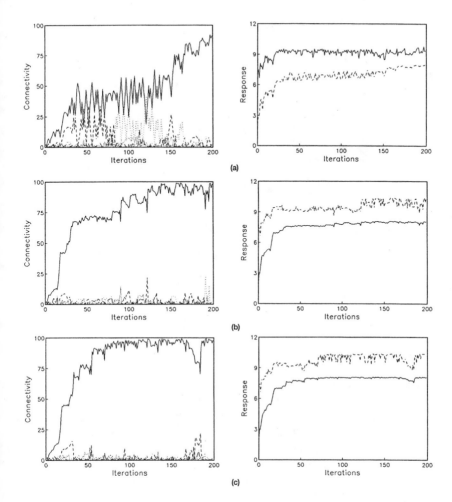

FIGURE 11.6   Connectivity strengths and responses when ALOPEX was conducted with constant connectivities $P_m(n)$, on the five templates of 21×21 elements. (a) Only moment response is used. (b) Responses from moments and connectivity differences are used "additively". (c) "Cooperatively". (d) "Alternately". (e) Only connectivity responses are activated.

the weight of the connectivity strengths. The model is then given by Equations 11.25–11.28, except for the response which is now expressed as

$$R_4(n) = \sum_{m=1}^{k+1} W'_m(n) P_m(n) \qquad (11.33)$$

where $W'_m(n)$ is the weight from the connectivity strengths. If the connectivity strengths initially start with small random numbers close to zero for the templates and 1.0 for the noise pattern, while negative (or zero) weights are given to the noise pattern, the system converges to one of the templates (Figure 11.9), i.e., only one

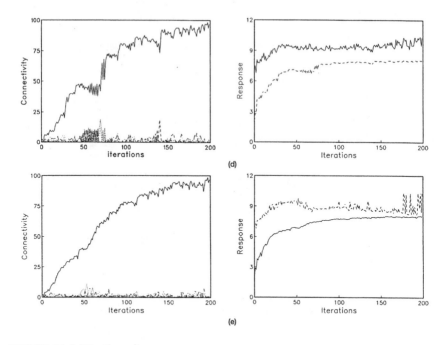

FIGURE 11.6 (Continued)

$P_m(n)$ becomes equal to 1.0 while the others are all equal to 0.0. This happens because the response term, which is actually the sum of the squares of connectivities $[W'_m(n)]$ proportional to $P_m(n)$, is affected by the differences in connectivities. Thus the sign of the response difference coincides with that of the maximum connectivity difference, which in turn has the advantage of increasing the response, since if the signs of the response and connectivity differences are the same, the calculated values are positive. The maximum is obtained when only one $P_m(n)$ is 1.0 and the rest are all 0.0.

If we assume that the system starts with small random numbers of the same amplitude as those of the template connectivities and ends up with only one template having the maximum values, this procedure can be thought of as the process of decreasing entropy while reaching the global extremum. If variations are allowed in connectivities within templates, the convergence is slowed down depending upon the degree of variations used.

This model is similar to other traditional neural network models in the sense that it uses connectivity strengths that are initially random numbers and adaptively converge to some values through an iterative procedure. However, this model has some different characteristics: (1) the connectivities are calculated from the previous two iterations (i.e., $P_{jm}(n-2)$, $P_{jm}(n-1)$) and determined by cross-correlations using response differences, (2) all parameters are changed simultaneously in each iteration, and (3) the connectivities have direct relationships to the templates.

Since many parameters are involved in the convergence of ALOPEX and the algorithm is not a linear process, it is not simple to analyze. Through simulations,

(a)

FIGURE 11.7   Examples of convergence when ALOPEX was conducted with $P_m(n)$ on real template images. (a) Templates. (b) Convergence when one template was used (iteration numbers are 10, 30, 60, and 90 from left to right at the top row and 120, 150, 200, and 300 at the bottom row). (c) Convergence when three templates were used (iteration numbers are same as (b) except the last one, which was 250). (d) Convergence on three templates when the response from the noise pattern had a negative effect (iteration numbers are 6, 8, 10, and 12 from left to right at the top row, and 14, 16, 18, and 20 at the bottom row).

however, image structures, amplitudes of noise in connectivity strengths and responses and their weights appear to influence the convergence. In a certain range, the high amplitude of the random noise in connectivities improves the speed of convergence but at the same time increases instabilities in connectivities.

One problem in using moment invariants in calculating responses is the initial distances between noise and templates. If the moment invariants of one template are distinctively closer to those of noise, that template appears to be favored for convergence. This phenomenon is being studied more closely at the present time. However, this effect indirectly confirms the fact that responses from moment invariants can direct the convergence. This is not a problem when we apply ALOPEX to the recognition of a "biased" starting image. Considering that moment invariants are sensitive to the overall structure of an image, a "biased" image would rapidly change the distances in favor of the template that has similar statistics with the "biased" image, that is an image that has been contaminated by a lot of noise and therefore is unrecognizable. In all cases, the responses from the moment invariants appear to "diverge" after the maximum connectivity (1.0) is reached. This is due to the fact that the sensitivity of moment invariants near the complete convergence is

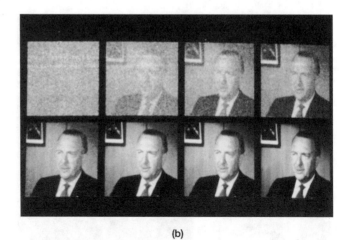

(b)

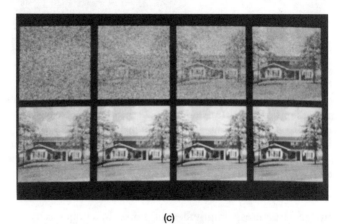

(c)

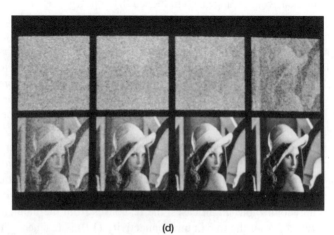

(d)

FIGURE 11.7 (Continued)

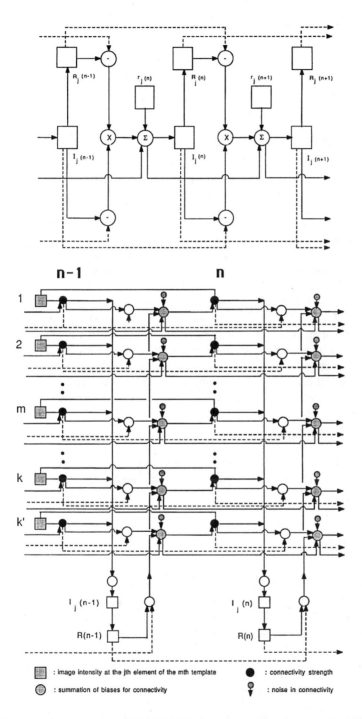

FIGURE 11.8   Flowcharts for ALOPEX. (a) On image intensities (traditional). (b) On connectivity strengths.

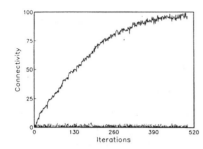 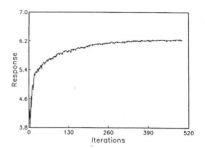

FIGURE 11.9 Convergence of connectivity strengths when ALOPEX was conducted with responses calculated only from connectivities and their weights.

very high. A slight connectivity change in another template due to noise or bias calculations will produce a new image which is in a "confusion state" between the templates (i.e., it is a mixture of the templates under consideration). These unexpected changes in the moment responses may be used as an indication of reaching complete convergence.

In that context, learning can be conceived as being the progression of convergence through time. The classical "weights" of neural networks are inherent in the connectivity curves. While in backpropagation algorithms only the final weights are saved and can be analyzed, the connectivity curves of the present algorithm show the history of changes, although no weights are needed for recognition. This algorithm can be and has been used with the traditional neural network architectures as well, applied directly to the templates as described in this chapter.

ACKNOWLEDGMENTS

The authors wish to thank T. Dasey and P. Hoeper for critically reviewing the manuscript and for helpful suggestions. Research supported by NSF Grant EET-8721327.

REFERENCES

1. Harth, E., Visual perception: a dynamic theory, *Biol. Cybern.*, 22, 169, 1976.
2. Harth, E., Unnikrishnan, K. P., and Pandya, A. S., The inversion of sensory processing by feedback pathways: a model of visual cognitive functions, *Science*, 237, 187, 1987.
3. Harth, E. and Unnikrishnan, K. P., Brainstem control of sensory information: a mechanism for perception, *Int. J. Psychophysiol.*, 3, 101, 1988.
4. Tzanakou, E. and Harth, E., Determination of visual receptive fields by stochastic methods, *Biophysical J.*, 15(42a), 1973.
5. Harth, E. and Tzanakou, E., ALOPEX: a stochastic method for determining visual receptive fields, *Vision Res.*, 14, 1475, 1974.
6. Tzanakou, E., Michalak, R., and Harth, E., The ALOPEX process: visual receptive fields with response feedback, *Biol. Cybern.* 35, 161, 1979.
7. Micheli-Tzanakou, E., Non-linear characteristics in the frog's visual system, *Biol. Cybern.*, 51, 53, 1984.

8. Deutsch, S. and Micheli-Tzanakou, E., *Neuroelectric Systems*, NYU Press, New York, 1987.

9. Dasey, T. J. and Micheli-Tzanakou, E., Neural fuzzy systems in handwritten digit recognition, *Industrial Electronics Handbook*, Irwin, D., Ed., CRC Press, Boca Raton, FL 1977, Ch. 95, 1231.

10. Zahner, D. and Micheli-Tzanakou, E., Artificial neural networks: definitions, methods and applications, *The Biomedical Engineering Handbook*, Bronzino, J., Ed., CRC Press, Boca Raton, FL, 1995, Ch. 184, 2689.

11. Micheli-Tzanakou, E., Uyeda, E., Ray, R., Sharma, A., Ramanujan, K. S., and Dong, J., Face recognition: comparison of neural network algorithm, *Simulation*, 64(1), 37, 1995.

12. Van Der Meer, S., Stochastic cooling and the accumulation of antiprotons, *Science*, 230, 900, 1985.

13. Kirkpatrick, S., Gelatt, C. D., and Vecchi, M. P., Optimization by simulated annealing, *Science*, 220, 671, 1983.

14. Hu, M.-K., Visual pattern recognition by moment invariants, *IRE Trans. Inf. Theor.*, IT-8, 179, 1962.

15. Smith, F. W. and Wright, M. H., Automatic ship photo interpretation by the method of moments, *IEEE Trans. Comput.*, C-20, 1089, 1971.

16. Dudani, S. A., Breeding, K. J., and McGhee, R. B., Aircraft identification by moment invariants, *IEEE Trans. Comput.*, C-26, 39, 1977.

17. Teague, M. R., Image analysis via the general theory of moments, *J. Opt. Soc. Am.*, 70, 920, 1980.

18. Reddi, J. J., Radial and angular moment invariants for image identification, *IEEE Trans. Patt. Anal. Machine Intell.*, PAMI-3, 351, 1981.

19. Abu-Mostafa, Y. S. and Psaltis, D., Recognitive aspects of moment invariants, *IEEE Trans. Patt. Anal. Machine Intell.*, PAMI-6, 698, 1984.

20. Teh, C. H. and Chin, R. T., On digital approximation of moment invariants, *Comput. Vision, Graph. Image Process.*, 33, 318, 1986.

21. Lo, C.-H. and Don, H.-S., 3-D moment forms: their construction and application to object identification and positioning, *IEEE Trans. Patt. Anal. Machine Intell.*, PAMI-11, 1053, 1989.

12 Multilayer Perceptrons with ALOPEX: 2D-Template Matching and VLSI Implementation

Daniel A. Zahner and
Evangelia Micheli-Tzanakou

12.1 INTRODUCTION

12.1.1 MULTILAYER PERCEPTRONS

Common network architectures consist of multiple layers of neurons, where each neuron in layer 1 is connected to all neurons in layer 2. A three-layer feed-forward architecture is shown in Figure 12.1. The input layer receives external stimulus, and the output layer generates the output of the network. The hidden layer and all the interconnections are responsible for the neurocomputation. The number of neurons in each layer and the number of layers necessary are problem dependent. Generally, as the number of nodes or neurons increases, problem complexity increases, as does the time to train the network. For linear activation functions, one layer is all that is necessary for linear separability. Additional layers are redundant,[1] when linear activation functions are employed. In the nonlinear cases, using two layers increases the nonlinear separability. Figure 12.1 shows a three-layer network, the simplest multilayer perceptron network.

The optimal number of hidden neurons needed to perform an arbitrary mapping is a subject of much debate. Methods used in practice are mainly intuitive determination or are found by trial and error. Mathematical derivation proves that a bound exists on the number of hidden nodes, m, needed to map a k element input set. The formulation is that $m = k - 1$ is an upper bound.[2] These results are consistent with the optimal number of hidden neurons, determined empirically in Reference 3. Others believe the number of hidden nodes necessary to be a function of the number of the separable regions needed as well as the dimension of the input vector.[4]

For most artificial neural networks there is an initial training phase in which the interconnection strengths are adjusted until the network has a desired output. Only after training is the network capable of performing the task it was designed to do. The training phase can be either supervised or unsupervised. In supervised learning,

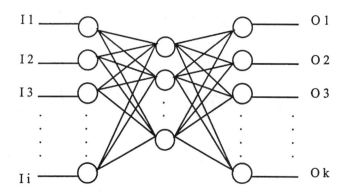

FIGURE 12.1   Multilayer perceptron network.

there exists information about the correct or desired output for each input training pattern presented.[5] In unsupervised learning no *a priori* information exists, and training depends on the properties of the patterns. Unsupervised learning is highly dependent on the training data, and information about the proper classification is often lacking.[5] For this reason, most neural network training is supervised.

It was not until the discovery of multilayer learning algorithms that interest in neural networks resurfaced. The most widely used training algorithm, called back-propagation, was initially discovered by Werbos,[6] although it went virtually unnoticed until 1985 when Parker rediscovered it.[7] In 1986, Rumelhart et al.[8] rediscovered the algorithm, and called it the delta rule. Their main contribution was not the discovery of the algorithm but their popularization of the algorithm, which has led to a renewed interest in neural networks. Another algorithm used for multilayerperceptron (MLP) training is the ALOPEX algorithm. ALOPEX was originally used for receptive field mapping by Tzanakou and Harth in 1973,[9] and has since been applied to a wide variety of optimization problems.[10–17] These two algorithms have been described in Chapter 2.

It should be stated that due to its stochastic nature, an efficient convergence for ALOPEX depends on the proper control of both the additive noise and the gain factor γ. Initially all parameters X_i are random, the additive noise is of Gaussian distribution with mean 0, and standard deviation, σ, initially large. The standard deviation, σ, decreases as the process converges to ensure a stable stopping point. Conversely, gamma, γ, increases with iterations. As the process converges ΔR becomes smaller and smaller, and an increase in γ is needed to compensate for this. Figures 12.2, 12.3, and 12.4 show the response, gamma, and sigma with iterations for a typical ALOPEX run. Additional constraints include a maximal change permitted for X_i, for one iteration. This bounded step size prevents the algorithm from drastic changes from one iteration to the next. These drastic changes often lead to long periods of oscillation, during which the algorithm fails to converge.

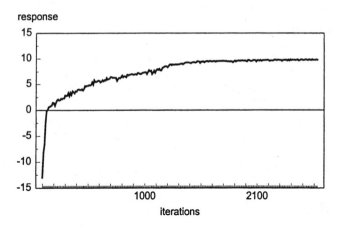

FIGURE 12.2   Response vs. iterations.

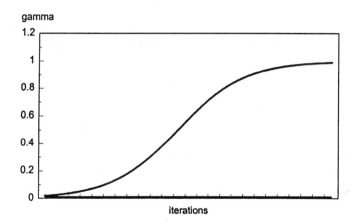

FIGURE 12.3   Gamma vs. iterations.

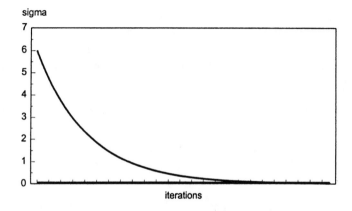

FIGURE 12.4   Gaussian noise standard deviation vs. iterations.

12.2 MULTILAYER PERCEPTRON AND TEMPLATE MATCHING

A three-layer perceptron is trained for pattern recognition using ALOPEX. The network is trained to recognize ten 5×7 templates corresponding to the ten digits, 0–9. Backpropagation training occurs as previously described. How ALOPEX is implemented in this application is described below.

For each 5×7 input pattern there exists a desired output vector O_k^{des}. The observed output, O_k^{obs}, is found by a single feed-forward pass through the fully interconnected layers of the network. Neurons or nodes in the hidden and output layers incorporate a nonlinear activation function, called a sigmoid.

As already has been discussed in Chapter 2, the response is calculated for the jth input pattern based on the observed and desired output

$$R_j(n) = O_k^{des} - \left(O_k^{obs}(n) - O_k^{des} \right)^2 \tag{12.1}$$

where O_k^{obs} and O_k^{des} are vectors corresponding to O_k for all k. The total response for iteration n is the sum of all the individual template responses, $R_j(n)$.

$$R(n) = \sum_{j=1}^{m} R_j(n) \tag{12.2}$$

In Equation 12.1, m is the number of templates used as inputs. ALOPEX iteratively updates the weights using both the global response information and local weight histories, according to the following:

$$W_{ij}(n) = r_i(n) + \gamma \Delta W_{ij}(n) \Delta R(n) + W_{ij}(n-1) \tag{12.3a}$$

$$W_{jk}(n) = r_i(n) + \gamma \Delta W_{jk}(n) \Delta R(n) + W_{ik}(n-1) \tag{12.3b}$$

where γ is an arbitrary scaling factor, $r_i(n)$ is an additive Gaussian noise, ΔW represents the local weight change, and ΔR represents the global response information. These values are calculated by

$$\Delta W_{ij}(n) = W_{ij}(n-1) - W_{ij}(n-2) \tag{12.4a}$$

$$\Delta W_{jk}(n) = W_{jk}(n-1) - W_{jk}(n-2) \tag{12.4b}$$

$$\Delta R(n) = R(n-1) - R(n-2) \tag{12.4c}$$

After training the network, it was tested for correct recognition using incomplete or noisy input patterns. The results show the robustness of the system to noise corrupted data. It should be noted that regardless of which training procedure was used, backpropagation or ALOPEX, the recognition ability of the system was the same. The only difference was in how the response grew with iterations. Two response curves are shown in Figure 12.5.

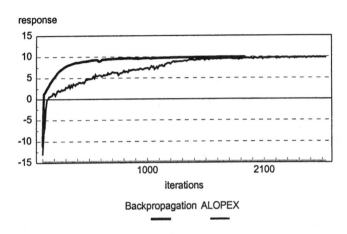

FIGURE 12.5 Response curves for ALOPEX and backpropagation.

It can be seen from Figure 12.5 that backpropagation converges faster than ALOPEX, particularly in the early periods of training. The networks were trained to 99% of maximal response; backpropagation converged in 1910 iterations, whereas ALOPEX took 2681 iterations to reach the same level.

The neural network's robustness is derived from its parallel architecture and depends on the network topology, not the learning scheme used to train. The network used was a three-layer feed-forward network with 35 input nodes, 20 hidden nodes, and 10 output nodes. The network's recognition ability was tested with noisy input patterns. Each 5×7 digit of the training set was subjected to noise of varying Gaussian distribution and tested for correct recognition. The original training templates were binary (0 or 1) images. The results, demonstrating the network's robustness, are shown in Figure 12.6. Note that even when the standard deviation approaches 1, the network correctly recognizes over 50% of the trained templates.

Artificial neural networks have shown a limited ability to solve problems, which conventional computers are unable to resolve. Image and speech recognition, motor control, and other such tasks which human brains perform well are stumbling blocks for the serial architecture. Artificial neural networks were derived from a conscious effort to mimic brain functions and are models of their biological counterparts. While ANN's are modeled after the human brain, they are far from repeating the brain's behavior. Severe limitations still exist, especially in terms of size and speed of the networks and in the understanding of the biological system.

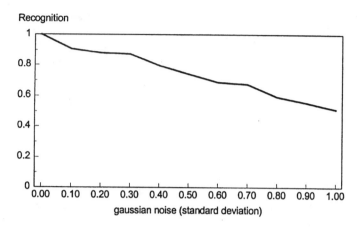

FIGURE 12.6   Recognition of noisy images by a trained MLP as a function of the standard deviation of the superimposed noise.

12.3   VLSI IMPLEMENTATION OF ALOPEX

Artificial neural networks (ANN) have existed for many years, yet because of recent advances in technology, they are again receiving much attention. Major obstacles in ANN, such as a lack of effective learning algorithms, have been overcome in recent years. Training algorithms have advanced considerably, and now Very Large Scale Integration (VLSI) technology may provide the means for building superior networks. In hardware, the networks have much greater speed, allowing for much larger architectures.

The tremendous advancement in technology during the past decades, particularly in Very Large Scale Integration (VLSI) technology, has renewed interest in artificial neural networks. Hardware implementation of neural networks is motivated by a dramatic increase in speed over software models. The emergence of VLSI technology has and will continue to lead neural network research in new directions. VLSI has advanced considerably over the last few years. Chips are now smaller, faster, contain larger memories, and are becoming cheaper and more reliable to fabricate.

Neural network architectures are varied, with over fifty different types being explored in research.[18] Hardware implementations can be either electronic, optical, or electro-optical in design. A major problem in hardware realization is often not due to the network architecture but to the physical realities of the hardware design. Optical computers, while they may eventually become commercially available, suffer far greater problems than do VLSI circuits. Thus, for the immediate and near future, neural network hardware designs will be dominated by VLSI.

Much debate exists as to whether digital or analog VLSI design is better suited for neural network applications. In general, digital designs are easier to implement and are a better understood methodology. Also, in digital designs, computational accuracy is only limited by the chosen word length. While analog VLSI circuits are less accurate, they are smaller, faster, and consume less power than digital circuits.[19]

For these reasons, applications that do not require great computational accuracy are dominated by analog designs.

Learning algorithms, especially backpropagation, require high precision and accuracy in modifying the weights of the network. This has led some to believe that analog circuits are not well suited for implementing learning algorithms.[20] Analog circuits can achieve high precision, at the cost of increasing the circuit size. Analog circuits with high precision (8 bits) tend to be as large as their digital counterpart.[21] Thus, high precision analog circuits lose their size advantage over digital circuits. Analog circuits are of greater interest in applications requiring only moderate precision.

Early studies show that analog circuits can realize learning algorithms, provided that the algorithm is tolerant of hardware imperfections such as low precision and inherent noise. In a paper by Macq et al., a fully analog implementation of a Kohonen map, one type of neural network, with on-chip learning is presented.[22] With analog circuits having been shown capable of the computational accuracy necessary for weight modification, they should continue to be the choice of neural network research.

Size, speed, and power consumption are areas in which analog circuits are far superior to digital circuits, and it is these areas that constrain most neural network applications. To achieve greater network performance, the size of the network must be increased. The ability to implement larger, faster networks is the major motivation for hardware implementation, and analog circuits are superior in these areas. Power consumption is also of major concern as networks become larger.[23] As the number of transistors per chip increases, power consumption becomes a major limitation. Analog circuits dissipate less power than digital circuits, thus permitting larger implementations.

Besides its universality to a wide variety of optimization procedures, the nature of the ALOPEX algorithm makes it suitable for VLSI implementation. ALOPEX is a biologically influenced optimization procedure that uses a single value global response feedback to guide weight movements toward their optimum. This single value feedback, as opposed to the extensive error propagation schemes of other neural network training algorithms, makes ALOPEX suitable for fast VLSI implementation.

Recently, a digital VLSI approach to implementing the ALOPEX algorithm was undertaken by Pandya et al.[24] Results of their study indicated that ALOPEX could be implemented using a Single Instruction Multiple Data (SIMD) architecture. A simulation of the design was carried out, in software, and good convergence for a 4×4 processor array was demonstrated.

The importance of VLSI to neural networks has been demonstrated. For neural networks to achieve greater abilities, larger and faster networks must be built. In addition to size and speed advantages, other reasons, including cost and reliability, make VLSI implementations the current trend in neural network research. The design of a fast analog optimization algorithm, ALOPEX, is covered below.

ALOPEX is an optimization procedure as already has been mentioned, in which the "best" value of a cost function or response is sought. The process uses a stochastic element (added Gaussian white noise) to avoid local extremes of the response. In other words, the added noise helps the procedure to find the global maximum or minimum value of the response. ALOPEX is an iterative procedure, where a large number of parameters are simultaneously changed by small amounts and then a new response is computed. The changes in the pixels of images are determined from the change in the response, the change in the pixel from the previous two iterations, plus the additive noise.

Let us assume that we have an array of 64 pixels which we call $I_i(n)$ where n represents the iteration. The additive Gaussian white noise is denoted by $r_i(n)$, and $R_j(n)$ is the response (or cost function) of the jth template at iteration n. The parameter $I_i(n)$ can then be found by the following equation:

$$I_i(n) = r_i(n) + \gamma \Delta I_i \Delta R + I_i(n-1) \qquad (12.5)$$

where γ is an arbitrary scaling constant and ΔI_i and ΔR are found from the following:

$$\Delta I_i(n) = I_i(n-1) - I_i(n-2) \qquad (12.6)$$

$$\Delta R(n) = R(n-1) - R(n-2) \qquad (12.7)$$

where $i = 1,2,3,\ldots,64$.

Let us assume that there are 16 templates to choose from, each with 64 pixels. The ALOPEX process is run on each of them, with the objective being to recognize (converge to) an input pattern. Due to the iterative behavior, if allowed to run long enough, ALOPEX will eventually converge to each of the templates. However, a "match" can be found by choosing that template which took the least amount of time to converge.

By convergence we mean finding either the global maximum or minimum of the response function. This response function can be calculated in many different ways, depending on the application. To allow this chip to be general enough to handle many applications, the response will be computed off the chip. A PROM can be used to compute the response based on the error between the input, $I_i(n)$, and the template. The PROM enables the response function to be changed to meet the needs of the application.

While the chip design is limited to only 64 ALOPEX subunits, the parallel nature of ALOPEX will enable many chips to be wired together for larger applications. Parallel implementations are made easy since each subunit receives a single global response feedback that governs its behavior. Backpropagation, on the other hand, requires dense interconnections and communication between each node. This flexibility is a tremendous advantage when it comes to hardwired implementations.

Originally the ALOPEX chip was designed using digital VLSI techniques. Digital circuitry was chosen over analog because it is easier to test and design. Floating point arithmetic was used to ensure a high degree of accuracy. The digital design consisted of shift registers, floating point adders, and floating point multipliers. However, after having done much work toward the digital design, it was abandoned in favor of an analog design. The performance of the digital design was estimated and was found to be much slower than an analog design. The chip area of the digital design was much larger than an analog design would be. Also, the ALOPEX algorithm would be tolerant of analog imperfections due to its stochastic nature. For these reasons, it seemed clear that a larger, faster network could be designed with analog circuitry.

The analog design needed components similar to the digital design to implement the algorithm. Mainly there needed to be an adder, multiplier, difference amplifier, a sample and hold mechanism, and a multiplexing scheme. These cells each perform a specific function and are wired together in a way that implements the ALOPEX process.

The chip is organized into 64 ALOPEX subunits, one for each pixel in the input image. They are stacked vertically, wiring by abutment. Each subunit is made from smaller components that are wired together horizontally and contains the following cells: a group selector, demultiplexor, follower aggregator, multiplier, transconductance amplifier, multiplexor, and another group selector.

The Gaussian white noise required for the ALOPEX process is added to the input before it reaches the chip. This will allow precise control of the noise, which is very important in controlling the stability of the algorithm. If there is too much noise, the system will not converge. If there is too little noise, the system will get stuck in local minima of the cost function. By controlling the noise during execution, using a method similar to simulated annealing[26] where the noise decays with time, it has been shown that the convergence time can be improved.[15] Also, by having direct control of the added noise, the component and functional testing can be done with no noise added, greatly simplifying the testing.

The addition, multiplication, and subtraction required by the ALOPEX algorithm are performed by the follower aggregator, Gilbert multiplier,[25] and transconductance amplifier, respectively. To understand how these units implement the equations of the ALOPEX process, let us rewrite the original ALOPEX equation as follows:

$$I_i(n) = r_i(n) + \text{bias}(n) \qquad (12.8)$$

where $r_i(n)$ is Gaussian white noise and bias(n) is defined as

$$\text{bias}(n) = \gamma \Delta I \Delta R + \text{bias}(n-1)) \qquad (12.9)$$

The follower aggregation circuit computes the weighted average of its inputs. By weighing the inputs equally, the circuit computes the average of the two inputs. The average is chosen instead of the sum since the circuit is more robust, in that the output never has to exceed the supply voltage. A straight summer is more difficult to design because voltages greater than the supply voltage could be needed. The output of the follower aggregator, $I_i(n)$ is sent to the multiplier where a C-switch acts as a sample and hold, to store the value of the previous iteration, $I_i(n-1)$. The difference between these signals is ΔI and is one input to the multiplier. The previous two responses, calculated off chip, are the other two inputs, representing ΔR. The output of the multiplier is $\gamma \Delta I \Delta R$, where γ is the gain of the multiplier and is controlled by the control signal gamma.

The output of the multiplier is a current equal to $\gamma\Delta I\Delta R$. The current can be either positive or negative, depending on the signs of ΔI and of ΔR. The output node acts as a capacitor that holds the bias voltage from Equation 12.9. This bias is then adjusted by an amount, $\gamma\Delta I\Delta R$, after each iteration. This bias is one of the inputs to the follower aggregator, the other being the input stimulus with added Gaussian white noise. The follower aggregator implements Equation 12.8, except that it computes the average or the sum divided by two.

The error signal is computed by the transconductance amplifier. The error is simply equal to the difference between $I_i(n)$ and the template to which you are matching. However, since $I_i(n)$ is equal to the sum divided by two, the template values must be halved before being multiplexed onto the chip. The error signal is computed, then multiplexed with $I_i(n)$ and sent off the chip. The error is used to compute the response $R(n)$. $I_i(n)$ is sent off the chip so that the operator can see the image as the algorithm converges, by sending the signal to some sort of display.

The power is supplied to the chip by four pins, two each for VDD and GND. The purpose of having two pins of the same signal is so that by placing them on opposite sides of the chip and by proper wiring, the resistive drop can be reduced.

In designing the chip, much effort was made in making it controllable and testable, while making the chip general enough that it could be used in a wide variety of applications. This is why the Gaussian white noise is added off chip, and also why the error signal is taken off chip for the computation of the response. This not only allows the response function to be changed to meet the requirements of the specific application, but it also provides the operator with accessible test points.

Despite the decrease in operating speed by a factor of four, due to time division multiplexing at both the input and outputs, the chip still operates at over 7,000,000 complete iterations per second. This speed may not even be attainable, given possible interfacing bottlenecks and much slower support hardware that is necessary for operation. Support hardware necessary for chip operation includes circuitry for the response calculation as well as memory to store templates. Depending on the application, A/D and D/A converters may be necessary. If this is the case, then 7 Mhz operation speed is more than adequate.

While backpropagation is the most widely used software tool for training neural networks, it is less suitable for VLSI hardware implementation than ALOPEX for many reasons. While backpropagation converges quickly, due to its gradient descent method it can often get stuck in local extrema. ALOPEX tends to avoid local extrema by incorporating a random noise component, at the expense of slightly longer convergence times.

The major differences arise when hardware implementation is discussed. Backpropagation is computationally taxing, due to the error computation needed for each node in the network. Each error is a function of many parameters (i.e., all the weights of the following layer). In hardware, very complex interconnections between all nodes are required to compute this error.

ALOPEX is ideal for VLSI implementation for a couple of reasons. First, the algorithm is tolerant of small amounts of noise; in fact, noise is incorporated to help convergence. Second, all parameters change based on their local history and a single value global response feedback. This single-valued feedback is much simpler to implement than the error propagation used in backpropagation.

REFERENCES

1. Minsky, M. and Papert, S., *Perceptrons: An Introduction to Computational Geometry*, M.I.T. Press, Cambridge, MA, 1969.
2. Huang, S. and Huang, Y., Bounds on the number of hidden neurons in multilayer perceptrons. *IEEE Trans. Neural Net.*, 2(1), 47, Jan. 1991.
3. Kung, S. Y., Hwang, J., and Sun, S., Efficient modeling for multilayer feedforward neural nets, Proc. IEEE Conf. on Acoustics, Speech Signal Processing, New York, April 1988, 2160.
4. Mirchandani, G., On hidden nodes for neural nets, *IEEE Trans. Circuits Syst.*, 36(5), 661, 1989.
5. Moore, K., Artificial neural networks: weighing the different ways to systemize thinking, *IEEE Potentials*, 23, Feb. 1992.
6. Werbos, P. J., Beyond Regression: New Tools for Prediction and Analysis in the Behavioral Sciences, Ph.D. Thesis, Harvard University, Cambridge, MA, 1974.
7. Parker, D. B., Learning Logic. Technical Report. Center for Computational Research in Economics and Management Science, M.I.T., Cambridge, MA, 1985.
8. Rumelhart, D.E., Hinton, G., and Williams, A., Learning internal representations by error propagation, in *Parallel Distributed Processing, Vol. 1, Foundations*, Rumelhart, D.E., and McClelland, J.L., Eds., MIT Press, Cambridge, MA, 1986.
9. Tzanakou, E. and Harth, E., Determination of visual receptive fields by stochastic methods, *Biophys. J.*, 15(42a), 1973.
10. Harth, E. and Tzanakou, E., ALOPEX: a stochastic method for determining visual receptive fields, *Vision Res.*, 14, 1475, 1974.
11. Tzanakou, E., Michalak, R., and Harth, E., The ALOPEX process: visual receptive fields by response feedback, *Biol. Cybern.*, 35, 161, 1979.
12. Holmstrom, L. and Koistinen, P., Using additive noise in backpropagation training, *IEEE Trans. Neural Net.*, 3(1), 24, Jan. 1992.
13. Venugopal, V., Sudhakar, R., and Pandya, A., An improved scheme for direct adaptive control of dynamical systems using backpropagation neural networks, *Circuits, Syst. Signal Proc.*, Sept. 1993.
14. Ciaccio, E. and Tzanakou, E., The ALOPEX process: application to real-time reduction of motion artifact, *Annu. Int. Conf. of IEEE EMBS*, 12(3), 1417, 1990.
15. Dasey, T. J. and Micheli-Tzanakou, E., A pattern recognition application of the ALOPEX process with hexagonal arrays, *Int. Joint Conf. Neural Net.*, Vol. II, 119, 1989.
16. Venugopal, K., Pandya, A., and Sudhakar, R., ALOPEX algorithm for adaptive control of dynamical systems, *Proc. IJCNN 1992*, II, 875, June 1992.
17. Micheli-Tzanakou, E., Non-linear characteristics in the frog's visual system, *Biol. Cybern.*, 51, 53, 1984.
18. Hecht-Nielsen, R., Neurocomputing: picking the human brain, *IEEE Spectrum*, 36, March 1988.
19. Mead, C. and Ismael, M., Eds., *Analog VLSI Implementation of Neural Systems*, Kluwer Academic Publishers, Norwell, MA, 1989.
20. Ramacher, U.and Ruckert, U., Eds., *VLSI Design of Neural Networks*, Kluwer Academic Publishers, Norwell, MA, 1991.

21. Graf, H. P. and Jackel, L. D., Analog electronic neural network circuits, *IEEE Circuits Devices Mag.*, 44, July 1989.

22. Macq, D., Verlcysen, M., Jespers, P., and Legat, J., Analog implementation of a Kohonen map with on-chip learning, *IEEE Trans. Neural Net.*, 4(3), 456, May 1993.

23. Andreou, A., et al., VLSI neural systems, *IEEE Trans. Neural Net.*, 2(2), 205, March 1991.

24. Pandya, A. S., Shandar, R., and Freytag, L., An SIMD architecture for the ALOPEX neural network, SPIE Vol. 1246, Parallel Architectures for Image Processing, 1990, 275.

25. Mead, C., *Analog VLSI and Neural Systems*, Addison-Wesley, New York, 1989.

26. Kirkpatrick, S., Gelatt, C. D., and Vecchi, M. P., Optimization by simulated annealing, *Science*, 220, 671, 1983.

13 Implementing Neural Networks in Silicon

Seth Wolpert and Evangelia Micheli-Tzanakou

13.1 INTRODUCTION

In spite of dramatic increases in the capacity and throughput of automated systems, there remain a number of descriptively simple yet highly desirable tasks that have remained elusive. These tasks are associated with the process known as pattern recognition. If machines were able to identify patterns in electrical, visual, mechanical, acoustic, or chemical signals as quickly and reliably as living systems, our world would be a very different place. A number of tedious operations could be performed tirelessly and accurately. We would no longer have the need for locks on our automobiles and homes or keyboards on our computers. For many years, engineers and mathematicians have worked to perform computer-based pattern recognition using geometric and statistical methods, but levels of accuracy commensurate with those of human operators have been difficult to obtain. To address the overwhelming utility to perform these tasks, engineers have begun to take cues from biological systems, the simplest of which are able to perform pattern recognition with relative ease and high reliability, as a matter of their very survival.

In order to deal with the sheer magnitude of living nervous systems, inroads have been taken historically to understand their workings by 'top-down' and 'bottom-up' approaches. Top-down approaches are based on outward observations of capability and behavior and have given rise to the field of Artificial Neural Networks, or ANNs. ANNs are based upon simplified models of individual neurons, which are highly interconnected via an array of variably coupled transmission units known as synapses. Such systems are generally implemented as computer models and have been most effective when configured and controlled in a manner tailored specific to a given pattern, method of assimilation and processing, and identification criteria. Generally implemented in the form of a computer simulation, these networks acquire data, train themselves, and evaluate possible solutions serially, and therefore require an inordinate amount of time and computational resources to function as well as traditional non-ANN pattern recognition methods. Clearly, these ANN methods have the potential to easily surpass conventional methods, but, in order to do so, they must be transplanted from the virtual environment within a serial computer to a dedicated hardware platform, where they may be implemented in a parallel and simultaneous manner. This would be consistent with theories of how living nerve circuits operate so quickly and reliably, and introduces the motivation for pursuing bottom-up approaches, based on observations of the structure and function of individual nerve cells.

13.2 THE LIVING NEURON

Living nerve cells have always been studied and modeled to the very limits of available electronic technology. Since as early as the 19th century, electrical models of processes observed in living nerve cells have undergone ardent development. The justification has been that machines had been previously unable to achieve the same tasks that living systems could so easily do; perhaps we could emulate them and put them to work in a number of endeavors. After all, the brain of an Einstein or a Shakespeare is not significantly different in structure or composition from the average human brain. The computational potential of the average human brain, then, must be remarkable, and an electronic model, which functions 100,000 times faster, would hold great potential as a computer for a variety of applications. The problem with this objective, however, has been in the sheer magnitude of the machinery. Consisting of 100 billion nerve cells, many of which have many thousands of interconnections, the human brain is a machine far beyond the analysis, design, and manufacturing capacities of any existing human technology. If such a "machine" or even a small part of it were to be replicated, however, such an effort must begin on the cellular level. In order to describe these efforts, a review of nerve cell structure and function is in order.

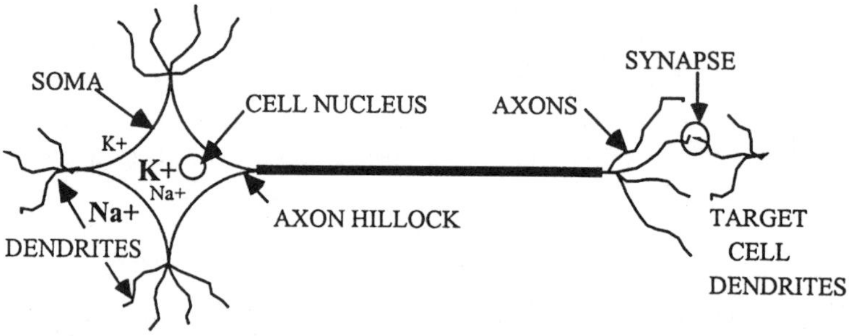

FIGURE 13.1   A typical living nerve cell.

A typical living nerve cell is depicted in Figure 13.1. Physically, it may be described as a tentacled elastic sac, whose interior and exterior are bathed in different conductive fluids separated from each other by the cell's outer material, known as *cell membrane*. This membrane draws upon the cell's metabolic processes to supply the energy it requires to accumulate specific ions against concentration gradients. Two key ions, potassium, which is accumulated inside the cell, and sodium, which is ejected to the exterior of the cell, have been identified as having the strongest role in the function of the nerve cell. The imbalance in distribution of these ions and several others forms the basis for an electrical potential within the cell relative to the conductive environment outside. This potential, known as *cell membrane potential*, rests nominally at 60–80 mV negative with respect to its external environment. Fluctuations in this cell membrane potential are the means by which neurons express their activity and communicate with sensory receptor cells, muscle cells, and other neurons.

The tentacles that branch off from the neuron's body, or *soma* are known as *dendrites*. Dendrites, which may number from zero to well into the thousands, branch off to other cells and collect sensory input signals based on those cells' levels of activity. These signals appear as electrical transients in membrane potential, which are accumulated over time and space, with the resultant sum appearing in the cell soma. Emanating from the soma is another singular tentacle known as the *axon*. Typically larger, longer, and better insulated than the dendrites, the axon conveys the output of its cell over long or short distances to target nerve and muscle cells. The point at which the axon attaches to the cell body is known as the *axon hillock*. There, the accumulated cell membrane potential is compared against a cellular *threshold* potential. When that threshold potential is exceeded, a separate mechanism in the axonal membrane gives rise to a single impulse, typically 80–100 mV in amplitude, and 1 ms in duration. This impulse is then propagated down the axon to its remote terminus, where individual fibers branch off and adjoin target nerve or muscle cells. While such an impulse is being generated, the cell enters a temporary state of total inexcitability, where no amount of stimulation can cause a second impulse to be superimposed over the first. This state soon elapses, and the cell gradually returns to an excitable condition. This phenomenon is known as *refraction*, and the interval of inexcitability is known as the *refractory period.*

Between the axonal terminus and the soma or dendrite of the target cell, a fluid gap forms a *synapse*, a physical discontinuity from one cell to the next. At the axonal terminal, a packet of chemicals is released into the synaptic gap, where it will migrate to the target cell and induce transient impulses in that target cell's membrane potential. These chemicals are known as neurotransmitters, of which over twenty different types have been identified. Neurotransmitters that induce negative transients in target cell membrane potential are known as inhibitory, while those that induce positive transients in target cell membrane potential are known as excitatory. Inhibitory stimuli suppress activity in target cells, while excitatory stimuli facilitate activity in target cells. Cells that induce large transients are said to have high synaptic weights, while those inducing little or no transients in target cells are said to have low synaptic weights. The magnitude or the duration of that transient may be affected by the synaptic weight, and changes in synaptic weight form the basis for training of ANNs, as well as learning in living nervous systems. The synapse also prevents reflection of impulses back to source cells, which would cause unbridled chaos to engulf the entire nervous system in a very short time.

Orchestrating the modification of synaptic weight in a network of cells learning to perform a new task or recognize a sensory image is the basis for top-down neuronal study. For bottom-up study, two other aspects of the operation of living neurons are of particular interest to those modeling its function: formulating the threshold of a nerve cell in terms of the spatial and temporal distribution of stimuli directed toward it and the relationship between conductivity of the membrane to the ions giving rise to membrane potential, present membrane potential magnitude, and time. These two aspects have been the bases for modeling individual nerve cells from two schools of thought.

13.3 NEUROMORPHIC MODELS

Since the era of the vacuum tube, a multitude of neuronal models composed of discrete components and off-the-shelf ICs have been published. Similar efforts in custom VLSI, however, are far fewer in number. A good introduction to a number of neuronal attributes, however, was presented by Linares-Barranco et al. of Texas A & M University.[1] CMOS-compatible circuits for approximating a number of mathematical models of cell behavior are described. In its simplest form, this model represents the cell membrane potential in the axon hillock as nothing more than a linear combination of an arbitrary number, n, of dendritic inputs, X, each of which is weighted by a unique multiplier, W, summed, and processed by a nonlinear range-limiting operator, f. The mathematical equation for this relationship is

$$Y_k = f\left\{\sum_{e=1}^{n} W_i X_i\right\} = f\{S_k\},\tag{13.1}$$

and this relationship is realized in the circuit model shown in Figure 13.2a and the CMOS circuit implementation in Figure 13.2b. This circuit is totally static and makes no provision for time-courses of changes in input or output signals, or intracellular relationships. In the implementation of Figure 13.2b, the *operational transconductance amplifier*, OTA, as described in Reference 16 and depicted in Figure 13.3, is used in lieu of operational amplifiers for this and most other VLSI neural network applications. Highly compatible with CMOS circuit technology, it is structurally simple and compact, realizable with only nine transistors, and provides reasonable performance. The only consideration it warrants is that its transfer function is a transconductance. As such, operations performed on its output signals must be oriented to its current rather than its voltage. When driving high load impedances, as is usually the case with CMOS circuitry, this is only a minor inconvenience, necessitating buffering for lower load impedances. In fact, under some circumstances, such as when algebraic summation is being performed, a current output may actually be an advantage, allowing output nodes to be simply tied together.

The nonlinear range-limiting operator, f, mentioned earlier, is necessitated by the observation that, for a given biological neuron, there are limits on the strength of the electrochemical gradients that the cell's ionic pumps can generate. This imposes limits on how positive and negative cell membrane potentials may go. Since a neuron may receive inputs from many other neurons, there is no such limit on the aggregate input voltage applied. As a result, an *activation function*, a nonlinearity of the relationship between aggregate input potential and output potential of a neuron, must be imposed. This is typically done in one of three different ways, the binary hard-limiter, which assumes one of only two possible states—active or inactive, the linear-graded threshold, which assumes a linear continuum of active states between

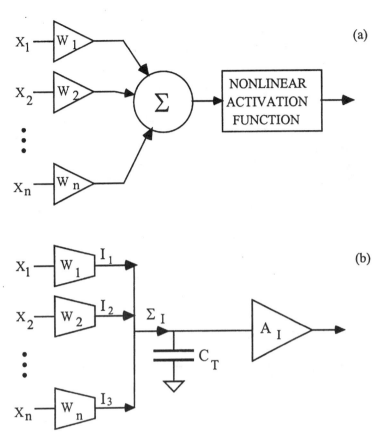

FIGURE 13.2   Circuit organization of a general purpose neuronal model (a), and a CMOS VLSI circuit implementation of such a model (b).

its minimal and maximal values, and the sigmoid, which assumes a sigmoidal distribution of values between its negative minimal and positive maximal output values. All three of these relationships are shown in the graphs of output potential vs. input potential of Figures 13.4a, b, and c, respectively.

Which type of activation function is employed depends on the type of artificial neuron and network in which it is implemented. In networks where cell outputs are all-or-none, such as McCullouch and Pitts models,[2] the binary threshold model is used. In networks where neurons are theorized to have variable output levels applied to distinctly designated excitatory and inhibitory inputs, such as Hopfield Networks, the linear threshold model is used. In networks where a synaptic connection must be both excitatory and inhibitory, depending on the level of activity, the sigmoid threshold is used. In either of the latter two activation functions, the slope of the overall characteristic can be varied to suit the sensitivity of the cell in question.

The basic neuron cell model shown in Figure 13.2a was designed for primitive neuronal models and learning algorithms. It performs linear summation of independently weighted synaptic inputs applied to a single node, and discriminates according

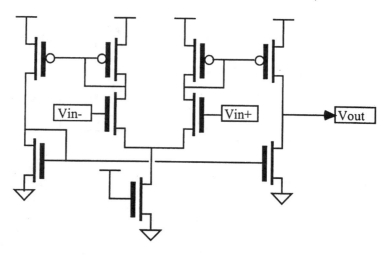

FIGURE 13.3   CMOS implementation of an Operational Transconductance Amplifier, OTA, widely used for realization models and networks in VLSI.

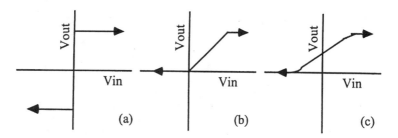

FIGURE 13.4   Nonlinear activation functions imposed on outputs of nerve cell models. Shown are the binary bipolar hard limiter (a), the linear graded potential (b), and the sigmoid potential (c).

to a binary threshold of zero, as shown in Figure 13.4a. Although the linear combination is an easy process to comprehend, its fidelity in the face of biological nerve behavior is restricted. In order to improve the applicability of such models, several improvements must be made to their mathematical descriptions. The first such improvement is the dynamic model. Like the model described by Equation 13.1, it includes linear combination, summation, and a nonlinear operator, in this case, the sigmoidal activation function. Consistent with transconductance amplifiers, however, its output is now expressed as a current in the form of CdS/dt. It also features I_B, which represents a fixed biasing current that determines a baseline level of activity, or threshold. This activity level represents a threshold that must be surpassed by the aggregate sum of weighted inputs in order for cell k to respond. Different cells may be assigned different thresholds, so that their responsiveness may be tuned to the demands of the network in which they reside. Finally, the dynamic model includes R, a self-relaxation term that insures that S, the cell output potential, will decay to zero when all dendritic inputs, X, are zero. The dynamic model is implemented using a "leaky integrator", which allows for the duration or persistence of input signals to be controlled. The equation for this behavior is

$$C\frac{dS_k}{dt} = I_{B\bar{k}}\frac{S_k}{R_k}\sum_{i=1}^{n}W_{ik}f(S_i)\qquad(13.2)$$

A mathematical model of this equation is given in Figure 13.5a, and the CMOS implementation in Figure 13.5b.

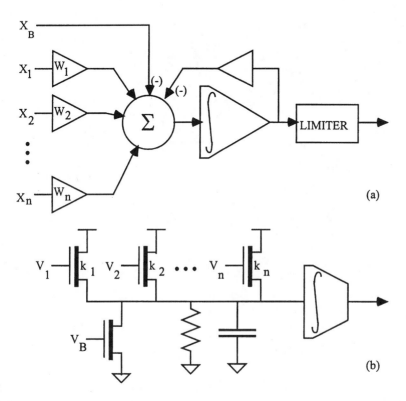

FIGURE 13.5   Circuit organization of a dynamic cell model (a) and a CMOS VLSI implementation of such a model (b).

More comprehensive features to facilitate functioning in a large population of nerve cells have been incorporated into the generalized model described by Carpenter and Grossberg.[3] This model includes the features of the dynamic model, as well as a more comprehensive facility for temporal summation with the self-forgetting, or persistence term, A_k. The H and L coefficients allow for the fixed and output voltage-dependent levels of activity in the network to be controlled. This keeps the network's signals from saturating at too high or low an overall level of activity. The E coefficient represents a fixed applied bias signal analogous to the I_B term of Equation 13.2. The Z coefficient represents a synaptic coupling analogous to W of Equation 13.2. Mathematically, the equation for the generalized model is given as

$$\frac{dS_k}{dt} = -A_k S_k + \left(H_k - L_k S_k\right)\left\{E_k + \sum_{i=1}^{n} Z_{ik} f\left(S_i\right)\right\}$$ (13.3)

and the model, along with a CMOS implementation are given in Figures 13.6a and 13.6b, respectively.

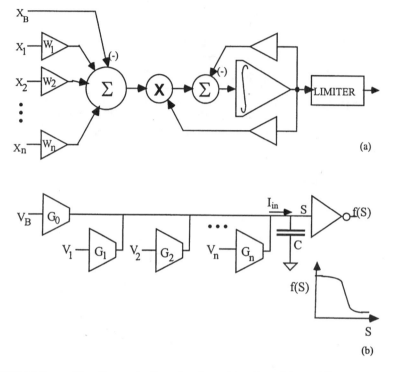

FIGURE 13.6   Circuit organization of a dynamic cell model providing unconstrained assimilation of excitatory and inhibitory inputs (a) and a CMOS implementation of that model (b).

This model is comprehensive enough to be appropriate for implementing artificial neural networks that realize Adaptive Resonance Theory (ART), as well as Hopfield Networks and McCullouch and Pitts networks, but still lacks one vital attribute of the living neuron. All of the cell models presented so far portray neurons as simplified cells whose output is expressed as a DC level that reflects some nonlinear function of the aggregate sum of input signals. This forms the basis for most ANN implementations. There is also a class of cell models whose output is a train of similar pulses whose frequency is varied, rather than a variable DC potential. For these frequency-modulated models there are also a series of circuit implementations.

Frequency-based neuronal models are similar to those already presented, in that they perform temporal and spatial summation of an arbitrary number of dendritic inputs, as well as their own current state of activity. They will also have activation functions assigned, depending on the type of cell and network. Unlike the activation functions of voltage-based models, these are imposed in recognition of the fact that, for a given biological neuron, an action potential cannot be elicited during the formation of its predecessor. This manifests itself as a limit on how close together in time two action potentials may occur from a given cell, and therefore, a limit on the maximal frequency at which a neuron can generate pulses. A neuron may receive inputs from many other neurons. While each of those inputs has a similar upper limit on its frequency, there is no such limit on the number of inputs, and therefore no limit on the overall input frequency. As a result, a non-linearity of the relationship between input frequency and output frequency of a neuron must be imposed. In most cases, this type of behavior may be brought about with the simple addition of a voltage-controlled oscillator, or VCO, to the output stage of one of the previously defined models with an activation function operator.

There are two CMOS VLSI implementations of oscillatory models of note, both of which are derived from the system of differential equations formulated by Hodgkin and Huxley[4] in 1952. In the course of producing an action potential, the neuronal cell membrane exhibits conductances to sodium and potassium ions that were found to be mathematical functions of time and of cell membrane potential. The Hodgkin-Huxley equations were derived to describe those time and voltage relationships. A popular circuit approach to realizing the oscillatory behavior required to synthesize a single pulse from one control input is to employ a hysteretic output stage. The organization of such a system is shown in Figure 13.7a, along with a CMOS circuit implementation in Figure 13.7b. It is apparent that this is a simple adaptation of the dynamic model shown in Figure 13.5.

The other approach to recreating such a circuit instability is in a Hodgkin-Huxley derivative known as the Fitzhugh-Nagumo model.[5] Based on a mutually antagonistic relationship between two cells and an I-V characteristic outwardly similar to that of a tunnel diode, the Fitzhugh-Nagumo model is somewhat more complex but still realizable in conventional CMOS VLSI subcircuits. The model for this circuit is given in Figure 13.8, and it formed the basis for one of the more successful CMOS VLSI implementations of single-neuron models.

In 1991, Bernabé Linares-Barranco et al.[6] fabricated and characterized a circuit whose behavior is based on the Fitzhugh and Nagumo equations. The variability membrane conductance characterized by the Hodgkin-Huxley equation was recreated as a piece-wise linear model, which was realized empirically using the circuit of Figure 13.8b. A series of OTAs whose transconductances correspond to the membrane ionic conductances over specified input voltage ranges was used to realize the transients in membrane conductance that give rise to the action potential. Fabricated prototypes were demonstrated to replicate several types of behavior commonly seen in living nerve cells, i.e., free-running sustained oscillation in a single cell and on-and-off, or *bursting* oscillation, as seen in a pair of mutually antagonistic cells. For both circuit configurations, oscilloscope photographs appear similar, albeit less noisy than intracellular recordings from live nerve cells.

Along similar lines, another CMOS implementation was developed by Mahowald and Douglas.[7] In this model, the time course of sodium and potassium currents

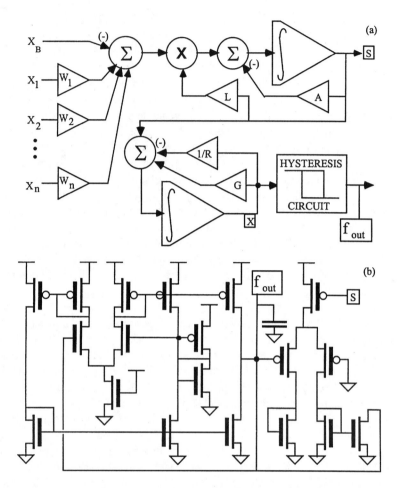

FIGURE 13.7   Circuit organization for a dynamic cell model producing variable-frequency output signals by the application of a hysteresis subcircuit (a), and a CMOS implementation of that model (b).

are recreated empirically, by virtue of fundamental similarities between ionic conductivity in neural membrane and that of appropriately biased MOSFETs. Structurally simple yet elegant circuits shown in Figure 13.9a, b, and c recreate the time and voltage courses of potassium activation, sodium activation, and sodium inactivation respectively in neural membrane. Rectangular current pulses of various amplitudes applied to the circuit show a striking similarity to similar impulses applied to living nerve cells. The circuit is highly compatible with larger scale applications, requiring minimal off-chip support, occupying under 0.1 mm² of chip area, consuming under 60 W of power, and able to operate a million times faster than their biological counterparts. With the incorporation of a dendritic array, networks of several hundred nerve cell analogs on a single chip have been envisioned.

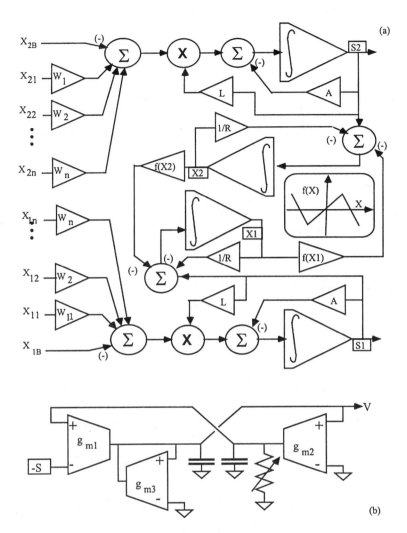

FIGURE 13.8   Circuit organization of a model of the Fitzhugh-Nagumo equation (a) and a CMOS implementation of that model (b).

Another well-executed implementation of VLSI-based nerve cells complements the Mahowald-Douglas model, concentrating less on overall nerve cell behavior and more specifically on how inputs to a neuron combine over time and space to affect a target cell.[8-9] Temporal and spatial summation and some topical applications have been modeled extensively in CMOS VLSI by Elias and Northmore. Recognizing that the strength, duration, and delay of a neuronal stimulus depend strongly on the physical location to which that stimulus is applied, Elias and Northmore recreate a linearly arrayed multicompartmental silicon dendrite, in which each segment or compartment has a specific capacitance to the cell's exterior, Cm, impedance of the internal fluid, or, cytoplasm, Ra, and impedance of a leakage path to the cell's exterior, Rm. Implemented using on-chip switched-capacitor analog networks, the

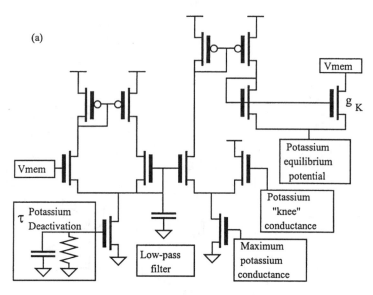

POTASSIUM ACTIVATION CIRCUIT

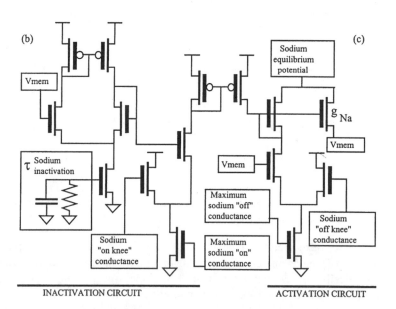

FIGURE 13.9   Circuits described by Mahowald and Douglas to realize sodium and potassium currents in active membrane. The potassium activation circuit is shown in part (a), the sodium activation is shown in part (b), and the sodium inactivation is shown in part (c).

authors demonstrate impulses that can persist millions of times longer that the impulses from which they originated. They also showed a mechanism by which a target cell's sensitivity may be keyed to any of a wide range of impulse shapes, durations, latencies, directional velocities, and repetition frequencies, as applied at various locations along a dendritic tree (topographic connection), or across a dendritic tree (laminar connection). The design of distributed compartments and their incorporation into a dendritic tree is shown in Figure 13.10a.

The facility of such networks to recognize specific spatial and temporal frequencies in arbitrary images was then applied to a VLSI-based system for recognition of binarized two-dimensional visual images. Due to the large number of possible input sites to a dendritic tree contained in a 40-pin IC package, a multiplexed approach was taken to transmission of data on- and off-chip. For the two-dimensional input images, one dimension is applied to topographic connections of the dendritic tree, and the other dimension is applied to laminar connections of the tree. As the image is scanned into the dendritic tree, spatial summation of the laminar inputs and temporal summation of the topological inputs results in a synchronized response unique to the pattern of the input image. Depictions of topographic and laminar connections to a dendritic tree are given in Figure 13.10b. The remainder of the circuitry in the implementation is associated with encoding and transferring data and synaptic coefficients. Dendritic trees of higher dimensions may be used to recognize images of higher dimension, and lateral inhibition and other real-time image processing operations are highly compatible with this method.

Another well-developed implementation of individual artificial nerve cells is the one by Wolpert and Micheli-Tzanakou.[10-11] While most neuromorphic models are based on the Hodgkin-Huxley equations, this one uses a sequencer to synthesize the action potential in three distinct phases. It also employs a different formulation for cell membrane and threshold potentials known as an integrate-and-fire model, presented and implemented in discrete components by French and Stein in 1974.[3] It makes use of the aforementioned leaky integrator and provides off-chip control over the response and persistence of stimuli assimilated into membrane potential. The model affords similar controls over the resting level and time constant of the cell threshold potential and allows for refraction to be recreated. This organization also affords control over the shape, resting level, and duration of the action potential and produces a TTL-compatible pulse in parallel with the action potential. These controls, all of which are continuously and precisely adjustable, make this model ideal for replicating the behavior of a wide variety of individual nerve cells, and it has been successfully applied as such. The organization for the French and Stein model is shown in Figure 13.11, and the Wolpert and Micheli-Tzanakou VLSI circuit was implemented as shown in Figure 13.12.

The Wolpert and Micheli-Tzanakou model is organized around three critical nodes, the somatic potential, the axonal potential, and the threshold potential. Each of these nodes is biased off-chip with an R-C network so that its resting level and time constant are independently and continuously controllable. Stimuli to the cell

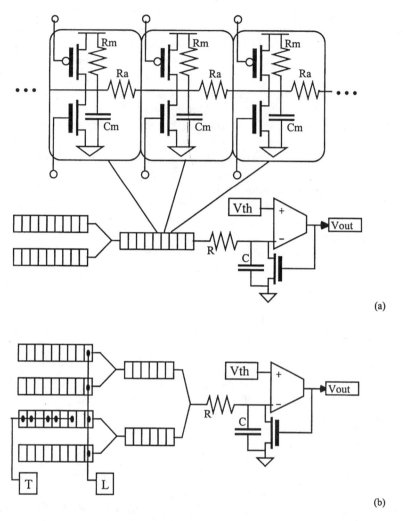

(a)

(b)

FIGURE 13.10   Circuits used to realize dendritic trees by Elias and Northmore. Dendritic compartment circuitry and the organization of compartments into a dendritic tree are shown in part (a), and the application of laminar and topographic summation in a dendritic tree are shown in part (b).

are buffered and standardized by truncation into $10\mu s$ impulses. Synaptic weight inputs on the excitatory and inhibitory pathways allow for this value to be increased or decreased from off-chip. The impulses are then applied to somatic potential by a push-pull MOSFET stage and compared to threshold potential by an OTA acting as a conventional voltage comparator. When threshold is exceeded, an action potential is synthesized and outputted. This waveform is then binarized and buffered to form a binary-compatible output pulse. Also at the same time, threshold is elevated to form the refractory period. The circuit consists of approximately 130 transistors

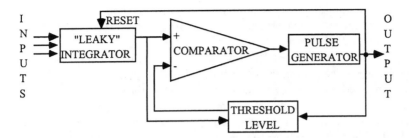

FIGURE 13.11   Organization of the "integrate and fire" model of neuronal behavior described by French and Stein.

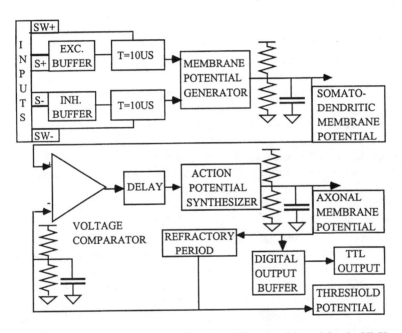

FIGURE 13.12   Implementation of the French and Stein model used for the VLSI prototype of the artificial nerve cell described by Wolpert and Micheli-Tzanakou.

plus a few on chip and discrete resistors and capacitors, and was implemented in a conventional CMOS technology, requiring a single-ended DC supply of 4-10 volts DC, and occupying 0.6 mm^2 of chip area.

With its critical nodes bonded out off-chip, the Wolpert-Micheli-Tzanakou neuromime's rate of operation may be accelerated from a biologically compatible time frame over several orders of magnitude. This model was first implemented in 1986 and is intended as a flexible and accurate aesthetic, rather than a mathematical model of cell behavior. In the time since then, it has been used to successfully recreate a number of networks from well-documented biological sources. Waveforms obtained in these recreations have shown a striking similarity to intracellular recordings taken from their biological counterparts. It has also been applied successfully to problems in robotics and rehabilitation.

Another well-conceived VLSI-based model of neuronal response is a hybrid neural processing element, PE, described by DeYong, Findley, and Fields.[12] Running at nominal CMOS VLSI speeds and having no need for internal nodes representing membrane and threshold potentials, this implementation requires far fewer components and is therefore much more appropriate for large-scale implementations in VLSI. In this model, each of the synaptic types, excitatory, inhibitory, and shunting, is implementable using seven transistors or less, and variability in synaptic weight costs an additional five transistors per synapse. The accumulated somatic potential is then applied to an axon hillock circuit, which performs threshold discrimination and generates an action potential pulse from under twenty transistors. This circuit has many of the features of the Wolpert-Micheli-Tzanakou model, including an arbitrary number of excitatory, inhibitory, and shunting inputs, a tangible threshold potential node, and biologically aesthetic waveforms, even though their durations and amplitudes are oriented to conventional analog and digital circuitry. The circuit is used to realize a one-by-four celled laterally inhibited winner-take-all network, which is of particular interest in pattern recognition operations, where the known pattern that is most similar to the unknown image is singled out over the remainder of less secure match candidates. Finally, models of neuronal function may be radically simplified to a voltage-controlled oscillator. This function may be realized in large quantity using a minimalist model known as the NTC, or Neural-Type Cell.

Recognizing that a neuron may be described as a voltage-driven pulse generator, Moon et al.[13] have been developing and applying NTC's to various problems in Artificial Neural Networks (ANN). The description of a neuron as a VCO is one that can be implemented as a small circuit of three MOSFETs, three resistors, and a capacitor, as shown in Figure 13.13. Although the circuit does not oscillate over a wide range of frequencies and its output frequency is not linearly related to its input level, its simplicity, small number of outward connections, and compact size make the NTC appropriate for implementation in large quantities. With the replacement of R6 with a voltage-controlled variable resistor, this circuit is able to assimilate variable synaptic weight as is manifest by a variable duty cycle on its output waveform. This circuit may also be tuned to function over a wide range of operating frequencies, as controlled by R6 and C. The NTC and the other VLSI circuits presented so far have all been conceived with the intent of replicating one or more aspects of nerve cell behavior. There are also many efforts directed at modeling cell-to-cell interactions, as theorized and observed in living nervous systems.

13.4 NEUROLOGICAL PROCESS MODELING

The modeling of interaction between nerve cells has been most widely pursued with respect to problems in image processing and computation. Image processing applications were mostly pioneered in VLSI form by Carver Mead, one of the world's leading educators and implementers of VLSI, and the models of vision and audition

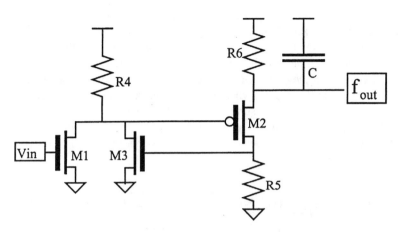

FIGURE 13.13   Schematic of the Neural-Type Cell described by Moon, Zaghloul, and Newcomb.[13]

he has developed focus on the simultaneous and immediate preprocessing of sensory images that is believed to take place before interpretation. The most common such processing step is known by the name of lateral, reciprocal, or mutual inhibition, and modeling of sensory processes that make use of lateral inhibition has been foremost in neural process modeling. Computation, on the other hand, encompasses system control, pattern recognition, clustering, and prediction. The latter three of these topics will be discussed in more detail in a future chapter. The former has several VLSI applications, one of which is as a general-purpose servo element.

DeWeerth and Mead, of Cal-Tech, and Nielsen and Astrom of Lund Institute of Technology in Sweden, have implemented a simple servo controller in custom VLSI.[14] The authors recognize that human tissues possess friction, elasticity, and internal damping, yet are capable of precise positioning and movement due to the presence of copious feedback and redundancy. Such a precise control system can surely provide excellent positional and motion resolution to electromechanical systems, as well. The OTA of Figure 13.3 was modified with the addition of a second, parallel output, whose current is the complement of the primary output. In addition, the biasing transistor, whose gate was depicted as being tied to V_{DD} in Figure 13.3, now has its gate tied to a DC reference input voltage, V_b. This input serves as an overall gain control for each OTA.

To implement the servo system, a number of these OTAs are connected with their corresponding outputs in parallel, as shown in Figure 13.14. As such, they represent a number of independently weighted synaptic inputs, whose outputs saturate as they approach maximal and minimal levels, forming a sigmoidal activation function. The aggregate complementary output currents are then pulse-width modulated. The complementary pair of variable duty cycle pulse trains that result are then buffered and applied directly to the terminals of a bidirectional DC motor. When system conditions demand motion in the positive direction, its synapses approach positive output currents, and the pulse train to the positive terminal of the motor approaches 100% while the duty cycle of the pulse train to the negative terminal approaches zero. This affects rotation in the positive direction. When full positive and negative motion are invoked, the motor turns rapidly, yet when the positive and negative pulse trains are roughly equal, there is a very fine resolution of motor control. In the servo system, the complementary outputs assume an agonistic/antagonistic relationship, where one signal exists at the expense of the other, and both signals cannot coexist simultaneously. This mutually inhibitory relationship is a frequently recurring theme in a wide variety of living nervous systems in a variety of organisms and has been modeled by a number of researchers.

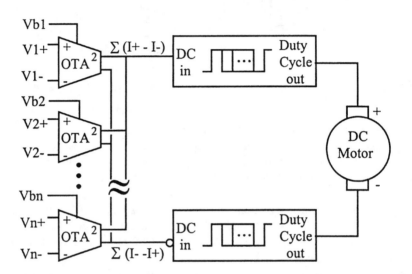

FIGURE 13.14   Organization of the VLSI neuron servo. System control directives are inputted to OTAs fitted with complementary outputs. Aggregate positive and negative motion directives are pulse-width modulated, buffered, and fed to the input terminals of a reversible DC motor.

Lateral inhibition is the process in which a cell containing some level of information encoded as its output level acts to inhibit and is inhibited by a similar adjoining cell, as depicted in Figure 13.15. For many years, this process has been observed with striking regularity in both one- and two-dimensional arrays of sensory receptors in a variety of systems, in a variety of organisms. In numerous morphological, mathematical, and circuit studies, it has been identified as a key image preprocessing step, which optimizes a sensory image in order to facilitate fast and accurate recognition in subsequent operations. Lateral inhibition accomplishes this by amplifying differences, enhancing image contrast, lending definition to its outward shape, and isolating the image from its background. While a digital computer would accomplish this process one pixel at a time, biological systems manage it in a manner that is both immediate and simultaneous.

Laterally inhibited behavior has been observed in pairs of cells implemented in hardware and software models by many researchers, but in dedicated VLSI by only a few. Notable among them, Nabet of Drexel University, and Pinter and Darling of

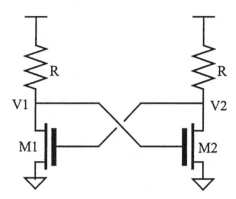

FIGURE 13.15   A pair of mutually inhibiting circuit nodes.

the University of Washington have extensively studied the stability and effectiveness of both pairs and linear strings of mutually inhibiting cells in CMOS VLSI and obtained results well-correlated with biological data.[15] This line of work has been explored in two dimensions in another series of VLSI-based models by Wolpert and Micheli-Tzanakou.[16] Arrays of mutually inhibiting cells that inhibit via continuously active connections and cells that inhibit by dynamic or strobed controls were both found to offer stable and variable control over the degree of inhibition. Arrays of hexagonally interconnected cells were found to be more stable than the square array, which tended to "checkerboard" when significant levels of inhibition were attempted. Feedback inhibition, where one array is used to store both the initial and inhibited images, was found to be as effective but less convenient to access than feed-forward inhibition, where separate input and inhibited images are maintained.

Characterization of lateral inhibition in the context of a more specific biological model has been pursued in another noteworthy effort by Andreou of Johns Hopkins University and Boahen of Cal-Tech. Multiple facets of cell-cell interactions, including both mutual inhibition and leakage of information between adjoining cells, were implemented in VLSI as a model of early visual processing in the mammalian retina.[17] There, adjacent cells on the photoreceptor layer intercommunicate through gap junctions, where their cell membrane potentials couple through a resistive path. Simultaneously, optical information from the photoreceptor cells are downloaded to corresponding cells of the horizontal layer, which have been shown to have mutually inhibitory connections. This interaction is illustrated in Figure 13.16. One-dimensional arrays, and subsequently, two-dimensional models of these relationships were implemented in analog VLSI and tested. Although little numerical data were published from these arrays, the two-dimensional array was demonstrated to produce a number of optical effects associated with the human visual system, including Mach bands, simultaneous contrast enhancement, and the Herman-Herring illusion, all of which are indicative of the real-time image processing known to occur in the mammalian retina.

Finally, the definitive VLSI implementation of a two-dimensional array is the well-known silicon retina devised by Carver Mead of Cal-Tech, and described in his text, "Analog VLSI and Neural Systems",[18] in addition to presenting a compre-

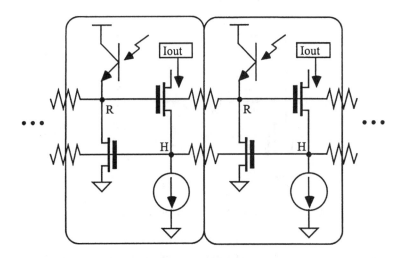

FIGURE 13.16   A mutually inhibitory pair, as implemented in analog VLSI by Andreou et al. This model represents the highly interconnected photoreceptor and horizontal cells of the retina, as indicated by the R and H nodes, respectively.

hensive treasury of analog VLSI circuits for a variety of mathematical operations necessary to implement neural networks in VLSI. The book then goes on to present several applications of analog ANNs, culminating in an auditory model of the cochlea and a visual model of the retina.

The 'silicon retina' is built around a 48 × 48 cell array of photosensors on a microchip. This array is then overlaid with a grid of resistors that replicates the gap junctions of the cells of the photoreceptor layer. Also incorporated into the array is a network of amplifiers, whose inputs are drawn from each adjoining node. The output of these amplifiers is an image that represents the Laplacian of the image, which replicates the mutual inhibition inherent in the horizontal cell layer. Because there are more pixels in the array than pins on the IC package that houses it, individual pixel data must be conveyed off-chip by an analog decoder/multiplexer. In tests, the circuit was shown to possess temporal and spatial response similar to those of living retinas, as evidenced by its recreation of a number of optical illusions associated with human vision. Since its initial description in 1988, many interesting modifications to the silicon retina have been implemented by Mead's students in the Computation and Neural Systems Laboratory at Cal-Tech.

An on-chip photoreceptor capable of transducing visual light over six orders of magnitude was implemented and published by Delbruck. They also developed a motion-sensitive silicon retina, which reacts to moving rather than stationary objects. This phenomenon has been observed many times in living retinas in a variety of organisms. Directional sensitivity was then applied to this principle by Delbruck and Benson. Velocity-sensitivity was later implemented by Delbruck, as well as the facility to optimize the focus of an image onto the surface of a chip by means of a distributed system of differentiators, a maximizer, and a servo mechanism to control positioning of an optical lens over the chip. This, along with intrinsic electronic control over contrast and brightness, constitutes a crucial first step in implementing a totally parallel visual system. This same objective has been brought to fruition in auditory system modeling, which has resulted in a number of commercial products now on the market.

A custom CMOS VLSI model of the human middle and inner ear has been implemented by Liu, Andreou, and Goldstein of Johns Hopkins University.[19] The eardrum and bones of the middle ear are modeled as a fifth order low pass filter with a second order pole at 15 kHz, and a third order pole at 100 kHz. The cochlea is modeled as a bank of thirty second-order band-pass filters, whose Q and center frequency are tuned by on-chip resistors and DC bias voltages. The hair cells of the cochlea perform nonlinear transformations and dynamic range compression, and are modeled by a series of 128 analog switches, voltage dividers, and voltage comparators.

The voltage comparison in this model is accomplished with the use of the OTA shown in Figure 13.3. The active filters, on the other hand, required an amplifier with a higher output impedance. This is because the circuits were operating in their subthreshold region. Subthreshold operation is used when the Vgs of a MOSFET is varied between zero and its specified threshold voltage. Transistors biased in this region pass small although coherently exponential drain-source currents. These currents are necessitated by the extremely long time constants of the human audio spectrum and the limited value of on-chip capacitors. A schematic for the modified subthreshold OTA with heightened output impedance is shown in Figure 13.17. Electrical signals in the VLSI implementation measured in response to sinusoidal tones correlate quite well to signals recorded under similar conditions in the auditory nerve of the cat. The overall organization of the system is shown in Figure 13.18. The technology for cochlear modeling has resulted in dramatic progress, not only in research efforts, but in commercial endeavors, as well.

A good deal of commercial success has been made in the area of cochlear implants. Deafness in humans is the result of a number of possible pathologies. In cases where deafness occurs due to physical damage to the structures of the outer and middle ear, those structures may be augmented by a surgically implantable microchip that decomposes incoming sounds into the fundamental and harmonic frequencies of which they are composed. The output of the device is then a multipolar electrode that applies various frequency signals to various points on the basilar membrane of the intact cochlea. In devices such as the implant manufactured by Cochlear Corp., a microphone is worn on the outside of the ear in much the same way as a hearing aid. The microphone then decomposes the sound wave into frequency bands, and transmits them through the skin to the implanted device in the form of a modulated radio frequency signal. The implanted device then rectifies the information signal to derive the power required to run its internal circuitry and to provide electrical stimulation of the appropriate region of the cochlear membrane. This eliminates the need for internal batteries, which may pose a health hazard due to their chemical contents or the surgical procedures required to install and replace them. While cochlear implants do not restore total hearing, they do impart the ability

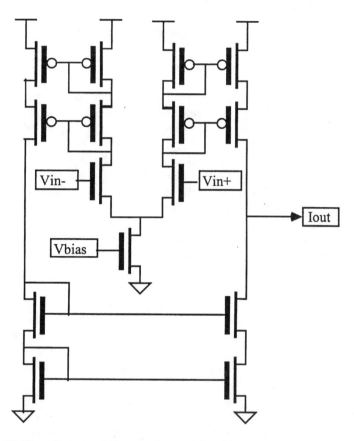

FIGURE 13.17   Operational Transconductance Amplifier adapted for use in subthreshold mode and elevated output impedance.

to receive sounds which aid in lipreading and overall awareness of the auditory environment. The cochlear implant also represents the leading edge in cybernetic implants, blazing the path for devices to augment hearing, vision, and sensation and movement, both visceral and somatic.

Clearly, the living nervous system has been the inspiration for a substantial amount of engineering development. Astounded and reassured by the utter throughput, reliability, and robustness of living sensory systems, engineers, mathematicians, neuroscientists, and computer scientists have doggedly endeavored to understand the workings of living nervous systems. Much progress has been made in understanding and recreating the structure and function of individual nerve cells. From morphological and electrophysiological study, progress has also been made in understanding the structure of small nerve circuits. Ahead lies the most profound frontier, that of the algorithms and control over activity in the brain.

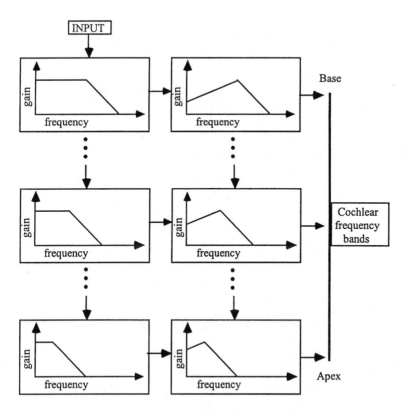

FIGURE 13.18 Organization of a silicon-based model of the auditory periphery, as described in Reference 19. Input sound waves are low-pass filtered to simulate the eardrum and bones of the middle ear, and then bank-filtered to simulate the function of the cochlea. Finally, their dynamic range is compressed to simulate the hair cells.

REFERENCES

1. Linares-Barranco, B., Sanchez-Sinencio, E., and Rodriguez-Vazquez, A., CMOS Circuit Implementations for Neuron Models, *Proc. IEEE Int. Symp. on Circuits and Systems,* New Orleans, LA, May 1990, vol. 3, 2421.
2. McCullouch, W. S. and Pitts, W., A logical calculus of the ideas imminent in nervous activity, *Bull. Math. Biophys.,* 5, 115, 1943.
3. Grossberg, S., Nonlinear neural networks: principles, mechanisms, and architectures, *Neural Net.,* 1, 17, 1988.
4. Hodgkin, A. L. and Huxley, A. F., A qualitative description of membrane current and its application to conduction and excitation in nerves, *J. Physiol.,* 177, 500, 1952.
5. Fitzhugh, R., Impulses and physiological states in theoretical models of nerve membrane, *Biophys. J.,* vol. 1, 445, 1961.

6. Linares-Barranco, B., Sanchez-Sinencio, E., Rodriguez-Vazquez, A., and Huertas, J. L., A CMOS implementation of FitzHugh-Nagumo model, *IEEE J. Solid-State Circuits*, 26(7), July, 956, 1991.

7. Mahowald, M. and Douglas, R., A silicon neuron, *Nature*, 354, 515, 1991.

8. Elias, J. G., Artificial dendritic trees, *Neural Comput.*, 5, 648, 1993.

9. Northmore, D. P. and Elias, J. G., Evolving Synaptic Connections for a Silicon Neuromorph, *Proc. IEEE World Congr. on Computational Intelligence*, Vol. 2, New York, 753, 1994.

10. Wolpert, S. and Micheli-Tzanakou, E., An Integrated Circuit Realization of a Neuronal Model, *Proc. IEEE Northeast Bioeng. Conf.*, New Haven, CT, March 13-14, 1986.

11. Wolpert, S. and Micheli-Tzanakou, E., A neuromime in VLSI, *IEEE Trans. Neural Net.*, 6(6), 1560, 1995.

12. DeYong, M. R., Findley, R. L., and Fields, C., The design, fabrication, and test of a new VLSI hybrid analog-digital neural processing element, *IEEE Trans. Neural Net.*, 3(3), May 1992.

13. Moon, G., Zaghloul, M. E., and Newcomb, R. W., VLSI implementation of synaptic weighting and summing in pulse coded neural-type cells, *IEEE Trans. Neural Net.*, 3(3), 394, May 1992.

14. DeWeerth, S. F., Nielsen, L., Mead, C. A., and Astrom, K. J., A simple neuron servo, *IEEE Trans. Neural Net.*, 2(2), March 1991.

15. Nabet, B. and Pinter, R. B., *Sensory Neural Networks: Lateral Inhibition*, CRC Press, Boca Raton, FL, 1991.

16. Wolpert, S. and Micheli-Tzanakou, E., Silicon models of lateral inhibition, *IEEE Trans. Neural Net.*, 4(6), 955, November 1993.

17. Andreou, A. G., Boahen, K. A., Pouliquen, P. O., Pavasovic, A., Jenkins, R. E., and Strohbehn, K., Current-mode subthreshold MOS circuits for analog VLSI neural systems, *IEEE Trans. Neural Sys.*, 2(2), March 1991.

18. Mead, C. A., *Analog VLSI and Neural Systems*, Addison-Wesley, New York, 1989.

19. Liu, W., Andreou, A. G., and Goldstein, M. H., Voiced-speech representation by an analog silicon model of the auditory periphery, *IEEE Trans. Neural Net.*, 3(3), May 1992.

14 Speaker Identification through Wavelet Multiresolution Decomposition and ALOPEX

Francis Phan and Evangelia Micheli-Tzanakou

14.1 INTRODUCTION

Speech intelligibility is most often corrupted by a noisy environment, which profoundly affects the hearing impaired. Hearing impairment ranks first in the United States among chronic disabilities. As communicative disorders go hand-in-hand with hearing impairment, there is an impetus to enhance speech intelligibility. Electronics miniaturization, along with digital signal processing techniques, has made speech enhancement more viable through digital-based hearing aids.[1]

Development of digital hearing aids for the hearing impaired has seen exciting growth over the last decade, due to the ongoing miniaturization and increasing power of microprocessor electronics. Digital hearing aids offer many significant advantages over older conventional analog hearing aids, in that they are programmable in nature.[2] This flexibility allows for the implementation of various signal processing techniques to enhance the intelligibility of speech or a particular sound source of interest. Computer simulation as a means of development for digital hearing aids facilitates design and is especially suitable for the evaluation of signal processing techniques. When developing working models of the digital hearing aid, the physiological properties of the ear canal and ear drum can be considered as well.[3]

The three primary operations of the digital hearing aid are amplification, filtering, and output limiting. Most research applied to the digital hearing aid lies within filtering in which two sources of noise have been delineated, 1) noise attributed to feedback and 2) noise attributed to the listening environment. An inherent problem with the structural design of conventional hearing aids is that of acoustic and electronic feedback, due to the microphone-receiver proximity and the acoustical dynamics of the ear canal. Feedback in the hearing aid degrades overall signal-to-noise ratio and thus limits the maximum usable gain of the instrument. Adaptive noise cancellation of feedback implementing LMS has been researched, simulated in depth, and described in References 4, 5, 6. The primary objective of our research relates to noise in the ambient listening environment, such as with competing speakers.

The benchmark for future generation digital hearing aids can be measured by how well it assists the hearing impaired in enhancing primary sounds of interest as well as the ability to suppress unwanted noise. The *cocktail party* effect refers to the mechanism by which humans can normally distinguish a particular sound source among many.[7] The notion of being able to differentiate one source among many sound sources for humans is a natural ability that is usually taken for granted. For the hearing impaired, the ability to achieve this effect is somewhat more difficult, as amplifying the sound also translates into amplifying noise, which conventional hearing aids normally do. Typically, most hearing aids implement static filtering that may be satisfactory for deterministic-based noise but is unsuccessful in filtering nondeterministic noise that is considered to degrade the primary sound source of interest, namely competing speech. As in the cocktail party effect, there may exist the problem of being able to differentiate one speaker from another, and thus the task of defining what is the primary sound source and what is not becomes more difficult. This complication is further compounded by frequency spectral overlaps among many sound sources. As shown in the following sections, it is rather difficult to differentiate one waveform from another once they have been combined.

As communicative disorders are a common result of hearing impairment, it seems reasonable that speech makes up the primary sound of interest for the hearing impaired. Our objective would be to implement a device that shall handle the cocktail party effect as a preprocessing element to the brain. With some assistance from the user, this device should be able to adjust its signal processing adaptively to focus in on a desired speaker and at the same time suppress the noise that may affect the speaker's intelligibility. Before signal processing can be applied to the sound signal arriving at the user's ears, a method must first be developed to differentiate speech from noise. This noise can be speech from a competing speaker. Although this objective is simple to describe, its actual implementation is not trivial. This problem breaks down into a speech and speaker recognition system in the presence of noise. Although this research was inspired by digital hearing aids applications, the principles of the system presented here can be applied to any speaker and speech recognition system.

Our research to develop this cocktail party preprocessor to the brain incorporates a variety of interdisciplinary paradigms, which include signal analysis through orthogonal wavelet transforms, feature extraction through the ALOPEX optimization method, and implementation of artificial neural networks. The first technique described is multiresolution decomposition through wavelets, a signal processing method that overcomes limitations imposed on signal processing techniques such as the Short Time Fourier Transform (STFT). The features that we used in this pattern recognition system are derived from the time-frequency coefficients generated by the wavelets. These features are implemented into ALOPEX, an optimization method inspired by a biological process. The methods for this research are based on the elements previously mentioned, which are integrated to form a speaker recognition system. A discussion follows which outlines the significance and limitations of the system and an outline for future work.

14.2 MULTIRESOLUTION ANALYSIS THROUGH WAVELET DECOMPOSITION

The phonemes of speech encompass a wide variety of characteristics in both the time and frequency domains. As an example, vowels and fricatives have almost complementary characteristics in which vowels are typically lower in frequency for longer time durations, whereas fricatives have a high frequency content for short durations.[8] Frequency analysis through conventional fixed window techniques such as the STFT are fixed window resolution operators in which the time duration of the analysis is inversely proportional to the bandwidth of the filters.[9] In other words, high frequency localization results in poor time resolution as high time resolution results in poor frequency localization. In extracting features from a sampled speech waveform, it would be useful to have a means to analyze the signal from a multiresolution perspective. Another motivation to pursue a multiresolution analysis of speech is that it somewhat models the cochlear mechanism of spectral decomposition during the initial stage of sound transduction in the ear, in which a time varying signal is spatially distributed in patterns along the basilar membrane. It has been shown that the nervous system processes spatially distributed patterns more efficiently than varying temporal signals.[10]

Wavelets are based on mathematical constructs that deal with the linear expansion of a signal into contiguous frequency bands. Instead of analyzing a signal with a single fixed window, as with short-time Fourier transform techniques, wavelets enable a signal analysis with multiple window durations that would allow for a coarse to fine multiresolution perspective of the signal.[11] Wavelets were popularized in the last decade after the detailed mathematical analysis by Grossman and Morlet in 1984.[12] Since then wavelets have been applied to all facets of signal processing, such as image processing and data compression. The multiresolution analysis implemented in this research is based on the wavelet decomposition algorithm developed by Mallat.[13]

Multiresolution analysis of a signal decomposes into a hierarchical system of subspaces that are one-dimensional and are square integrable. Each resolution of the decomposition consists of a multiresolution subspace and an orthogonal subspace. These subspaces can also be respectively referred to as the *discrete approximation* and the *detail signal* at a particular resolution. Orthogonality implies that no correlation exists between subspaces of different resolutions. Each subspace is spanned by basis functions that have scaling characteristics of either dilation or compression, depending on the resolution. The implementation of these basis functions is incorporated in a recursive pyramidal algorithm in which the discrete approximation of a current resolution is convolved with quadrature mirror filters in the subsequent resolution. Quadrature mirror filters (QMFs) are a pair of filters whose frequency responses are complementary.[14] Essentially, they are high and low pass filters that define the bandwidth for a particular resolution. A particular resolution in the decomposition process can also be referred to as an octave.

Figure 14.1 displays plots for the detail signal and discrete approximation. In each plot, the original sampled waveform is shown first. Beneath the original waveform are the multiresolution decompositions represented in four octaves. Notice that for each subsequent octave there exists a down sampling by a factor of two. Due to

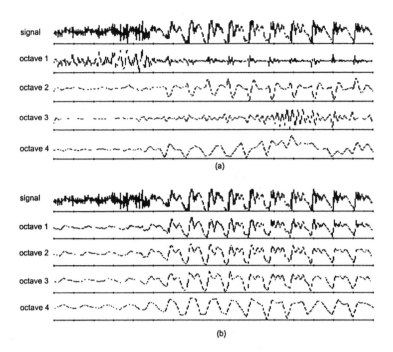

FIGURE 14.1   Wavelet representation of a speech waveform (a) signal detail, (b) discrete approximation.

convolution with the QMFs, the discrete approximations indeed appear as low pass filtering where conversely, the detail signal reflects high pass filtering.

The quadrature mirror filters implemented are based on the 23 tap FIR filter defined by Mallat[13] which we shall denote as $h(n)$, which represents the low band of the QMF. The high band of the QMF can be derived from $h(n)$ as follows:

$$g(n) = (-1)^{1-n} h(1-n) \qquad (14.1)$$

Figure 14.2 is a graph of these filters.

The original speech waveform sampled at 8 kHz is convolved with the high pass and low pass QMFs. The resulting coefficients of the filter convolutions are subsampled or decimated by a factor of two and represent the coefficients for that octave. The signal details correspond to the coefficients generated from the high pass filter convolution of $g(n)$. The signal detail coefficients will eventually be used as the feature space for this particular octave. The approximate signal corresponds to the coefficients generated from the low pass filter convolution of $h(n)$. The approximate signal coefficients are then fed in as input to the next successive stage of QMFs. Figure 14.3 is a schematic representation of the process described above.

The subsampling that proceeds the QMF convolutions provides for time dilation to a more coarse resolution. In subsampling, every other sample is dropped so that the original number of coefficients is down-sampled by a factor of two. For each

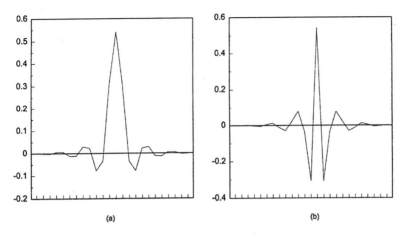

FIGURE 14.2   Plot of the (a) high pass filter coefficients and (b) low pass filter coefficients.

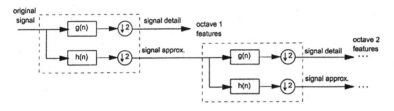

FIGURE 14.3   Successive recursive stages of multiresolution decomposition using Mallat QMFs.

successive octave, the filter lengths of the QMFs remain static; thus, the decimation of the approximated and detail signals is responsible for the time dilation.

The frequency bands on to which wavelet coefficients are projected are contiguous in nature. Thus there are no overlaps or notches in the frequency response before the Nyquist criterion. For a speech waveform sampled at 8 kHz, the frequency bandwidth that corresponds to each of four octaves are is shown in Table 14.1.

TABLE 14.1
Frequency Bandwidths for Four
Successive Octaves of a Waveform
Sampled at 8 kHz

Original Signal Sampled at 8 kHz	
First octave	2 kHz – 4 kHz
Second octave	1 kHz – 2 kHz
Third octave	500 Hz – 1 kHz
Fourth octave	250 Hz – 500 Hz

14.3   PATTERN RECOGNITION WITH ALOPEX

Paradigms in pattern recognition usually rely on templates in which a closest match or correlation is determined for a given input pattern. Artificial neural networks provide a nondeterministic means for pattern recognition and rely on algorithms to update connectivity weights between nodes of successive layers. Pattern recognition can be implemented by more intuitive approaches, such as by direct template comparisons to an input pattern and then determining the least squares error among the templates.

For the problem at hand, once the time-frequency features have been obtained for the various speech waveforms, the objective is to use these features in a pattern recognition scheme and in a manner that avoids the undesirability of local convergence. How then is it possible to devise an algorithm which effectively finds the global maxima and minima without local convergence yet in a manner that does not require inefficient scanning for the solution? The ALOPEX algorithm is a method developed to do just that and avoid these complications. The overall optimization of a function takes into consideration a global component as well as a local one. The ALOPEX paradigm incorporates a stochastic element to prevent convergence to local minima and inverts the typical pattern recognition process by generating features of a pattern as opposed to extracting them.

Through the iterative process of ALOPEX, the input pattern $X(n)$ should eventually transform into the template that has the maximal response $R(n)$ of all the templates. It is in this way that ALOPEX can be said to be a pattern generator. The template in which the input most closely converged to is determined by the template with the maximal response value.

14.4   METHODS

The speech processing of this research involving the implementation of the tools outlined earlier was implemented as software routines developed in C on a PC. The data processing windowing scheme was developed in part to memory limitations imposed on this platform. This memory restriction, however, is a typical design complication associated with embedded control instrumentation, such as with digital hearing aids, and is treated as a "real world" problem in this research. Thus, the software routines used in this research have been developed to circumvent memory restrictions, but at the cost of computational speed. As with any practical signal processing application, there is a trade off between memory and speed. However, as VLSI technology is continually developing faster microprocessors, speed in the overall perspective of this research takes a lesser precedence to memory overhead. Figure 14.4 is a block diagram of the overall system. Presently, speaker identification operates off line as data preprocessing, and links between each of the components are not automated yet.

14.4.1   DATA ACQUISITION

Subjects were asked to speak into a unidirectional microphone leaving about a 1-cm distance between the mouth and the microphone. The sampling rate and analog to

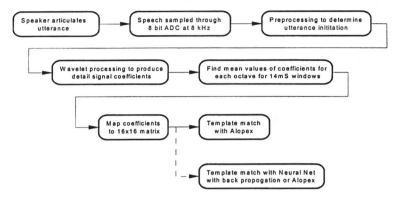

FIGURE 14.4   Block diagram of speaker identification system.

digital conversion resolution are 8 kHz at 8 bits, respectively. Although careful studies have shown that speech sounds such as vowels and fricatives are inherently not band-limited, it has been observed that high frequencies above 4 kHz fall off appreciably to that more than 40 dB below the peak spectra, which is typically between 1 and 2.5 kHz.[20] It seems reasonable then to define the Nyquist frequency for speech to be 4 kHz, which is the same for "telephone speech", enabling a minimal sampling rate of 8 kHz for intelligible speech. For this speech processing application, 8-bit AD resolution is chosen over 16-bit resolution because the added overhead required in memory does not justify the increased dynamic amplitude range.

14.4.2 DATA PREPROCESSING

Six speakers were used as subjects; three were male and three were female. The subject comprised speakers with American, Chinese, and European accents. Each speaker was asked to articulate a series of 10 words three times each. These words were chosen to represent a variety of vowel phonemes and are listed below.

TABLE 14.2
Test Utterances Articulated Three Times Each by Six Subjects and Arpabet Representation

Utterance	Vowel	Utterance	Vowel
beet	IY	hot	AA
bit	IH	bought	AO
bet	EH	foot	UH
bat	AE	boot	UW
but	AH	bird	ER

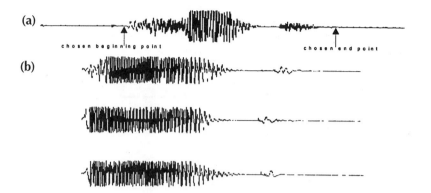

FIGURE 14.5   (a) Example of determining endpoint for an utterance of "hot". (b) Three separate utterances of the word "bird".

Each utterance is assumed to have a duration of 0.50 s, and thus the speech samples have been edited to truncate any trailing space in the utterance.

Once the waveform is on the editing screen, the beginning of the waveform is visually determined in which samples before this beginning point are deleted. In most cases, detecting the beginning of the utterance is obvious. However, for some utterances that begin with low energy fricatives, as in *foot* and *hot*, the beginning of the waveform is more ambiguous. For utterances that have beginning points that are hard to delineate, the three waveforms are approximately lined up in time, and the endpoints are determined by centering the waveform in a 0.50-s window. Thus, this preprocessing stage renders all the waveforms to a 0.50-s duration, in which the beginning of the sampled waveform corresponds to the approximate beginning of the utterance. It is also assumed that multiple utterances of a word possess similar envelope characteristics with respect to one another in time. However, in the figure shown above, different articulations of the same utterance show some variability.

14.4.3   REPRESENTING THE WAVELET COEFFICIENTS FOR TEMPLATE MATCHING

Once the input speech waveform has been preprocessed, it is made available to the wavelet transform. The signal detail coefficients, which are the time-frequency features, are generated for four octaves. For the pattern recognition input configuration, these coefficients are mapped to a vector whose length is a power of two. In the case of ALOPEX, this shall be a vector of 256 parameters. For an artificial neural network, the input would consist of 256 nodes.

Figure 14.6 illustrates how the coefficients from each of the four octaves are mapped into a matrix form. Each octave of signal detail in the wavelet decomposition is divided into 14 ms bins. Each bin contains the mean value of the coefficients that fall within that bin. Each octave has a maximum of 64 bins spanning 896 ms. Signals that are shorter than 896 ms are zero padded to that duration. For four octaves, the number of bins totals 256 and is used to form the pattern vector.

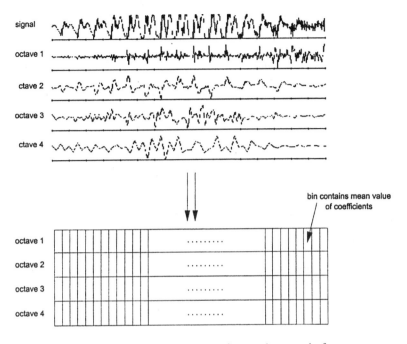

FIGURE 14.6   Deriving the 256 time-frequency features into matrix form.

As described earlier, the wavelet transform consists of convolutions of quadrature mirror filters. Because linear convolution can be computationally inefficient, frequency-based convolution was implemented instead. A problem with frequency convolution, however, is that the FFT algorithm processes its input as if it was a periodic signal resulting in side lobe artifacts at the boundaries of the signal. Another important problem is that too much memory overhead is required in processing the entire length of the waveform at one time. To circumvent these limitations, an overlapping windowing technique was devised in which the sampled waveform is processed in windows of 256 bytes with overlaps of 32 bytes, in which each 8-bit sample is stored in a byte.

Before the next window is processed, the resulting wavelet coefficients for the current window are swapped to disk. The current window that is read is processed entirely, but not all the coefficients are written to disk, as the coefficients closest to the end boundary contain the artifacts associated with the FFT. The coefficients from the next window overwrite the artifact area of the previous window. The choice of the 32-byte overlap for the signal is not an arbitrary value. A system of overlapping coefficients for the respective octaves must be taken into consideration as well. Another important consideration is that for the decimated waveform, subsequent octave lengths are not static and continually divide by a factor of two. Wavelets with orthogonal basis functions, as with this application, require that the signal processed be a power two. Thus for windowed portion of a signal it is 256 bytes long, as shown in Figure 14.7. Table 14.3 shows the byte overlap based on the frequency band and the signal size.

TABLE 14.3
Relative Frequency Band and Length Characteristics
for Wavelet Octaves

Octave	Freq. Band	Size in Bytes	Byte Overlap
Windowed waveform	Sampled at 8 kHz	256	32
First octave	2 – 4 kHz	128	16
Second octave	1 – 2 kHz	64	8
Third octave	500 Hz – 1 kHz	32	4
Fourth octave	250 Hz – 500 Hz	16	2

The overlap values are determined as a ratio of windows overlapping each other by a factor of 1/8. As described earlier, the mean values of the coefficients are taken for a window subdivided by two. To calculate the time resolution that these subdivisions represent, the actual size of the window is truncated by the overlap of the proceeding window. From the table above, the processed window can be calculated to be a length of 224 samples. When subdivided by two and considering an 8-kHz sampling rate, each subdivision represents mean values taken for a 14 ms duration. Thus for a 500-ms sample, 18 of 64 subdivisions are required to represent the waveform. The remainder of the subdivisions are zero padded. The maximum signal length that can be processed is 896 ms.

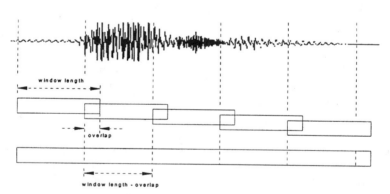

FIGURE 14.7   Processing the waveform by overlapping windows.

14.5 RESULTS

Three versions of the utterance "bat" were sampled from six speakers and were preprocessed as described earlier. One waveform of the three versions from each of the speakers was used as templates where the remaining two versions were used as test inputs. Speaker recognition for clean and corrupted input waveforms was performed using the ALOPEX template matching scheme. The cocktail party effect was simulated by digitally mixing in with the input waveform another speaker uttering the word "vision". Tests were also conducted with input waveforms corrupted with –20 dB white noise. Training data are not used as testing data for the results presented. Figure 14.8 displays plots of an input waveform before and after it was altered. The speaker recognition scheme is based upon the coefficients of the wavelet representation of the speech waveform.

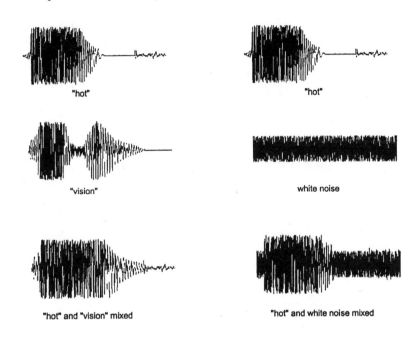

FIGURE 14.8   Input speech patterns before and after waveform altering.

Table 14.4 shows results for speaker recognition for the three test scenarios described above. The column headings are the initials of the speaker templates where row headings are initials of speaker inputs. A "thumbs up" or "thumbs down" indicates if the expected classifications were made successfully. The boxed "x" indicates misclassification. For the *no noise background* case, there was one misclassfication in which speaker MZ was misidentified as FP. The other five speakers were correctly identified. For the –20 dB white noise case, misclassfication occurred for three of the six speakers. For the cocktail party effect in which the word *vision* was mixed in with the *bat* utterance, the results were identical to that of the no noise test case.

A more rigorous application of the cocktail party effect was simulated for speaker recognition between two speakers simultaneously articulating the same utterance. The utterance *bat* from speaker FP was mixed in with the *bat* utterances of the speaker test data set. For this situation, the speaker recognition system correctly classified three of the five speakers.

Speech recognition was tested for one speaker articulating three words. In all, three utterances for each word were sampled. For a particular word, one utterance

TABLE 14.4
Case 1. Speaker Recognition Against Competing Noise and Speakers

	No noise background						−20 dB white noise						Cocktail Party Effect					
	FP	MZ	KL	BL	YA	DZ	FP	MZ	KL	BL	YA	DZ	FP	MZ	KL	BL	YA	DZ
FP	◊	–	–	–	–	–	◊	–	–	–	–	–	◊	–	–	–	–	–
MZ	☒	♪	–	–	–	–	–	♪	–	–	–	☒	☒	♪	–	–	–	–
KL	–	–	◊	–	–	–	–	☒	♪	–	–	–	–	–	◊	–	–	–
BL	–	–	–	◊	–	–	–	–	–	◊	–	–	–	–	–	◊	–	–
YA	–	–	–	–	◊	–	–	–	–	–	♪	☒	–	–	–	–	◊	–
DZ	–	–	–	–	–	◊	–	–	–	–	–	◊	–	–	–	–	–	◊

TABLE 14.5
Speaker Recognition for Simultaneous Utterance of "bat" Between Two Speakers

	MZ	KL	BL	YA	DZ
MZ	♪	–	–	–	☒
KL	–	◊	–	–	–
BL	–	–	◊	–	–
YA	–	–	–	◊	–
DZ	–	–	–	☒	♪

was trained as a template, while the two other utterances were used as test inputs. Training data were not used as testing data. The methods of preprocessing and wavelet representation are the same as with the speaker recognition system. Table 14.5 column headings represent the template utterances as the row headings represent the test utterances. The number suffix appended to the listed utterance is intended to differentiate multiple articulations of the same utterance. Three methods of pattern recognition were implemented: ALOPEX template matching (see Chapters 12 and 17), artificial neural network using an ALOPEX training algorithm, and artificial neural network using a backpropagation learning algorithm. The neural network topology used in each neural network implementation consisted of 256 nodes in the input, 10 nodes in the hidden layer, and 3 nodes in the output layer. Each node of the output layer represents one of the three templates (speakers).

The utterances used have similar time alignment but possess variable prosodics. *Beet2* is slightly higher in pitch as compared to *beet3*. Template matching with ALOPEX correctly identified the utterance where the neural networks failed to classify. The ANN with ALOPEX converged halfway with all the templates where the ANN with backpropagation misclassified. *Boot3* has an exaggerated "t" stop consonant sound, which was misclassified by ALOPEX and the ANN with back-propagation but correctly identified with ALOPEX template matching. *Bought*, which sounded similar for all utterances, was correctly classified by each paradigm.

TABLE 14.6
Interspeaker Speech Recognition for Three Words of Two Utterances Each

	ALOPEX			ANN & ALOPEX			ANN & Backprop		
	beet1	beet1	bought1	beet1	boot1	bought1	beet1	boot1	bought1
beet2	◆	–	–	◆.5	.5	.5	◆.13	☒.89	.01
beet3	◆	–	–	◆1.0	0.0	0.0	◆.89	.09	.04
boot2	–	◆	–	0.0	◆1.0	0.0	.63	◆.23	.30
boot3	–	◆	☒	.19	◆1.0	0.0	☒.89	◆.09	.04
bought2	–	–	◆	0.0	.01	◆1.0	.04	.01	◆.98
bought3	–	–	◆	0.0	0.0	◆1.0	.04	.01	◆.98

For the neural network topology used, ALOPEX as a training algorithm showed stronger convergence as compared to backpropagation, as shown in Figure 14.9.

14.6 DISCUSSION

The results that have been presented in our study show promise of using wavelet representation of discrete short speech utterances for speaker recognition in the presence of noise or a competing speaker for a cocktail party effect. The current implementation of our system, however, is somewhat limited considering the small number of templates used in the pattern recognition schemes. However, as the template and vocabulary size increase, so does computational and memory overhead required for the pattern recognition paradigms. Thus, the primary focus was in establishing the wavelet representation of speech for time-frequency tokens in a multispeaker environment.

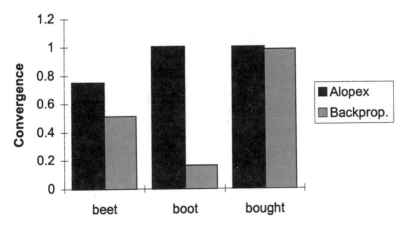

FIGURE 14.9   Average convergence comparisons for ALOPEX and backpropagation training algorithms.

Although a certain amount of speaker cooperation is assumed, variability in multiple speech samples is expected to exist. Since wavelets produce frequency features localized in time, preprocessing can be greatly enhanced though a time alignment technique such as Dynamic Time Warping. Prosody of an utterance, however, provides more of a spectral variability. Thus, as discussed earlier, wavelets provide a multiresolution analysis of a signal that Short Time Fourier Transforms cannot provide. Low frequency features associated with vowels and high frequency features of consonants can be distinguished more readily with multiple filter lengths as opposed to a static one.[16]

The subsampling characteristic of wavelet processing for multiple octaves and use of QMFs have made it possible for wavelets to be used in data compression and reconstruction. In association with the feature mapping algorithm implemented in this study, the actual amount of data used for speaker recognition represents a *90% data reduction* from the original sample size. This is significant in that it minimizes memory and computational overhead for the pattern recognition algorithms.

The ALOPEX template-matching paradigm was demonstrated to be fairly robust to white noise or multiple speaker corruption. This technique works to generate the features of the template to which it converges, in this case the wavelet coefficients of the speech waveform. Future work would involve resynthesizing the speech signal without its noise components based on the ALOPEX-generated features.

The speech input is taken from a monaural sound source. Speaker recognition for the cocktail party effect in the system is based on time-frequency characteristics of speech stored as templates and does not rely on binaural phase information. However, sound localization is an important factor in the cocktail party effect and has been addressed with techniques in adaptive beamforming.

Due to computational overhead, the implementation of the cocktail party pre-processor into a digital hearing aid is presently unlikely. However, a feasible method of speaker identification in the presence of competing noise has been demonstrated, which is a complication to which all speaker and speech recognition systems are susceptible.

ACKNOWLEDGMENTS

It is with great pleasure that the authors acknowledge the financial support and trust of the Albert and Ethel Herzstein Foundation, Houston, Texas. The authors also thank the speakers in this study for their participation.

REFERENCES

1. Levitt, H. L., Speech processing aids for the deaf: an overview, *IEEE Trans. Audio Electroacoust.*, 21, 269, 1973.
2. Levitt, H. L., Digital hearing aids: a tutorial review, *J. Rehabilitation Res. Dev.*, 24, 7, 1987.
3. Kates, J. M., A time-domain digital simulation of hearing aid response, *J. Rehabilitation Res.*, 27, 279, 1990.

4. Kates, J. M., Feedback cancellation in hearing aids: results from a computer simulation, *IEEE Trans. Signal Proc.*, 39, 553, 1991.

5. Chabries, D. M., Christiansen, R. W., and Brey, R. H., Application of the LMS filter to improve speech communication in the presence of noise, *IEEE ICASSP*, 148, 1982.

6. Widrow, B., Grover, J. R., and McCool, J. M., Adaptive noise canceling: principles and applications, *Proc. IEEE*, 63, 1692, 1975.

7. Mitchell, O. M., Ross, C. A., and Yates, G. H., Signal processing for a cocktail party effect, *J. Acoust. Soc. Am.*, 50, 656, 1971.

8. Deller, J. R., Proakis, J. G., and Hansen, J. H. L., *Discrete-Time Processing of Speech Signals*, Macmillan, New York, 1993.

9. Cody, M. A., The fast wavelet transform, *Dr. Dobb's J.*, 16, April 1992.

10. Gulick, W. L., Gescheider, G. A., and Frisna, R. D., *Hearing: Physiological Acoustics, Neural Coding, and Psychoacoustics*, Oxford University Press, 1989.

11. Rioul, O. and Vetterli, M., Wavelets and signal processing, *IEEE Signal Proc. Mag.*, 14, October 1991.

12. Grossmann, A. and Morlet, J., Decomposition of hardy functions into square integrable wavelets of constant shape, *SIAM J. Math. Anal.*, 15, 723, 1984.

13. Mallat, S. G., A theory of multiresolution signal decomposition: the wavelet representation, *IEEE Trans. Patt. Anal. Machine Intell.*, 11, 674, 1989.

14. Rabiner, L. R. and Juang, B. H., *Fundamentals of Speech Recognition*, Prentice Hall, Englewood Cliffs, 1993.

15. Micheli-Tzanakou, E., Non-linear characteristics in the frog's visual system, *Biol. Cybern.*, 51, 53, 1984.

16. Tzanakou, E. and Harth, E., Determination of visual receptive fields by stochastic methods, *Biophys. J.*, 15, 42a, 1973.

17. Tzanakou, E., Michalak, R., and Harth, E., The ALOPEX process: visual receptive fields with response feedback, *Biol. Cybern.*, 35, 161, 1979.

18. Deutch, S. and Micheli-Tzanakou, E., *Neuroelectric Systems*, NYU Press, New York, 1987.

19. Dasey, T. J. and Micheli-Tzanakou, E., A pattern recognition application of the ALOPEX process using hexagonal arrays, *Int. Joint Conf. Neural Net.*, 2, 119, 1989.

20. Rabiner, L. R. and Schafer, R. W., *Digital Signal Processing of Speech Signals*, Prentice Hall, Englewood Cliffs, 1978.

15 Face Recognition in Alzheimer's Disease: A Simulation

Evangelia Micheli-Tzanakou

15.1 INTRODUCTION

"Visual agnosia" is an impairment of the nervous system in recognizing visual stimuli such as faces, words, and other objects. These manifestations occur although the subjects might exhibit signs of intact basic perceptual performance in discriminating brightness. The defect of not recognizing faces is also called prosopagnosia, and it appears to be exceedingly impaired, relative to the impairment in recognizing other types of objects. Cases like this have been reported in the literature.[1] There are yet other cases where the recognition of certain objects other than faces is also defected. One such category is animals.[2] There the recognition of the class animals was intact but the species was totally or almost totally mistaken. The same is true for plants,[3] as well as buildings and monuments[3-4] and clothing articles.[5-6] Damasio et al.[5-6] also exhibited some evidence that food is another category yet. Feinberg et al.[7] and McCarthy and Warington[8] have reported cases with object recognition difficulties. Yet the level of face recognition was reported as satisfactory. A number of other studies have suggested the presence of face-selective cells in the temporal cortex of the monkey.[9-11] The response of these cells was reported to be invariant to rotation and partial occlusion of the face. However, cells were not significantly or consistently responsive to line-drawn faces. It has also been reported that some of the face-selective cells were not facial-feature selective.

These studies motivated us to develop a neural network that would recognize faces in a similar fashion. We used commonly available feature extraction techniques for image compression to isolate "features" from a face. In addition, the aging process, both normal and abnormal, was simulated by adding noise to the weights of a trained network in order to simulate memory decay with age.

15.2 METHODS

The gray scale images of two individuals, a male and a female, were acquired using a JVC camcorder connected to a TARGA board inside a PC. Different versions of the faces were obtained: full smile, no smile, partially occluded profiles, etc. The video frames were converted into eight 16×16 digital images per person using standard data compression techniques.

A circular mask was used in order to eliminate backgrounds from the pictures. The mask sets all pixel intensities outside a given radius (user selected) to zero. The number of inputs to the neural network depends on the number of features selected by the different methods.

Two different training algorithms are used, backpropagation (BP) and ALOPEX (both described in detail elsewhere in this book). In the paragraphs that follow, we will describe the methods used for extraction of features.

Feature extraction was achieved by different methods, such as moments, edges, wavelet and Fourier coefficients (F-CORE), all described in Chapter 4.

The masked images were convolved with a 3×3 Prewitt operator to extract edge information from these images. The edge image was created by thresholding the convolved image. If the pixel value of the convolved image was less than the set threshold, then it was set to zero; otherwise it was set to 255.

The wavelet transform provides a multiresolution representation of signals or images and has been used extensively in the last few years, both in signal and image processing.[13–14] The orthogonal wavelet transform can be composed of a set of quadrature mirror filters. In this study 256 wavelet coefficients were obtained from each image, but only the eight highest in amplitude were used as inputs to the neural network.

The same approach was taken with F-CORE, which is a Fourier-based method.[15] After finding the Fourier spectrum of an image, the method sorts the frequency coefficients from max to min. The user can select a percentage of these coefficients in decreasing order. The selected highest in magnitude coefficients are used as inputs to the neural network. This way most of the energy of an image is used.

15.3 RESULTS

Once the features from the different methods are obtained, they are used as inputs to the neural network, and both BP and ALOPEX are used for comparison. The two methods exhibited a large difference in the number of iterations needed for training. BP reached the maximum number of iterations (30,000) allowed by the program before reaching the optimal convergence. At that point the program is stopped. ALOPEX, on the other hand, converged at approximately 4,000 iterations in the worst case. In the testing mode the results are comparable to the training level for each method of training.

Figure 15.1 shows curves of the system response vs. iterations for the selected features from each of the methods. We tested the effect of noise added to the inputs with different standard deviations of the Gaussian noise. As it is shown in Figure 15.2, the response of the network decreased exponentially as the noise level on the inputs increased. In this case no noise was added to the weights. That means that we assumed there is no damage to the "brain." In Figure 15.3, the percent of recognition vs. standard deviation of the noise applied on the weights (assumed to be damage to the connectivity of neurons in the "brain") is shown. No noise was added to the inputs. Notice that the response drops exponentially with increasing

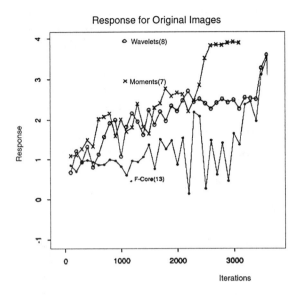

FIGURE 15.1 Features selected using different methods required different numbers of iteration to converge to 99% recognition during training.

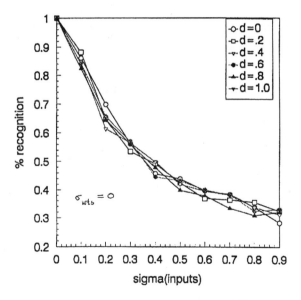

FIGURE 15.2 The response of the network decays exponentially as the noise level on the inputs increases. No noise applied to the weights.

standard deviation of the Gaussian noise applied on the weights. Also notice that the greater the sigma, the lower the starting point of recognition (y-axis intercept). This should be compared with Figure 15.2, where all responses start from 100% and deteriorate as the inputs decay. Finally, Figure 15.4 shows the effects of noise (or damage) to the "brain" as well as on the inputs. Responses decayed exponentially with increasing weight decay and increased damage to the input image. The increased noise to the input image, simulates the changes of a face due to aging.

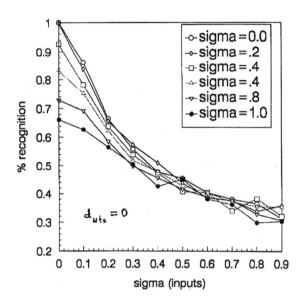

FIGURE 15.3   The response drops exponentially with increasing noise component. Noise applied to the weights.

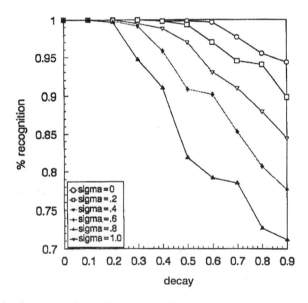

FIGURE 15.4   Response decayed exponentially with increasing neuron decay. Noise applied to the inputs.

15.4 DISCUSSION

Responses of the network to different images were tested under a variety of conditions that systematically validated the robustness of the neural network. Features from faces (looking straight ahead, rotated 45°, profiles, and edge images) were used as inputs to the network. The network was trained mainly on the frontal views of the face, to capture the most important features. The network was tested with various noisy faces as well. This was done to compare the recognition of highly degraded images by networks trained on noiseless or low noise images. In another effort to test the robustness of the network and to determine the extent to which responses to parts of the faces could account for a response to the entire face, parts of the faces were occluded.

We simulated the natural aging by using a decaying mechanism on the weights, and the Alzheimer's abrupt changes by using a Gaussian noise added to the weights. The robustness of the neural network was tested by adding noise to the images. This way a comparison could be made between the simulated normal aging and Alzheimer's disease.

The results indicate that the percent recognition and number of iterations needed are closely related to the choice of features. The ALOPEX algorithm had a much faster convergence as compared to the BP method. Wavelet coefficients and moments served as good feature extractors. Rotation and occlusion of parts of the face did not decrease the recognition to any appreciable degree. The response decreased exponentially though, with an increasing noise on the input images as well as with an increase in simulated decay of neuronal function, thus indicating that similar symptoms may be observed under different pathological conditions and different causes.

REFERENCES

1. DeRenzi, E., Current issues in prospagnosia, in *Aspects of Face Processing*, Ellis, H. D., Jeeves, M.A., Newcombe, F., and Young, A., Eds., Martinis Nijhoff, Dordrecht.

2. Shuttleworth, E. C., Syring, V. I., and Allen, N., Further observations on the nature of prospagnosia, *Brain Cognition*, 1, 302, 1982.

3. Gomori, A. J. and Hawryluk, G. A., Visual agnosia without alexia, *Neurologica*, 34, 947, 1984.

4. Assal, G., Favre, C., and Anderes, J., Nonrecognition of familiar animals by a farmer: zooagnosia or prosopagnosia for animals, *Rev. Neurolog.*, 140, 580, 1984.

5. Damasio, A. R., Damasio, H., and Van Hoesen, G. W., Prosopagnosia: anatomic basis and behavioural mechanisms, *Neurology*, 32, 331, 1982.

6. Desimone, R., Albright, T. D., Gross, C. D., and Bruce, C., Stimulus-selective responses of inferior temporal neurons in the macaque, *J. Neurosci.*, 4, 2051, 1984.

7. Feinberg, T. E., Gonzalez-Rothi, L. J., and Heilman, K. M., Multimodal agnosia after unilateral left hemisphere lesion, *Neurology*, 36, 864, 1986.

8. McCarthy, R. A. and Warrington, E. K., Visual associative agnosia: a clinico-anatomical study of a single case, *J. Neurol. Neurosurg. Psych.*, 49, 1233, 1986.

9. Perret, D. I., Rolls, E. T., and Cann, W., Visual neurons responsive to faces in the monkey temporal cortex, *Exp. Brain Res.*, 47, 329, 1982.

10. Desimone, R., Face-selective cells in the temporal cortex of monkeys, *J. Cogn. Neurosci.*, vol. 3, 1991.

11. Desimone, R., Albright, T.D., Gross, C.G., and Bruce, C., Stimulus-selective properties of inferior temporal neurons in the macaque, *J. Neurosci.*, 4(8), 2051.

12. Hu, M. K., Visual pattern recognition with moment invariants, *IRE Trans. Inf. Theor.*, 8, 173, 1962.

13. Mallat, S., A theory for multiresolution signal decomposition: the wavelet representation, *IEEE Trans. Patt. Anal. Mach. Intell.*, 11(7), 674. 1989.

14. Weiss, L. G., Wavelets and wideband correlation processing, *IEEE Signal Proc. Mag.*, 13, 1994.

15. Micheli-Tzanakou, E. and Binge, G. M., F-CORE: a Fourier-based image compression and reconstruction technique, *SPIE Proc. Visual Comm. Image Proc. IV*, 1199, 1563, 1989.

16. Tzanakou, E. and Harth, E., Determination of visual receptive fields by stochastic methods, *Biophys. J.*, 14(42a), 1973.

17. Harth, E. and Micheli-Tzanakou, E., ALOPEX: a stochastic method for determining visual receptive fields, *Vis. Res.*, 14, 1475, 1974.

16 Self-Learning Layered Neural Network

Faiq A. Fazal and Evangelia Micheli-Tzanakou

16.1 INTRODUCTION

The basic task of any pattern recognition system is to decide on the class membership of the current input pattern to the system. One approach is to make use of decision functions, if the input pattern has *n* items Euclidean space. Consider, for example, the two-dimensional cases depicted in Figure 16.1. We note that in Figure 16.1a the input patterns can be put into two different cases, *c1* and *c2*, and a linear decision function *d1* exists such that for any pattern, *p, d1(p)>0* if *p* belongs to *c1* and *d1(p)<0* if *p* belongs to *c2*. Figure 16.1b shows a more complicated case of clustering, which requires three decision functions to establish a pattern's membership.

For more involved classification schemes, one may have to turn to a nonlinear decision surface. For example, Figure 16.1c shows pattern classes separated by a circle. A detailed and mathematically rigorous discussion on this topic can be found in Tou and Gonzalez.[1] It is apparent that the success of this scheme depends on two factors: (a) the form of the decision function and (b) the ability to determine its coefficients.

Often decision functions are not prewired into pattern classifiers but heuristically develop as the classifier experiences input patterns during the training period. This is referred to as *clustering*. Several methods of clustering exist[1] and have found a variety of applications.[2,9] For example, the first of the input patterns during the training period forms a class of its own and becomes the initial prototype for the class. If the second pattern is *similar* to the first pattern, it is put in the first class, and the prototype for the class is adjusted so that the difference between it and the two patterns in the class is minimized. If the second pattern is not similar, then it forms a new class of its own and becomes the initial prototype for that class. This process is repeated for each of the patterns, forming a new class only if the pattern does not match the prototypes of the existing classes. Several measures exist for similarity, which include finding the following two: 1) the minimum distance between the prototype and the pattern and 2) the dot product between the prototype and the pattern. A pattern belongs to a class represented by the prototype, if this measure is less than or exceeds a specified threshold, respectively.

Numerous questions can be asked regarding the quality of the clustering mechanism, such as

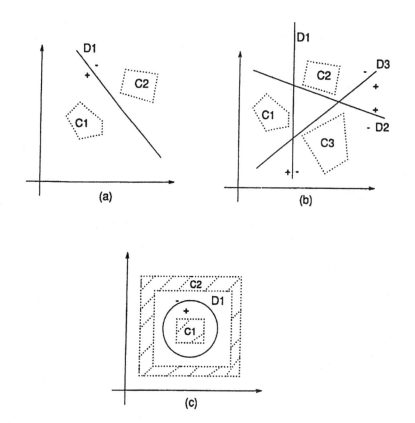

FIGURE 16.1 Different types of clustering: (a) Simple separation of classes, (b)more complex clustering, (c) clusters separated by a circle.

- How distinct or redundant are the prototypes for the various classes? For this estimation, the dot product between prototype pairs could be used.
- How much of an "overlap" exists between the classes?
- How correlated are the samples within each class? This can serve as measure of the selectivity of the clustering procedure.
- Are there lots of prototypes with very few samples in them? This reflects on the sensitivity of the clustering mechanism to noise.
- Is the clustering mechanism dependent on the order in which the patterns are applied?
- Is the clustering mechanism dependent on the rate at which clusters are formed?

In this chapter, we study the pattern clustering performance of a well-known neural network (NN) model, namely, the layered NN (Neocognitron) of Fukushima.[3] This model is a self-organizing (implying unsupervised learning) classifier of input patterns, which is capable of tolerating shifts in position and a certain degree of deformity of the input pattern. The following section reviews the neocognitron model in terms of the classical pattern recognition techniques without getting into the details. This will help provide insight into the underlying mechanisms of the neocognitron, suggest quantitative measures for its performance, and encourage experimentation with techniques not discussed by Fukushima.[3] A simplified version of the neocognitron is also described in Deutsch and Micheli-Tzanakou.[4]

Note that the present study excludes aspects of the neocognitron which deal with tolerance to deformity and shifts in position. This is because the underlying mechanisms for the functionality are not necessary to or explicitly integrated into the more difficult task of unsupervised pattern classification. In addition, the functionality is hard wired and does not involve learning.

16.2 NEOCOGNITRON AND PATTERN CLASSIFICATION

Figure 16.2 conceptualizes the pattern classification model embodied in the neocognitron. The first thing we note is the *distribution* of the *decision functions* involved in pattern classification. Instead of having a set of decision functions which operate over the entire input field, the neocognitron architecture distributes the decision mechanism over several levels. The decision functions at the first level work over very small portions of the input representation and, accordingly, decide over the existence of low-level features in the various parts of the input field. Thus, given the pixel-input representation of Figure 16.3a, the first level decision function may collectively map it into a representation involving corners and line-segments, as depicted in Figure 16.3b. The mapped representation, which must preserve the *spatial relationship* of the higher-level features, now serves as input to the next level of decision functions. This level, in turn, produces a topographic map of the primary input in terms of more complex features. The process continues to the top-most level whose decision functions collectively decide on the correct classification of the entire input to the network. Thus, in Figure 16.3c the top-most level of a two-level network will put the topographic map of Figure 16.3b into a class which could be labeled A.

This approach to classification can be viewed as a *divide-and-conquer* technique to solving the problem. However, this distribution implies that at the lower level decisions are made on a local basis without taking the entire picture into account. In the presence of noise in the input pattern this scheme may not perform as well as a single-level scheme would have done. This is also true even when such distributed schemes take top-down expectations into account. After all, top-down expectation can only be initiated after receiving some initial evidence from bottom-up. Also, in image reconstruction through top-down excitation, this scheme may cause convergence problems.

The basic decision mechanism is the same at all levels. Each level has a set of decision functions which work in *parallel* and in *competition* in order to decide on the features present in the different sections of the input-field. This process is conceptualized in Figure 16.4.

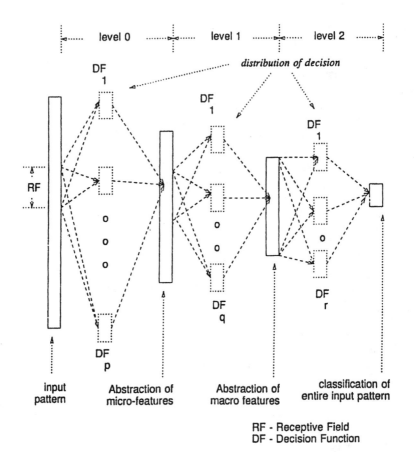

Figure 16.2 Pattern classification model embodied in Neocognitron. It includes three levels, each containing a decision function (DF).

The decision functions for level l work parallel on their respective receptive field in order to decide which feature is present in it. The decision functions also compete with one another to decide on the *winning* feature for each section (also referred to as *competition area*). The decision for each competition section collectively forms the input to the next level of decision functions. It should be noted that the output of each decision function, which is shown as a single point in Figure 16.4, is typically represented by the states of a collection of units in the actual neural model.

In the neocognitron the decision functions at each level are implemented by the a vector associated with each plane in the feature-detecting layer (known as the S-layer) of that level. The number of planes in each S-layer thus places an upper limit on the number of classifications that can be made at the corresponding level. Each unit of a given S-plane attempts to decide if the prototype feature

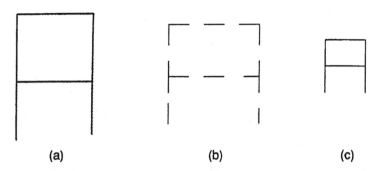

(a)	(b)	(c)

FIGURE 16.3 Input representation of a pattern to each level of the Neocognitron in Figure 16.2. (a) Pixel input (level 0), (b) first level representation of a decision function involving corners and line segments (level1), (c) classification stage (level 2).

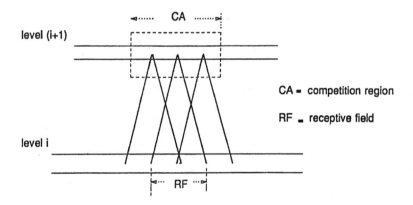

FIGURE 16.4 Parallel and competitive execution of decision functions.

represented by the plane's a vector is presented in the unit's receptive field in the previous layer.

Mathematically, this decision is specified by the following discriminant:

$$df = r\Phi\left[\frac{\sum_i a_i u_i - kb\sqrt{\sum_i c_i u_i^2}}{1 + kb\sqrt{\sum_i c_i u_i^2}}\right]$$

where

$$\Phi(n) = \begin{bmatrix} 0, \, x \le 0 \\ x, \, x \ge 0 \end{bmatrix}$$

and

$$k = \frac{r}{r+1}; \; (r = Inhibition \; Factor)$$

While the vector a is the same for each of the units in a plane, vector u, which represents the feature currently present in a unit's receptive field, may be different for each unit. Vector c and scalar b are used to compute the average excitation in a unit's receptive field.

From Equation 16.1 it can be seen that df is a decision function of the quadratic form, based on the equations below:

$$\sum_i a_i u_i > kb \sqrt{\sum_i c_i u_i^2}$$

$$\left(\sum_i a_i u_i \right)^2 - k^2 b^2 \left(\sum_i c_i u_i^2 \right) > 0$$

In the neocognitron the values of a and b are developed during the training period, which c is a constant vector associated with each plane. All planes at a given level have the same c. In the neocognitron, c follows an exponentially decreasing function over the receptive field with the constraint that

$$\sum_i c_i = 1$$

The learning (or clustering) mechanism of the neocognitron can be described by the procedure given below.

16.2.1 Training Algorithm

1. apply the next training pattern
2. for each level in the network perform the following, bottom-up for each section in the input to this level:
   a. determine the plane whose a vector has the closest match with the feature contained in this section,
   b. update a and b as follows:

$$\begin{bmatrix} \Delta a_i = qc_iu_i \\ \Delta b = qv \end{bmatrix}$$

where q is the learning rate and v is the average inhibitory excitation computed u and c.

16.3 OBJECTIVES

The objectives of the simulation experiments include an attempt to understand the issues about the clustering mechanism that were raised in the introduction. Specifically, we are interested in the dependence of these issues on the form and the parameters of Equation 16.1 and Equation 16.4. The varied parameters are the following:

- inhibition-factor, r, from Equation 16.1
- learning-rate, q, from Equation 16.4
- the form of vector c, i.e., exponentially decreasing vs. uniform
- the initial values for vector a, i.e., random vs. primed
- thresholding the selection of winning units in the competition area with a *threshold factor*, ϑ. Experimentation has shown that such threshold can reduce the development of noisy or redundant features. Essentially, only those units whose activation exceeds $\vartheta*$ *average-activation* are selected for a weight update.

In relation to the distributed nature of clustering and the fact that the neocognitron has a prefixed limit on the number of clusters that can be formed at each level, the following issues are also investigated:

- How should the inhibition-factor and learning rate vary from one level to another?
- How many applications of the training patterns are necessary for learning to develop?
- Does it help to intermix the patterns from the different classes?

16.4 METHODS

The applied input stimulus consisted of an array of pixels whose values are set to 0 or 1 in order to create different types of patterns. A facility provided by the simulation environment allowed creation of noisy patterns from the originals. The noise introduced by this facility is random. However, the user can control the Hamming distance* of the noisy pattern from the original one. In the two studies reported in the next section, the mix of the original (i.e. non-noisy) pattern to noisy

---

* If the pixels composing a pattern are viewed as elements of a vector, then the Hamming distance between two patterns is equal to the number of pixels in which they differ.

patterns was 2:1:1, where the three numbers refer to the proportion of original, 1-Hamming-noisy and 2-Hamming-noisy patterns, respectively.

In these simulations, there exist several fixed and variable parameters as listed below.

The *fixed* parameters are the number of levels, layers, planes, and units, and the size of the receptive field. The *variable* parameters are the learning rate, q, the inhibition factor, r, and the form of the vector c.

Two cases are considered for vector c, namely, a uniform distribution of connection weights and an exponential distribution of connection weights. In addition, the initial value for vector a assigned to each of the learning planes is a variable parameter. For a, two cases of initial values are considered, namely, (a) *random* assignment of the weights and (b) primed assignments, mixed with random assignments. *Primed* assignment of the initial value to an a vector gives it a slight bias to certain types of patterns in its receptive field. The pattern types included horizontal, vertical, and diagonal lines.

Another variable parameter is ϑ, the threshold factor used in deciding the winning unit in each competition area.

The *network performance* is evaluated in terms of its capability to *learn* and *recall* after learning is over. *Learning* is evaluated in terms of (a) the number of planes used at each level and (b) how orthogonal are the a vectors for the planes that are used in learning. The *dot-product* between the a vectors is used for this purpose. After learning is completed for each level, the dot-product between the a vectors is computed, along with the minimum, average, and maximum values.

In the results presented here (Figures 16.6 through 16.10 and 16.14 through 16.16), the learned a vector for each *plane* is shown in a two-dimensional grid format to make obvious the correspondence between the vector and the feature that it detects. Each number within the grid represents the relative sensitivity of the learned pattern to excitation occurring at the location of the number. For clarity, zero sensitivity is represented by blank space. *Recall* is evaluated in terms of the activation states of the units at the various levels.

Two different studies, *Study A* and *Study B,* were performed. The results and their implications are presented next.

16.5   STUDY A

Study A was made with a two-level network having an input level and a recognition level.

16.5.1   NETWORK DESCRIPTION

The table below summarizes the description of the network used in this study. The entire *NA* stands for Not Applicable.

The input patterns consist of a 5×5 array of pixels. Figure 16.5 shows the patterns that are used in Study A. If we consider patterns 5 and 6 as being similar to 4 and 1, respectively, then we expect the network to cluster them into four classes. This should result in four planes being used in level 1.

TABLE 16.1
Description of Network Used in Study A

	Level 0	Level 1	
	layer INP	layer Vc	layer S
# of planes	1	1	10
Units per plane	5×5	1	1
Receptive field size	NA	5×5	5×5

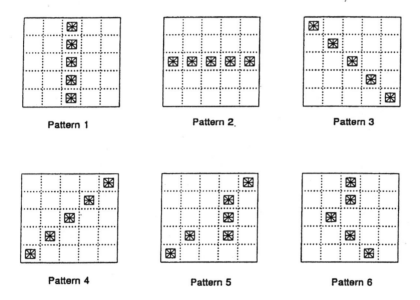

Pattern 1	Pattern 2	Pattern 3

Pattern 4	Pattern 5	Pattern 6

FIGURE 16.5   Input patterns for Study A.

16.5.2   RESULTS FROM STUDY A

Several observations are made, and are listed below:

A1 *The larger the inhibition factor, the more discriminatory is the clustering process which, in turn, results in larger number of clusters.* This can be seen in Figures 16.6a, 16.7a, and 16.8a for different inhibition factors. These figures show that there is an increase in the number of planes used up in the clustering process as the inhibition factor increases.

A2 *The greater discrimination resulting from increased inhibition may cause the development of redundant planes.* This can be seen by comparing the connections in Figures 16.6a, 16.7a, and 16.8a. For example, in Figure 16.8a, which shows the results for the highest inhibition, we see similar patterns (4 and 7, 2 and 6) represented by different planes. However, in Figure 16.6a, which corresponds to the least inhibition, only four planes develop. This is about the number that we would expect based on a human inspection of the stimulus shown in Figure 16.5. This redundancy is also reflected in the increase of the dot-product of the a vectors of the developed planes, as seen in Figures 16.6b, 16.7b, and 16.8b.

A3 *Decreasing the learning rate has little effect (actually, negative, if any) on the results for the one level classification attempted in Study A.* This can be inferred by comparing the results in Figure 16.8 with those in Figure 16.9. In the latter the learning rate was decreased to 0.75, and the number of vectors increased appropriately so that the same amount of learning occurred. As a result of a decrease in the learning rate, q, we note that the number of clusters increased by 1 and the dot-product also increased slightly.

A4 *Using the exponential form of c only seems to worsen the classification* in the one level case. This can be seen by comparing the results shown in Figure 16.6 with those shown in Figure 16.10. They differ in the form of vector c. The exponential form of c results in fewer clusters than are actually required. For the exponential case, we also note an increase in the value of the dot-products between clusters.

A5 *A high inhibition factor causes a sharp drop in the response of a feature detecting plane, even with a single missing element in the feature. This could be a problem if the input consists of a macro-feature which contains many instances of this feature.* Figure 16.11 illustrates this phenomenon for a plane detecting a diagonal line. With a high I inhibition-factor (in this case equal to 4), the decrease in activation is significantly greater than with a low inhibition factor (in this case equal to 1). Figure 16.12 illustrates the problem this could cause in the network's response to a macro-feature containing contiguous instances of a micro-feature (feature 1). With a high inhibition factor, the one missing pixel may cause the activations of units at level $l + 1$ to decrease significantly, which would prevent the feature's detection at that level.

16.6 STUDY B

This study was made with a three-level network: an input level, a micro-feature recognition level, and a level recognizing the total input pattern. Table 16.2 summarizes the description of the network used in this study. The entry *NA* stands for Not Applicable. The input patterns consisted of a 9×9 array of pixels. The patterns are shown in Figure 16.13.

16.6.1 RESULTS FROM STUDY B

Several observations are made, and are listed below:

TABLE 16.2
Description of Network used in Study B

	level 0 layer INP	level 1 layer Vc	level 1 layer S	level 2 layer Vc	level 2 layer S
# of planes	1	1	12	1	12
Units per plane	9×9	7×7	7×7	1	1
Receptive field size	NA	3×3	3×3	7×7	7×7
Competition area	NA	NA	2×2	NA	3×3

B1 *With appropriate values for the inhibition factor and the learning rate, the neocognitron seems to extract appropriate micro-features at the first level, which are then used in recognizing the different letters at the second level.* This observation is substantiated by noticing that the a vectors developed for level 1 (Figure 16.14a) correspond to features that are apparent through visual inspection of the letters (Figure 16.13). Also, after learning has occurred, each letter is associated with a response from only one plane in level 2 (Figure 16.14c). The responding plane is unique to that letter, thus signifying the recognition of the input letters by the trained network. For the experiment on which this observation is based, the variable parameters are set as follows:
   a. for pass 1, in which the network is trained with four instances of each pattern:
      • the inhibition factor is set at 5 for level 1, and 8 for level 2,
      • the learning rate is set for 0.5 for level 1, and 1 for level 2.
   b. for pass 2, in which the network is trained with seven instances of each pattern:
      • the inhibition factor is set to 5 for level 1 and 8 for level 2,
      • the learning rate is set to 2.0 for level 1 and 9 for level 2.
B2 *The high inhibition factor required to distinguish between letter A and R resulted in the network being very sensitive to missing features in the input.*
B3 *Lowering the inhibition factor to reduce this sensitivity resulted in failure to distinguish between A and R.*
B4 *Not thresholding the selection of winning features in each competition area resulted in the development of redundant micro features.* This is evidenced by the development of connections for plane 8 in level 1, as shown in Figure 16.15a.
B5 *Primed, instead of totally random initialization of a resulted in a better clustering at level 1.* This is evidenced by comparing the connection tables for level 1 in Figure 16.14a and Figure 16.16a (the unprimed case). We note that the unprimed case resulted in more clusters, with increased dot-product between them. However, this does not seem to affect the capacity

of the network to distinguish between the letters, as evidenced by the distinct activation of the planes in level 2 (Figure 16.16c).

16.7 SUMMARY AND DISCUSSION

The neocognitron is analyzed in terms of classical pattern recognition techniques. Its ability to recognize characters is demonstrated through simulations. Useful observations are made about the performance of this task. The most critical factors for the process appear to be the selection of the learning rate and the inhibition factor. The neocognitron seems to provide a viable approach for optical character recognition. Several copies of the type of network used in study B could be used in parallel to recognize items like zip-codes or social security numbers.

Future work will consider the hardware implementation of this type of neocognitron. The local decision functions at each level could be implemented with simple processors with a small amount of local memory that could store the values of a, c, b and other parameters. Finally, it would be interesting to perform studies of the type reported in this chapter on some of the other (References 5–8) pattern clustering approaches that are based on neuro-computing paradigms.

PLANE 0 (5×5)

49				11
	49			
		48		11
	11	49		
			49	

PLANE 1 — empty

PLANE 2 (5×5)

11		86		
		73	11	
11	48	48		
		99		
		48	48	

PLANE 3 — empty

PLANE 4 — empty

PLANE 5 — empty

PLANE 6 — empty

PLANE 7 (5×5)

11		11		98
				98
		49	48	
11	86	11	49	
98	11			

PLANE 8 (5×5)

.	11			
48	36	49	49	49
			11	

PLANE 9 — first cell: .

PLANE 10 — empty

PLANE 11 — empty

(a)

PLANE_NO	0	1	2	3	4	5	6	7	8	9
0	1.00		0.21					0.29	0.31	
1										
2			1.00					0.16	0.31	
3										
4										
5										
6										
7								1.00	0.23	
8									1.00	
9										

(b)

** minimum = 0.16 maximum = 0.31 average = 0.25

FIGURE 16.6   Results from Study A with INPUT_STIMULUS: stimulus 1, IF: 1.0, LR: 3.0, type of **c**: UNIFORM, initialization of **a**: RANDOM. (a) CONNECTIONS for Level 1 plane. (b) DOT_PRODUCT of CONNECTIONS after LEARNING.

PLANE 0					**PLANE 1**					**PLANE 2**					**PLANE 3**			
56				13	13				13	13	56						41	
56							13		41	13						41	
	56	13			13				56			13	56				
	13	56							56							56	
		56		13					56								56

PLANE 4					**PLANE 5**					**PLANE 6**					**PLANE 7**			
																13	98
																	98
																41	56
														13	98	13	56
														99	13		

PLANE 8					**PLANE 9**					**PLANE 10**					**PLANE 11**			
	13																
56	41	56	56	56														
		13															

(a)

PLANE_NO	0	1	2	3	4	5	6	7	8	9
0	1.00	0.45	0.31	0.06				0.26	0.31	
1		1.00	0.32	0.01				0.70	0.21	
2			1.00	0.52				0.20	0.27	
3				1.00				0.06	0.27	
4										
5										
6										
7								1.00	0.22	
8									1.00	
9										

(b)

** minimum = 0.01 maximum = 0.70 average = 0.28

FIGURE 16.7   Results from Study A with INPUT_STIMULUS: stimulus 1, IF: 2.0, LR: 3.0, type of **c**: UNIFORM, initialization of **a**: RANDOM. (a) CONNECTIONS for Level 1 plane. (b) DOT_PRODUCT of CONNECTIONS after LEARNING.

PLANE 0

99				24
	98			
		98	24	
		24	98	
			99	98 24

PLANE 1

		24		98
				99
				98
		98	24	98
98	24			

PLANE 2

74			
73	23		
73			
74			
73			

PLANE 3

		49	
		74	
73			
73			
		74	

PLANE 4

23				24
			24	
	24			
24				

PLANE 5

PLANE 6

24	24		
23			
24		24	73
24		74	

PLANE 7

				73
				73
			74	

PLANE 8

	24					
98	73	98	98	98	24	24
			24			

PLANE 9

		24	
24	24		
		24	
		24	

PLANE 10

PLANE 11

(a)

PLANE_NO	0	1	2	3	4	5	6	7	8	9
0	1.00	0.24	0.25	0.06	0.45		0.44	0.25	0.31	0.06
1		1.00	0.15	0.08	0.54		0.10	0.71	0.20	0.10
2			1.00	0.56	0.27		0.79	0.26	0.26	0.40
3				1.00	0.01		0.36	0.00	0.20	0.78
4					1.00		0.41	0.79	0.21	0.01
5										
6							1.00	0.21	0.27	0.41
7								1.00	0.21	0.01
8									1.00	0.42
9										1.00

(b)

** minimum = 0.00 maximum = 0.79 average = 0.30

FIGURE 16.8 Results from Study A with INPUT_STIMULUS: stimulus 1, IF: 4.0, LR: 3.0, type of **c**: UNIFORM, initialization of **a**: RANDOM. (a) CONNECTIONS for Level 1 plane. (b) DOT_PRODUCT of CONNECTIONS after LEARNING.

Supervised and Unsupervised Pattern Recognition

PLANE 0

86	5	5	5
	86	5	
5	86		5
	5	86	
			80

PLANE 1

5		11	5	92
				99
5		98	5	
5	98	11	98	
98	5	5		

PLANE 2

5	92	5	
5	86	11	5
	92		
5	92	5	
	92		

PLANE 3

86		11
92	5	5
86	11	
92	5	
5	92	

PLANE 4

5			5
			5
		5	
5			

PLANE 5

11	11	5	11	11
				11
		5		

PLANE 6

5	5		
	5		
	5		
	5		

PLANE 7

5		5	86
	11	86	
86			5
5	86		
86	5		5

PLANE 8

5		5		5
				5
86	80	86	86	86
5				
5			5	

PLANE 9

	5	
5	5	
	5	
		5

PLANE 10

PLANE 11

(a)

PLANE_NO	0	1	2	3	4	5	6	7	8	9
0	1.00	0.23	0.28	0.08	0.45	0.14	0.44	0.24	0.27	0.07
1		1.00	0.11	0.10	0.57	0.22	0.12	0.72	0.24	0.10
2			1.00	0.62	0.26	0.11	0.81	0.27	0.22	0.42
3				1.00	0.10	0.31	0.45	0.11	0.23	0.80
4					1.00	0.13	0.42	0.81	0.27	0.05
5						1.00	0.14	0.13	0.85	0.50
6							1.00	0.23	0.26	0.44
7								1.00	0.25	0.05
8									1.00	0.42
9										1.00

(b)

** minimum = 0.05 maximum = 0.85 average = 0.31

FIGURE 16.9 Results from Study A with INPUT_STIMULUS: stimulus 1, IF: 4.0, LR: 0.75, type of **c**: UNIFORM, initialization of **a**: RANDOM. (a) CONNECTIONS for Level 1 plane. (b) DOT_PRODUCT of CONNECTIONS after LEARNING.

(a)

PLANE 0 — (empty grid)

PLANE 1

			5	
			16	
	5	1	5	
1				

PLANE 2

	12
	29
14	37
	33
5	2

PLANE 3 — (empty grid)

PLANE 4 — (empty grid)

PLANE 5 — (empty grid)

PLANE 6 — (empty grid)

PLANE 7 — (empty grid)

PLANE 8

1			1	
5	1	6		
5	13	99	14	6
5	1	5		
1				

PLANE 9 — (empty grid)

PLANE 10 — (empty grid)

PLANE 11 — (empty grid)

(a)

(b)

PLANE_NO	0	1	2	3	4	5	6	7	8	9
0										
1		1.00	0.08	0.55					0.20	
2			1.00	0.18					0.64	
3				1.00					0.23	
4										
5										
6										
7										
8									1.00	
9										

(b)

** minimum = 0.08 maximum = 0.64 average = 0.31

FIGURE 16.10 Results from Study A with INPUT_STIMULUS: stimulus 1, IF: 1.0, LR: 3.0, type of **c**: EXPONENTIAL, initialization of **a**: RANDOM. (a) CONNECTIONS for Level 1 plane. (b) DOT_PRODUCT of CONNECTIONS after LEARNING.

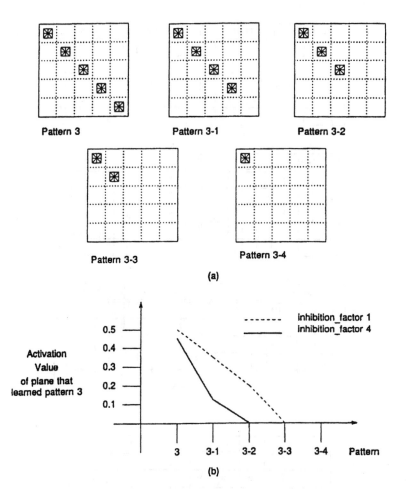

FIGURE 16.11   Neocognitron's sensitivity to distortion in learned features. (a) An example pattern (Pattern 3) with distortions (3-1 to 3-4). (b) Activation as a function of distortion.

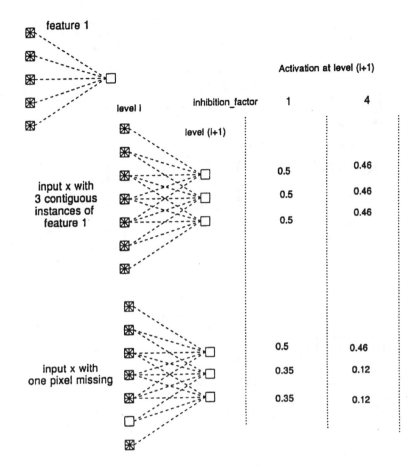

FIGURE 16.12   Adverse effect of high inhibition.

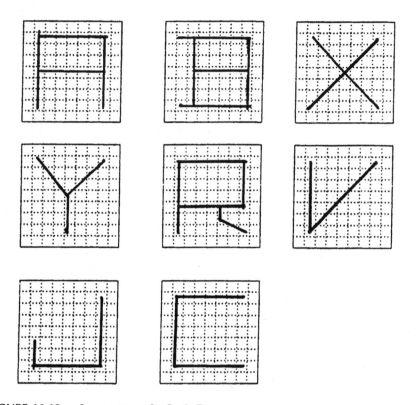

FIGURE 16.13   Input patterns for Study B.

PLANE 0			PLANE 1			PLANE 2			PLANE 3			PLANE 4			PLANE 5		
				99		3	1	4	49								48
1	1	1		99	67	4				49			48			48	
		1		99			4	4			49			48			

PLANE 6			PLANE 7			PLANE 8			PLANE 9			PLANE 10			PLANE 11		
32	32	32			65	71			48	48	49				60		
					63	71				48		77	77	77	60	60	
			65	65	65	71				49					29		

(a)

PLANE_NO	0	1	2	3	4	5	6	7	8	9	10	11
0	1.00	0.45	0.53	0.56		0.40	0.11	0.52	0.29	0.52	0.85	0.55
1		1.00	0.34	0.31		0.31	0.31	0.40	0.00	0.40	0.52	0.75
2			1.00	0.75		0.78	0.60	0.61	0.49	0.66	0.26	0.35
3				1.00		0.33	0.33	0.26	0.33	0.52	0.33	0.32
4												
5						1.00	0.33	0.52	0.33	0.26	0.33	0.32
6							1.00	0.26	0.33	0.77	0.00	0.32
7								1.00	0.26	0.60	0.26	0.13
8									1.00	0.26	0.33	0.32
9										1.00	0.26	0.25
10											1.00	0.64
11												1.00

(b)

** minimum = 0.00 maximum = 0.85 average = 0.40

	PLANES										
PATTERNS	0	1	2	3	4	5	6	7	8	9	10
A				.31							
B		.27									
X					.18						
Y						.20					
R										.30	
V	.21										
J											.15
C					.34						

(c)

FIGURE 16.14 Results from Study B. (a) CONNECTIONS for Level 1 learning. (b) DOT_PRODUCT of CONNECTIONS for Level 1. (c) ACTIVATION of PLANES in Level 2.

PLANE 0		PLANE 1			PLANE 2			PLANE 3			PLANE 4			PLANE 5		
29	98	50	38	50			58	27	15	27	42	42	42			8
	99						56	27					42	68	68	68
	98				58	58	58	13		27			42			

PLANE 6			PLANE 7			PLANE 8			PLANE 9			PLANE 10			PLANE 11	
70			15		15						42	58	58	56		37
70	70	70	13			42			42	27		58				24
42			15			42				42		58				37

(a)

PLANE_NO	0	1	2	3	4	5	6	7	8	9	10	11
0	1.00	0.38	0.26	0.50	0.33	0.33	0.36	0.64	0.40	0.82	0.33	0.00
1		1.00	0.28	0.71	0.77	0.05	0.30	0.64	0.00	0.26	0.77	0.40
2			1.00	0.53	0.60	0.29	0.34	0.46	0.32	0.26	0.40	0.76
3				1.00	0.75	0.31	0.52	0.69	0.00	0.32	0.64	0.60
4					1.00	0.29	0.43	0.46	0.00	0.24	0.60	0.76
5						1.00	0.83	0.30	0.41	0.52	0.30	0.30
6							1.00	0.46	0.34	0.44	0.56	0.20
7								1.00	0.37	0.44	0.46	0.33
8									1.00	0.76	0.32	0.00
9										1.00	0.50	0.00
10											1.00	0.28
11												1.00

(b)

** minimum = 0.00   maximum = 0.83   average = 0.41

PATTERNS	PLANES										
	0	1	2	3	4	5	6	7	8	9	10
A					.38						
B								.41			
X							.19				
Y						.12					
R				.42							
V									.12		
J		.26									
C							.44				

(c)

FIGURE 16.15   Results from Study B referred to in observation. (a) CONNECTIONS for planes in Level 1 after learning. (b) DOT_PRODUCT of CONNECTIONS for Level 1. (c) ACTIVATION of PLANES in Level 2.

PLANE 0		PLANE 1			PLANE 2			PLANE 3		PLANE 4			PLANE 5	
28	98	43	43	43			56	1		1	41	41	41	
	99						56	1			41	41		
	98				56	56	56	1		1	41		41	

PLANE 6			PLANE 7			PLANE 8			PLANE 9		PLANE 10			PLANE 11
69			15					41	26		57	57	56	41
69	69	70	15		15	41			26	25	57			41
41			15	15		41			26		57			41

(a)

PLANE_NO	0	1	2	3	4	5	6	7	8	9	10	11
0	1.00	0.43	0.26	0.33	0.33	0.40	0.35	0.33	0.33	0.85	0.33	0.43
1		1.00	0.26	0.49	0.77	0.00	0.28	0.26	0.33	0.29	0.77	0.33
2			1.00	0.62	0.60	0.32	0.34	0.62	0.52	0.23	0.40	0.26
3				1.00	0.59	0.00	0.54	0.41	0.80	0.23	0.58	0.76
4					1.00	0.00	0.43	0.42	0.26	0.23	0.60	0.52
5						1.00	0.34	0.63	0.00	0.71	0.32	0.00
6							1.00	0.77	0.44	0.48	0.56	0.55
7								1.00	0.26	0.45	0.60	0.29
8									1.00	0.28	0.51	0.33
9										1.00	0.45	0.28
10											1.00	0.26
11												1.00

(b)

** minimum = 0.00 maximum = 0.85 average = 0.41

PATTERNS	PLANES										
	0	1	2	3	4	5	6	7	8	9	10
A					.45						
B								.45			
X			.36								
Y	.32										
R						.34					
V											.28
J							.17				
C			.34								

(c)

FIGURE 16.16   Results from Study B referred to in observation. (a) CONNECTIONS for planes in Level 1 after learning. (b) DOT_PRODUCT of CONNECTIONS for Level 1. (c) ACTIVATION of PLANES in Level 2.

REFERENCES

1. Tou, J. T. and Gonzalez, R. C., Pattern Recognition Principles, Addison Wesley, Reading, MA, 1974.
2. Micheli-Tzanakou, E., Visual receptive fields and clustering, *Behav. Res.-Meth. Instrument.*, 15(6) 553, 1983.
3. Fukushima, K., Neocognitron: a new algorithm for pattern recognition tolerant of deformations and shifts in position, *Patt. Recogn.*, 15(6), 455, 1982.
4. Deutsch, S. and Micheli-Tzanakou, E., *Neuroelectric Systems*, New York Univ. Press, New York, 1987.
5. Kohonen, T., *Self-Organization and Associative Memory*, Springer-Verlag, Berlin, Heidelberg, New York, Tokyo, 1984.
6. Widrow, B. and Winter, R., Neural nets for adaptive filtering and adaptive pattern recognition, *IEEE Comput.*, 21(3), 25, 1988.
7. Carpenter, G. A. and Grossberg, S., A massively parallel architecture for a self-organizing neural pattern recognition machine, *Comput. Vis. Graph. Image Proc.*, 37, 54, 1987.
8. Linsker, R., Self-organization in a perceptual network, *IEEE Comput.*, 21(3), 105, 1988.
9. Chon, T.-S. and Micheli-Tzanakou, E., Pattern and Feature Extraction, *Proc. IASTED — Int. Symp. Machine Learning and Neural Networks*, 1990, 14.

17 Biological and Machine Vision

Evangelia Micheli-Tzanakou
and Raymond Iezzi, Jr.

17.1 INTRODUCTION

Feature detectors, as well as feature generators, have been the subject of many papers. Since Receptive Fields (RFs) are considered to be the trigger features of cells; they can be considered to be critical *tokens*, which are representative of a stimulus. Stimuli are presented in a variety of ways. The RFs can be determined as the sum of points in the visual space where the *optimal stimulus* excites them all at a rate larger than a specific threshold. This threshold is larger than the cell's spontaneous activity. This way an image can be segmented into a large number of RFs of neighboring cells. More sophisticated methods have also been used, such as the reverse correlation method[8] and the ALOPEX method[14] that uses the cell's response as a feedback in order to find the optimum two-dimensional pattern of its RF. The ALOPEX method reverses the pattern recognition process and makes a pattern extractor become a pattern generator. Marcella[9] and Daugman,[5] simultaneously proposed a model for simple cortical RFs. Their hypothesis was tested experimentally and was found to be valid for every simple cell tested.[12]

The basic assumption for the model was that the response of a simple cortical cell is strongly localized in both the space and frequency domains. In the space domain, such a localization defines the RF profile of a cell, while in the frequency domain, it defines the cell's spatial frequency tuning curve. Therefore, the representation of images in the visual cortex involves features from both domains. Furthermore, it suggests a mathematical description where the product of the *localization* in these two domains is minimum.

In this chapter we propose a neural network approach to the notions described above, and we attempt to map RFs in the human cortex using the optimization technique ALOPEX.

17.2 DISTRIBUTED REPRESENTATION

A representation in which the features can be used for an effective further processing of information and in which they can be represented by combinations of activities of elements rather than the activities of neurons that are sensitive to these features is called a *distributed representation*. Local mechanisms can estimate how often features (and any existing combinations of them) and any reinforcements are available or are taking place. One can imagine the reinforcement signal being available at all elements that carry information about a feature having occurred.

17.3 THE MODEL

Most of vision problems can be formulated as an optimization of some cost function. Some of them are intrinsically global types of computation, like template matching or pattern recognition. Others, like the computation of intrinsic images of a scene[11] or image reconstruction,[6] can be restricted to local computations. The first kind requires processes with global support, while the second kind can be performed by processes with local support.[1] The problem with global processes is that they require a vast number of connections to even start approaching the goal. With local processes, a small number of connections is needed, and the number of iterations usually does not depend on the input array.

In this model, we use local processes to solve visual tasks which require a global kind of computation. Any optimization problem requires a search over the configuration space for an optimal set of variables to be found. The cost function depends on those variables. Convergence can be improved by parallel processing.[4] One can divide the process of searching for an optimal solution into several phases. We might start with a global optimum in some very coarse scale and then keep refining the scale until a more precise solution is found. This method, known as the *pyramidal processing,* has been extensively used in the past.[13] The often-encountered problem with this approach is how to combine the results from the different levels of the pyramid. In order to prevent such a problem from occurring, we introduce a novel method where all processes work on the same data set in each scale.

The neural network used is based on the same method used in the past.[10] It consists of an input array where the data are presented and several levels of optimizing processes. Every process computes its local cost function. The processes of the upper levels have larger RFs than those of the lower levels. The determination of the RF size is problem dependent and must be done heuristically (Figure 17.1). The RFs of the processes at any level are overlapping to allow for spreading of a "*temperature*". The temperature is defined as the reciprocal of the response strength. In addition to overlapping, any process is laterally connected to its nearest neighbors at its own level of the pyramid. All these connections are mutually inhibitory. This is necessary in order to ensure temperature transfers from "warmer" to "colder" regions. The competition is gradual rather than winner-takes-all (WTA).

This type of neural architecture is shown in Figure 17.2 and is based on a method presented in Marsic and Micheli-Tzanakou.[10]

Every process (A_k) computes its local cost function. In addition to overlapping, any process is laterally connected with its nearest neighbors at its own pyramid level.

17.4 A MODIFIED ALOPEX ALGORITHM

The calculation of the response R_k (cost function) is problem specific. Examples of cost functions for low-level vision tasks are given in Reference 11, while an example

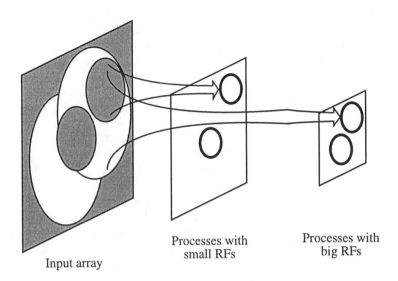

FIGURE 17.1   Pyramidal structure of the neural network. The optimization processes are organized into levels according to their RF sizes.

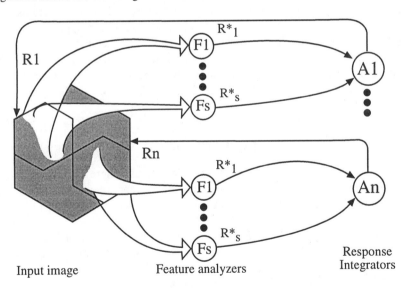

FIGURE 17.2   Neural network architecture. Each feature cluster consists of s feature analyzers F_i. The response integrators A_k compute the overall response R_k of the cluster k.

of template matching is given in the next section. The goal is to maximize the response R_k for each optimization process. Each process is running on its own data subset. These subsets are overlapping. One iteration in the process consists of (a) calculation of all responses for all processes using problem-specific cost function equations and (b) updating the corresponding subset. Updating any variable in the nth iteration is computed as in the original ALOPEX algorithms.[14]

The degree of the mutual inhibition among the response integrators of the same size RFs is computed in the following way. The response integrator A_k with the lowest response should afford the biggest change. Accordingly, it should inhibit its neighbors by the greatest amount. This is done by the following equation:

$$x_i(n) = x_i(n-1) + \Delta x_i^k(n-1) / M^k(n-1) \qquad (17.1)$$

where $M^k(n)$ is the amount of inhibition of the response integrator A_k. The $M^k(n)$ is calculated in the following way: in the beginning, we set it equal to 1 for all A_k. If some of them have responses greater than their neighbors for some threshold value, its $M^k(n)$ is increased by one. So, the highest level of inhibition of a pixel is equal to the number of its closest neighbors. At equilibrium, all response integrators will have inhibition levels equal to one, and this serves as a *stopping condition*.

Setting different thresholds for the different RF sizes generates a pyramidal processing. We are always checking whether all processes have the same response. Therefore, the value of the threshold determines the coarseness of the scale on which we search for an optimum at the particular level of the pyramid. If a large threshold is used, all processes will soon achieve an equilibrium, and processing at that scale is terminated. Conversely, a low threshold will initiate a very precise search for an optimum. All levels of the pyramid are synchronized so that processes with the largest RFs have to achieve an equilibrium, in order to allow for processes with smaller RFs to commence their work. Only when processes at an upper level whose RFs overlap with the RF of a lower level process are inactivated is a particular lower level process activated. However, higher order processes continually check for an equilibrium, and if it is not there any more, they are immediately activated, and lower level processes inactivated. This continues until all levels of the pyramid are at equilibrium.

17.5 APPLICATION TO TEMPLATE MATCHING

In this application we used hexagonal arrays as inputs. These hexagonal lattices contain the image that should converge to one of the stored templates. Each optimizing process consists of two layers of "neurons", as shown in Figure 17.2.

1. *Feature analyzers (FAs)*, F_j, with RFs connected to the input array for a feature cluster. Every feature cluster has as many neurons (feature analyzers) as the number of learned templates.
2. *Response integrators (RIs)*, A_k, compute local cost functions by combining the responses of all FAs within the feature cluster. Lateral inhibitory connections in between processes are implemented at this layer.

The response of an FA is computed as a combination of pseudo-χ^2 expressions for each feature F_j within a cluster of features:

$$R_j^*(n) = \Sigma_I \left[x_i(n) - F_{ij} \right]^2 \qquad (17.2)$$

The FA responses are then combined in a nonlinear way to calculate the response (cost function) of the whole feature cluster k:

$$R_k(n) = \Sigma_j \left(1 / R_j^*(n)\right)^2 / \Sigma_j \left(1 / R_j^*(n)\right) \qquad (17.3)$$

The goal here is to get the input image to converge to one of the templates. The number of hexels (hexagonal pixels) in such a hexagonal image (i.e., in one of the RFs) is $H = 3N^2 - 3N + 1$, where N is the number of hexels per side. Uniformly distributed noise in the range of 0–30 is added to the input image. The global ALOPEX needs about 3500 iterations to erode the excessive line, whereas the distributed one, with 19 RFs of $N = 9$, needs 1350–1750 iterations (Figure 17.3), depending on the parameter γ. Introducing the next level of the pyramid with smaller RFs reduces the number of iterations to less than 200. This depends on the choice of the parameters γ and σ and the thresholds of both levels of the pyramid.

Although a global ALOPEX takes much less than 146 parallel ALOPEXes on a serial computer, since, most of the time, most of them are inactive and convergence takes much fewer iterations, the whole task takes less time overall than the global ALOPEX. Evidently, implementation on a parallel computer will result in a much faster process. The presented architecture of the network can be considered as a pyramid or processing cone,[13] the main difference being that there is no difficulty combining results from the different layers, since all levels work on the same data. Another advantage is the stopping condition that forces the process to stop when equilibrium is reached everywhere. It may be applied to problems where the combining of state variables into a cost function is done in a "homogeneous" way, and where all local cost functions have approximately the same optimal value.

In all cases we have three memorized or learned templates consisting of the hexagonal representation of the numbers "EIGHT", "NINE", and a damaged "SIX", while the input image is a complete "SIX". The upper left line of the input pattern "SIX" should disappear upon convergence, as is clearly the case shown in Figure 17.3.

17.6 BRAIN TO COMPUTER LINK

17.6.1 GLOBAL RECEPTIVE FIELDS IN THE HUMAN VISUAL SYSTEM

Since a very large portion of vision can be based on the properties of the RF profiles, it would be very interesting to find a way to study human cortical RFs (visual or auditory). Since we cannot perform single unit recordings in the human brain, another noninvasive method should be applied. As already has been mentioned, ALOPEX was originally used as a RF mapping technique. The underlying assumption was based on the fact that a cell would fire most rigorously when the stimuli best match its RF's spatial characteristics. Therefore, by presenting a stimulus and then modifying it using the cell's response as a guide, one can eventually converge to a pattern that is a global RF. The so-found RF has a very high correlation with

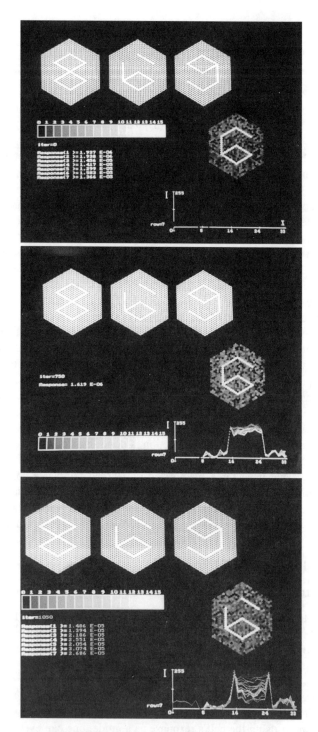

FIGURE 17.3 The ALOPEX process at work. After 1750 iterations the complete "SIX" in the input image has converged to the learned one (damaged "SIX").

the RF mapped with the classical methods and yet it is more specific in defining the spatial arrangement of the emerging pattern pixel intensities. For humans, instead of the response of a single cell, we can measure an aggregate response of thousands of cells located in the visual cortex. Smaller contributions from other areas are contributing as well. The measured signal in this case is the visual evoked potential (VEP). The problem that arises is what to use as a response function. One would like a response that is maximized when ordered patterns are used as stimuli and is minimized when random patterns are used. In a series of experiments carried out in our laboratory, it was found that, when a sequence of equi-luminant patterns is presented and the only variable that changes is the amount of order in successive patterns, then there exists a large correlation between the change in the N_2 peak of the VEP waveform and the bi-directional transition from disordered patterns to ordered ones. Furthermore, if a biasing orthogonal to the sequence pattern is presented as an OFF pattern, the previous measure changes direction completely.[2,7] These results show that there is a way to perform an ALOPEX process on humans, by establishing a BRAIN-TO-COMPUTER LINK, in order to find an optimum pattern for any human visual system. If the N_2 peak characteristic (amplitude or latency) change and/or changes in the neighboring peaks are used as response feedback, then it is possible that the process will eventually converge to an ordered pattern. This pattern can then be assumed to be a global human RF.

17.6.2 THE BLACK BOX APPROACH

A Fourier analysis of the stimulus patterns can also be done. Fifteen patterns that ranged from totally random converging to a bar were used as stimuli. When these patterns were analyzed with a two-dimensional Fast Fourier transform (2D-FFT), they showed a gradual sweep of spatial frequencies in the very low spatial frequency range (0.2–0.5 cycles per degree). Using a modified perceptron neural network as shown in Figure 17.4, two-dimensional spectra were correlated to VEP amplitude responses using a novel blackbox approach.

This perceptron has one input layer with as many input nodes as the number in the power spectrum. Each input node receives one and only one input, corresponding to one point from the power spectrum. An activation function given by

$$Y_j(n) = \sum_j X_{ij} * W_i(n) \qquad (17.4)$$

is applied to the single output node. As noted from Equation 17.4, this is a linear function. For each stimulus image, a power spectrum plot was generated that contained 128 frequency points; therefore, 128 input nodes were used, no hidden layers, and only one output node. Subscript i indexes each one of these frequencies in a given power spectrum, while j identifies which template in the set of 15 patterns used as stimuli is presented to the network. The idea then is that given a power spectrum, indexed by $j = 1,...m$, Y_j should output the VEP voltage amplitude, corresponding to the stimulus image, j. The weights W_i, when optimized correctly, should provide a single global solution for all input spectra, X_{ij}, and all amplitudes,

Y_j. Notice that there is no index j for the weights. This *single set* of weights must provide the simultaneous solution *for all input power spectra and all output VEP amplitudes*. In providing the simultaneous solution for all input spectra and all output amplitudes, the neural network weights actually describe the spatial frequency tuning sensitivity curve for the human visual system under study, for the set of patterns and responses studied. A schematic representation of what is described above is given in Figure 17.5.

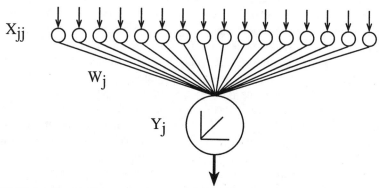

FIGURE 17.4   The black box approach of the ALOPEX process for the human brain.

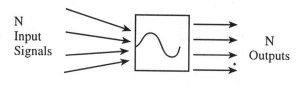

Feedforward Mode

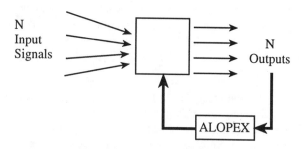

Supervised Learning Mode

FIGURE 17.5   An example of how an input sinusoidal function is reproduced by ALOPEX using spatial frequencies.

Using this technique, it is possible to predict which spatial frequencies of the stimuli were most important in producing any response trends noted. Those spatial frequencies that enhance the VEP response should have strong positive weights and vice versa. Network training was performed, using ALOPEX. Figure 17.6 demonstrates the use of the ALOPEX black box solution technique in a real time ALOPEX-VEP experiment where the subject was instructed to concentrate on mentally producing an ordered pattern, starting with a totally random pattern of intensities. As the process is iterative, in each iteration the subject gets closer to the order pattern that the subject tries to produce. On the left of the figure, we see the computed spatial frequency sensitivity tuning curves for P_{100}, N_2, and P_{100}/N_2 peak-to-peak amplitudes.

These curves, when convolved with the power spectrum of an input image, will produce the images corresponding to VEP response amplitude, shown to the right of the figure. For comparison, the desired outputs, which are the physiologically measured VEP amplitudes, are superimposed. It is important to notice that the fundamental spatial frequency for all of these ALOPEX images can be computed on the basis of their intrinsic check size. This results in a fundamental spatial frequency of 0.35cpd, the same as that found from the peak in the figures on the left. Most importantly, we also see that there is another strong peak approximately at 5cpd. Another such case is shown in Figure 17.7. Similar results were obtained from all other subjects tested in this study. Adjacent to these peaks, strong inhibitory regions were also observed. It is critical to note that these spatial frequency tuning curves are *not* pattern dependent, since the images produced by these subjects were totally different for each subject—each one had their own preferred pattern evolution. They resulted from independent, subject-specific, ALOPEX *feedback optimization of stimulus and brain response*. As such, the patterns generated by each subject are totally different and unique for each subject. Nevertheless, they all support the concept of facilitatory and inhibitory spatial frequency channels and the fact that stronger VEP amplitudes are associated with stronger more narrow energy peaks in the 2-D FFT. This technique, therefore, allows us to study human visual RFs, as well as facilitatory and inhibitory interactions in the human visual system.

17.7 DISCUSSION

The ALOPEX optimization techniques described in this work have allowed us to examine spatial frequency analysis systems intrinsic to the biological brain, by either using a visual evoked potential or other types of evoked responses for different types of stimuli.

The premise of the black box technique was that the visual system analyzes spatial frequency information, and that the VEP amplitudes are the result of some complex membership function, which weighs the response for each spatial frequency. That was the case as well presented in Figure 17.2, with the feature analyzers and the response integrators. The single weighting function corresponds to the input weights of a single node filter or neural spectrum for any given ALOPEX image (consisting of 128 frequency spindles, or points), to a single VEP output voltage that corresponds to either the P100 amplitude, or the N2 amplitude, or the P100/N2

Spatial-Contrast Analysis for Subject #1, MV

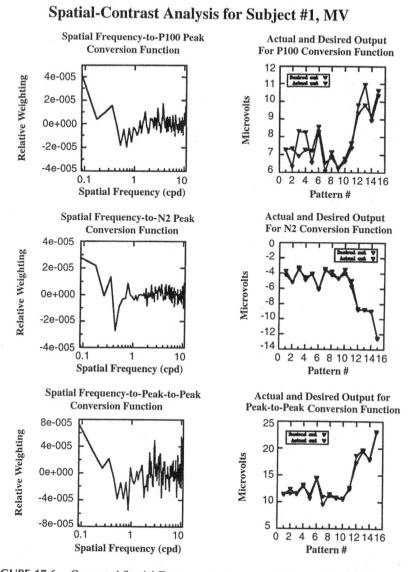

FIGURE 17.6   Computed Spatial Frequency tuning curves for an on-line ALOPEX experiment (subject MV).

peak-to-peak amplitude. This amplitude was recorded in response to that ALOPEX image when presented as a stimulus. In order to arrive to a single global solution for the node input weights, the entire set of images FFTs for each image FFT in a set was used, and a VEP voltage was computed and compared to the actual VEP amplitude measured from the subject. An error term was computed by summing the difference between the actual subject's VEP amplitude for the given pattern and

Spatial-Contrast Analysis for Subject #2, LM

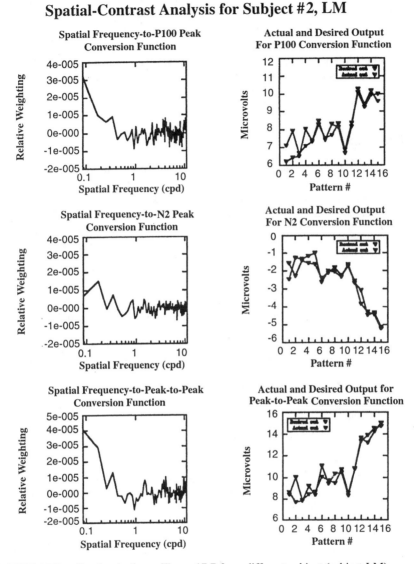

FIGURE 17.7   Graphs similar to Figure 17.7 for a different subject (subject LM).

computer-generated VEP amplitude, produced by the node when presented with the 2D-FFT of that pattern, over all images generated in a set. The ALOPEX algorithm optimized the single set of input weights, so as to minimize the global error term for all VEP responses. The ultimate single set of weights provided the computed Transfer Function (TF) for spatial frequency sensitivity of the VEP parameter in question. Using this technique, one may examine the relationship between specific VEP parameters and their spatial tuning.

The novel application of these techniques has allowed us to relate VEP response trends to the information content of the ALOPEX-generated stimuli, using a response feedback mechanism from the subject's brain, in this unique *brain-to-computer link!* Thus, the pattern recognizers in the human brain became pattern generators. The brain is the ultimate neural network that can work in both a supervised and an unsupervised manner, in order to extract features from signals and image patterns to make sense of situations and the environment.

REFERENCES

1. Abelson, H., Towards a theory of local and global support in computation, *J. Theor. Comp. Sci.*, 6, 41, 1978.
2. Cottaris, N. P., Iezzi, R., and Micheli-Tzanakou, E., VEP as a response of the visual system to pattern convergence, *IEEE Conf. Eng. Med. Biol.*, 12, 895, 1990.
3. Dasey, T. J. and Micheli-Tzanakou, E., A Pattern Recognition Application of the ALOPEX Process on Hexagonal Arrays, *Proc. Int. Joint Conf. Neural Net.*, II, 119, 1989.
4. Dasey, T. J. and Micheli-Tzanakou, E., The Unsupervised Alternative to Pattern Recognition I. Classification of Handwritten Digits, *Proc. Workshop on Neural Networks*,1992, Auburn University, 328.
5. Daugman, J. G., Two dimensional spectral analysis of cortical receptive field profiles, *Vision Res.*, 20, 847, 1980.
6. Geman, S. and Geman, D., Stochastic relaxation, Gibbs distributions, and the Bayesian restoration of images, *IEEE Trans. Patt. Anal. Mach. Intell.*, PAMI-6 (6), 712, 1984.
7. Iezzi, R., Micheli-Tzanakou, E., and Cottaris, N., Effects of pattern convergence and orthogonality on visual evoked potentials, *IEEE Conf. Eng. Med. Biol.*, 12, 897, 1990.
8. Jones, J. P. and Palmer, L. A., The two dimensional spatial structure of simple receptive fields in cat striate cortex, *J. Neuroph.*, 58(6), 1187, 1987.
9. Marcella, S., Mathematical description of the responses of simple cortical cells, *J. Opt. Soc. Am.*, 70(11), 1297, 1980.
10. Marsic, I. and Micheli-Tzanakou, E., Distributed optimization with the ALOPEX process, *Proc. 12th Conf. IEEE Eng. Med. Biol. Soc.*, 12, 1415, 1990.
11. Poggio, T., Torre, V., and Koch, C., Computational vision and regularization theory, *Nature*, 317(6035), 314, 1985.
12. Pollen, D. A. and Ronner, S. F., Phase relationships between adjacent simple cells in the visual cortex, *Science*, 212, 1409, 1981.
13. Rosenfeld, A., *Multiresolution Image Processing and Analysis*, Springer-Verlag, Berlin, 1984.
14. Tzanakou, E., Michalak, R., and Harth, E., The Alopex process: visual receptive fields by response feedback, *Biol. Cybern.*, 35, 161, 1978.

Index

A

Activation function, 62, 66, 236, 280–282,
285
Adaptive beamforming, 314
Adaptive Linear Element (ADALINE), 64
Adaptive Resonance Theory (ART), 5, 20,
165, 284
Additive Model (AM), 47
Additivity and Variance Stabilization
(AVAS), 47
Aging-associated visual agnosia, 317, 320,
321
Ahmad, S., 215
ALgorithm Of Pattern EXtraction, See
ALOPEX
Ali, F., 165
ALOPEX, 26, 27–28, 65, 69–72,
141–144, 241–246
brain to computer link, 351–353, 358
comparison with backpropagation, 26
connectivity strengths and convergence,
242, 243–246, 249–262
converged connection vectors, 148, 149
face recognition system, 318, 321
FCM (fuzzy C-means) algorithm and,
156–157
FCM-based handwritten digit recognition
system, 171–182
classification errors, 175, 178–179
cluster validity measures, 177–178
comparison with supervised system,
180
computational demand, 174, 177
generalization capability, 180, 181
hardware resources, 180
testing using backpropagation,
175–176, 179–180
general pattern recognition system design,
144
general updating equation, 70

generalized algorithms, 243–246
inter-network feedback and, 149–150
mammogram classification system,
203–206, 209–214, 217–218
moment invariants and convergence,
246–249, 252–255, 259
multilayer perceptron training, 71–72,
265–275
convergence rate vs. backpropagation,
269
template matching, 268–269
output variance maximization, 148–151
Parkinson's disease model, 237
receptive field mapping, 70, 232–233,
241, 244, 347–353
reinforcement rules, 242–243
retinal damage classification system, 223
simulated annealing and, 143
speech recognition system, 302, 308,
310–314
three-dimensional neural network, 232
update equation, 142, 147, 148, 150
utility of, 144, 181–182, 191–192
visual evoked potentials system,
187–193, 353-358
black box approach, 353-355
clustering analysis and, 188–191
generalization capability, 193
usefulness of, 191–192
VLSI implementation, 72, 270–275,
See also Very large scale integration
Alternate Conditional Expectation (ACE),
47
Alzheimer's-associated visual agnosia, 317,
321
Analog design, 72, 270–273
Andreou, A. G., 295, 297
A-norm distance, 155
Artificial neural networks (general considerations), 19, 61–75, 277–278, See
also Neural networks
Associative memory, 20, 27

Astrom, K. J., 293
Auditory system modeling, 297–298, See
    also Speaker recognition system
Autoassociative memory, 20
Average squared residual (ASR), 38–40,
    44
Axon, 279
Axon hillock, 279, 280, 292

B

Backpropagation, 21–27, 65–69, 165,
    266, 268
    comment on terminology, 25
    comparison with ALOPEX, 26
    convergence enhancement methods, 68,
        69, See also Simulated annealing
    face recognition system, 318, 321
    local extrema problem, 69
    Madaline Rule, 25
    mammography applications, 73
    optimization in multilayer perceptron
        structures, 25
    optimization machine, 25–26
    slow convergence rate, 29–30
    speaker recognition system, 312–313
    testing of ALOPEX-trained handwritten
        digit recognition system, 175–176,
        179
    through time, 69
    VLSI suitability, vs. ALOPEX, 274
Bartlett statistic, 85–87
Baum, E., 29
Bayesian classifiers, 4, 7, 10–18
    equivalence of LDM and minimum TPM
        classifier, 14
    learning vector quantization, 14–16
    linear discriminant analysis and
        classification, 12–13
    linear discriminant score, 11–12
    Nearest Neighbor classifier, 18
    probabilistic neural nets and, 74
    quadratic discriminant score, 11
    self-organizing map, 15
Bezdek, J. C., 153–155
BFGS algorithm, 27, 48, 104
Biased random walk, 141
Binary threshold model, 281
Black box approach, 353–355

Blurring, 170
Boahen, K., 295
Boltzmann distribution, 140
Boltzmann Machine Learning, 27
Bootstrapping estimation, 52, 54–55,
    100–105
Borland Database Engine (BDE), 222
Bottom-up neuronal analysis, 279
Boundary samples, 215
Brain structure and processes, 61–62,
    278–280
Brain to computer link, 351–353, 358
Broyden, Fletcher, Goldfarb, and Shannon
    (BFGS) algorithm, 27, 48, 104
BRUTO, 34, 48, 49
Bryson, A. E., 22
B-spline wavelet transform, 112–114
Buja, A., 3, 37, 48
Bursting oscillation, 285

C

Canonical correlation analysis (CCA), 3, 37,
    40–44, See also Optimal scoring
Casey, R., 164
Cell membrane potential, 278–279, 280
Center of mass, 168, 223
Centering, 81–82
Cepstrum analysis, 122–123
Chaining, 138
Chan, H., 73
Character recognition, See Handwritten
    digit recognition
Chaudhuri, B. B., 124–125
Chinese alphabet, 164
Chromosome classification, 74–75
Classif library, 48
Classification and regression tree (CART),
    6–7, 34, 48, 49
Classifiers, 3–56, 136, See also Canonical
    correlation analysis; Flexible
    discriminant analysis; Linear
    discriminant analysis; Optical
    scoring
    Bayesian, See Bayesian classifiers
    clustering, See Clustering
    comparison of experimental results,
        48–49
    cost for misclassification, 7

divide-and-conquer technique, 325
entropy values, 122
exemplar, 5–6
Expected Cost for Misclassification
(ECM), 3–4, 8–9
feature extractor design considerations,
137
K-nearest neighbors (KNN) classifier, 5,
16, 18, 48, 49
layered neural network, See Neocognitron
Maximum A Posteriori (MAP), 4, 9, 10,
32
multi-class optimal classifiers, 9–11
neural networks, 7, 19–49
nonparametric regression approaches, 3
optimal design criteria, 3–4
optimal scoring, 36–40
regression methods, 34–40
flexible discriminant analysis via
optimal scoring, 37–40
linear discriminant analysis via optical
scoring, 37, 41–46
optimal canonical correlation analysis,
37, 40–41
software resources, 48
space partition methods, 6
system performance assessment, 50–53,
See also Prediction error evaluation
bootstrapping estimation, 52
hold-out method, 51
jackknife estimation, 53, 73
K-fold cross-validation, 51–52
prediction error, 50
Total Probability of Misclassification, 4, 9
Clinical diagnosis, See Medical diagnosis
Clustering, 32–34, 137, 138
Adaptive Resonance Theory (ART)
networks, 20
cluster validity pattern, 155–156
FCM algorithm, See Fuzzy C-means
(FCM) algorithm
ISODATA method, 153–154
CMOS implementation, 280–292, 295,
297, See also Very large scale
integration
Cocktail party effect, 302, 310–311, 314,
See also Speaker recognition system
Codebook vectors, 14–18
Color vision tests, 185
Combinatorial optimization, 181

Competitive learning, 14–15
Computer-aided diagnosis (CAD), 72–74,
197, See also Medical diagnosis
Conditional density functions, Bayesian
classifiers and, 4, 10–11
Connection weights, 147
Connectionist models, 18, 62
Connectivity strength, 64, 242, 243–246,
249–262
Contents-addressable memory, 20
Contrast enhancement, 81
Convergence
defined, 68
enhancement, 30–34, 68
Kullback-Leibler distance, 32
Quickprop, 31–32
suggested heuristics, 30–31
weight decay, 33–34
gradient-based methods, 28–29
Cost function, 138, 241
ALOPEX, 141–142
mean field annealing, 141
vision model, 348
CRIMCOORD, 13, 100
Curse of dimensionality, 4

D

Damasio, A., 317
Data compression, wavelet applications, 314
Daugman, J. G., 347
Decision function, 323
neocognitron, 325–328
DeLong, M. R., 238
Delta rule, 24, 266
Dendrites, 279
Deutsch, S., 325
DeWeerth, S. F., 293
DeYong,, M. R., 292
Diagnosis, See Medical diagnosis
Digit recognition, see Handwritten digit
recognition
Digital VLSI, 270–271
Dilation, 84–85
Dimensionality, curse of, 4
Dimensionality reduction, 96–100, 136,
144, See also Feature extraction
methods
discriminant analysis, 96, 98–100

mammogram feature extraction system, 201
principal component analysis, 96–98
Discrete B-spline wavelet transform, 114
Discrete Fourier transform (DFT), 122
Discrete wavelet series, 111
Discrete wavelet transform (DWT), 112
Discriminant analysis, 96, 98–100, See also Flexible discriminant analysis; Linear discriminant analysis
Douglas, R., 285–288
Downes, P., 74
Dubes, R. C., 123
Dubuisson, M-P., 123
Dynamic model, 282, 283
Dynamic time warping, 314

E

Eden, M., 164
Elias, J. G., 287
Energy function, 138
Entropy, 121–122
    fuzzy C-means routine, 156
    Karhunen-Loéve expansion and, 146
    mammogram classification system, 201, 212
    Maximum A Posteriori, 32
    SGLD matrix, 128
Epoch learning, 23
Erosion, 84–85
Error Backpropagation, 21, 22–27, See also Backpropagation
Expected Cost for Misclassification (ECM), 3–4
    Minimum ECM classifier, 8–9
    optimal ECM classifier, 12
Expert systems, 72–73
Eye related diseases, 221–227

F

Face recognition, 317–321
Fahlmann, S. E., 68
Farber, R., 28
Fast algorithm, 117–119

Fast Fourier transform (FFT), 353
F-CORE decomposition, 223, 224, 226, 318
Feature analyzers (FAs), 350–351
Feature extraction methods, 109–130, 136, 137–138
    cepstrum analysis, 122–123
    classifier accuracy and, 137
    entropy, 121–122
    face recognition system, 318, 321
    fractal dimension, 123–125
    general pattern recognition system design, 144–153
    invariant moments and, 119–121, See also Moment invariants
    Karhunen-Loéve (K-L) expansion, 137, 145–147
    mammogram classification system, 200–201
    moment invariants, 137
    neural network application, 147–153
    SGLD texture features, 126–130
    unsupervised handwritten digit recognition system, 171–175
    wavelet transform, 109–119, See also Wavelet analysis
Feature-map classifier, 5
Feedback inhibition, 295
Feedback networks, 19–20, 63–64, 149–150
Feed-forward inhibition, 295
Feed-forward networks, 19, 20–22, 63–64, 265, See also Multilayer perceptrons training methods for, 27–28
Feinberg, T., 317
Fields, C., 292
Findley, R. L., 292
Fisher's Discriminant function, 12
Fitzhugh-Nagumo model, 285
Flexible discriminant analysis (FDA), 3, 34, 37
    classification application via optical scoring, 46–48
    multi-response regression and, 36
    software resource, 48
Floyd, C., Jr., 73
Follower aggregator, 273
Fractal analysis, 74, 123–125

Frequency analysis, 303
Frequency based neuronal models,
  284–285
Fukushima, K., 165, 324
Fuzzy clustering, 138
Fuzzy C-means (FCM) algorithm,
  153–159
  handwritten digit recognition system,
    171–182, See also under ALOPEX
  locally optimal solutions, 156–157, 174
  pattern labeling, 157–159
  visual evoked potentials system, 188
Fuzzy covariance matrix, 155, 174, 177
Fuzzy logic, 153, 154, 178

G

Gabor transform, 109, 112, See also Short-
  time Fourier transform
Gagnepain, 124–125
Galar, R., 141
Gallant, A. R., 28
Gaussian Pyramid, 223
Generalization performance, 30–34
  modular neural network mammography
    application, 213–218
  unsupervised handwritten digit recogni-
    tion system, 180, 181
  unsupervised system for visual evoked
    potentials, 193
Generalized delta rule, 24
Genetic sequence classification, 74–75
Giger, M., 74
Gilbert multiplier, 273–274
Goldstein, M. H., 297
Gonzalez, R. C., 165, 323
Gradient descent methods, 22–27, 65, 69,
  139, See also Backpropagation;
  Steepest descent gradient step
  optimization
  enhancing convergence rate, 30–34
  justification for, 26
  slow convergence, 29–30
Gray-tone spatial-dependence matrix, 126
Grimsdale, R. L., 164
Grossberg, S., 283
Grossman, A., 303

H

Hamiltonian, 138
Hamming distance, 329
Handwritten digit recognition, 79–105,
  163–182
  ALOPEX trained FCM scheme,
    171–182
  Chinese alphabet, 164
  classification errors, 175, 178–179
  cluster validity measures, 177–178
  commercial applications, 163
  comparison of unsupervised vs.
    supervised system, 180
  computational demand, 174, 177
  data collection, 165–166
  dimensionality reduction, 96–100
    discriminant analysis, 96, 98–100
    principal component analysis, 96–98
  fuzzy neural network system, 157
  generalization capability, 180, 181
  hardware resources, 180
  methodology categorization, 164
  model-free strategy, 79
  prediction error evaluation, 100–105
  preprocessing, 166–170
    Bartlett statistic, 85–87
    contrast enhancement, 81
    dilation, 84–85
    segmentation, 81
    size normalization, 85, 87
    skeletonization, 82–84
    smoothing by median filter, 81
    translation normalization ("centering"),
      81–82
  results, 170–176
  testing using backpropagation,
    175–176, 179–180
  Zernike moments, 87–96, See also
    Zernike moments
    features from, 92–96
    reconstruction by, 90–92
Haralick, R. M., 126
Hardlimiter, 62
Hardware design, 270, See also Very large
  scale integration
Harth, E., 65, 70, 266
Hastie, T., 3, 37, 48

Haussler, D., 29
Haykin, Simon, 202
Hearing aids, 301
    cochlear implants, 297–298
    cocktail party effect and, 302
    feedback in, 301–302
    Hebb, Donald, 64
    Hebbian training, 148–149
    feature cell output value, 151
    three-dimensional neural network, 230,
        232, 238
Heisenberg's uncertainty principle, 109
Hessian matrix update (BFGS), 27, 48, 104
Heteroscedacity, 97
Hidden layers, 21, 24, 28–29, 64, 265, 268
Highleyman, W. H., 165
Hilbert's 13th problem, 28
Hinging hyperplanes, 47
Hinton, G., 21, 22, 31
Histogram equalization, 81, 223
Ho, Y. C., 22
Hodgkin-Huxley equations, 285
Hold-out method, 51
Hopfield networks, 27, 182, 281, 284
Hornik, K., 28, 29
Hudson, D., 72
Hu invariants, 119–121, 201, See also
        Moment invariants
Hybrid neural processing element, 292

I

IBM, 164
ID3, 6
IEEE, 62
Image data management system, 221
Image power spectrum, 223
Image preprocessing, See Preprocessing
Inter Quartile Range (IQR), 55
Interaction spline method, 47
Interset distance, 98–99
Intraset distance, 98
Invariant moments, See Moment invariants
ISODATA, 153–154

J

Jackknife estimation, 53, 73
Jacobs, R. A., 30

K

Karhunen-Loéve (K-L) expansion,
    137, 145–147
ALOPEX system for visual evoked
    potentials, 188
converged connection vectors, 148, 149
Fuzzy C-means (FCM) algorithm,
    158–159
Keller, J. M., 123, 124, 125
K-fold cross-validation, 51–52
K-means algorithm, 6
K-nearest neighbors (KNN) classifier, 5,
    16, 18, 48, 49
Kohonen map, 271
Kolmogorov, A. N., 28
Kullback-Leibler distance, 27, 32

L

Lapedes, A., 28
Lateral inhibition, 293, 294–295
Layered neural network, 323–342, See
    also Multilayer perceptrons;
    Neocognitron
Le Cun, Y., 28, 179
Leaky integrator, 282
Learning Vector Quantization (LVQ), 5,
    14–18
Least mean square (LMS) algorithm, 65
Leave-one-out method, 51
Linares-Barranco, B., 280
Line thinning ("skeletonization"), 82–84,
    168
Linear discriminant analysis (LDA), 3,
    12–13, 49
    classification application, 37–44
    via optical scoring, 44–46
    equivalence of LDF and minimum TPM
        classifier, 14
    equivalence with canonical correlation
        analysis and optical scoring, 37,
        42–44, 48
    generalization to nonlinear flexible
        discriminant analysis, 37
    performance assessment, 55
    two-group regression and, 34–36
Linear discriminant classifier, 5
Linear discriminant score, 11–12
Linear threshold model, 281

Lippmann, R. P., 3, 7
Liu, W., 297
LMS algorithm, 6
LREG, 49, 55

M

Madaline rule, 25
Mahalanobis distance, 6, 18, 155
Mahowald, M., 285–288
Mallat, S., 112, 303, 304
Mammography, 73–74, 197–218
 data acquisition and preprocessing,
  199–200
 distribution of entropy values, 122
 feature extraction, 200–201
 modular neural networks, 202–203
 ALOPEX parameters, 204–206,
  209–214, 217–218
 classification results, 203–207
 data normalization, 207–209, 216
 generalization, 213–218
 misclassification error, 208–209
 system sensitivity and specificity,
  206–207
 training percentages, 213
MANOVA model, 85
Marcella, S., 347
Marsic, I., 348
Maximum A Posteriori (MAP) classifier, 4,
 9, 10, 32
Maximum growth factor, 68
McCarthy, R. A., 317
McCulloch and Pitts model, 64, 281, 284
Mead, Carver, 292, 293, 295–296
Mean field annealing, 141
Median filtering, 81
Medical axis transform (MAT), 74–75
Medical diagnosis, 72–74, See also
 specific applications
 mammography, See Mammography
 retinal damage classification, 221–227
 visual evoked potentials, 185–193
Metropolis algorithm, 139–140
Micheli-Tzanakou, Evangelia, 65, 70, 266,
 289–292, 325, 348
Minimum TPM decision rule, 11
Minsky, M., 21, 28, 65
Misclassification rate, 50

Modular neural networks, 202–203
 mammogram classification system,
  203–218, See under Mammography
 retinal damage classification system,
  223–227
Moment invariants, 119–121, 137
 ALOPEX and, 246–249, 252–255,
  259
 mammogram feature extraction system,
  201
 retinal damage classification system, 223
Moon, G., 292
Morlet, J., 303
Multi-class optimal classifiers, 9–11
Multilayer perceptrons, 19, 21–22, 64,
 265–275, See also Neocognitron
 ALOPEX alternative to backpropagation
  training, 182
 approximation capability, 28
 convergence rate, backpropagation vs.
  ALOPEX, 269
 optimization, 25–26
 template matching, 268–269
 training with ALOPEX, 71–72
Multiple Sclerosis, visual evoked potentials
 and, 185–193
Multiresolution decomposition, 34, 223,
 302–314, See also Wavelet analysis,
 multiresolution decomposition
Multi-response regression, flexible discriminant analysis and, 36
Multivariate Adaptive Regression Splines
 (MARS), 34, 47, 49
 software, 48
Multivariate function, approximation using
 superposition theory, 28
Multivariate regression, See Regression
 methods

N

Nabet, B., 294
Nagy, G., 164
Nearest neighbor classifier, 5, 15–16, 18
Neocognitron, 165, 324–342
 distribution of decision functions,
  325–328

high inhibition effects, 341
pattern classification model, 325–329
sensitivity to distortion in learned
  features, 340
simulation experiment
  methods, 329–330
  network performance evaluation, 330
  objectives, 329
  parameters, 330
  three-level network, 332–334, 342
  two-level network, 330–332,
    335–339
training algorithm, 328–329
Neural networks, 7, 19–48, 61–75, See
  also ALOPEX; Backpropagation
brain to computer link, 351–353, 358
classifier applications, 19–20,
data sets for testing, 165
definitions and terminology, 62–64
FCM algorithm, See Fuzzy C-means
  (FCM) algorithm
feed-forward networks, 19, 20–22, 63
general considerations of artificial neural
  networks, 19, 61–75, 277–278
gradient methods and, See Gradient
  descent methods
hybrid symbolic machine learning
  algorithm, 74
justification for gradient method for nonlinear function approximation, 26
layered, 323–342, See also Multilayer
  perceptrons; Neocognitron
modular, 202–218, 223–227
multivariate regression and classification
  properties, 7, See also Regression
  methods
pyramidal processing, 348–351
shift invariant, 73
symbolic processing system, 72–73
temperature, 348
three-dimensional architecture,
  229–238, See also Three-dimensional neural network architecture
topology, 19, 20–21, 63–64, See also
  Feedback networks; Feed-forward
  networks
training methods, 27–28, 147–153,
  See ALOPEX; Backpropagation
universal approximation, 28–29
VLSI, See Very large scale integration

Neural networks, applications, 19–20, See
  also Pattern recognition; specific
  applications
chromosome and genetic sequence
  classification, 74–75
expert systems, 72–73
handwritten digit recognition, 79–105,
  163–182
medical diagnosis, 72–74
mammography, See Mammography
retinal damage classification,
  221–227
visual evoked potentials, 185–193
speaker recognition, 302–314
vision models, 347–358
visual evoked potentials, 185–193,
  353–358
Neural processing element, 292
Neural-type cell, 292
Neurological process modeling, 292–298
Neuromorphic models for VLSI, 280–292,
  See also under Very large scale
  integration
Neuromorphic systems, 62, See also Neural
  networks
Neurons, 19, 61, 62, 278–280
Neurosarcoidosis, 186
Neurotransmitters, 279
Newton's method, 25, 27, 31, 48
Nguyen, D., 69
Nielsen, L., 293
Nnet, 48, 49
Nodes, 62
Noise thresholding, 166–167
Nonlinear morphological processing, 84
Noordewier, M., 74
Northmore, D. P., 287

O

Objective function, 138
Oja, E., 148–149, 142
OLVQ1, 49
On-line learning, 23, 26
Operational transconductance amplifier
  (OTA), 280, 282, 285, 290, 293
Optical character recognition (OCR), 164,
  See Handwritten digit recognition
Optical computers, 270

Optimal scaling (OS), 3
Optimal scoring (OS), 36, 37–40, 42–48
  equivalence with linear discrimi-
      nant analysis and canonical correla-
      tion analysis, 37, 42–44, 48
  linear discriminant analysis and, 44–46
  software, 48
  translation of dimensions into discrimi-
      nant coordinates, 42–43
Optimization, 27–28, 69, 138–144, 241,
      See also ALOPEX; Backpropaga-
      tion; Training
  enhancing convergence rate and general-
      ization, 30–34
  feed-forward multilayer perceptrons and,
      25–26
  Galar's biased random walk, 141
  justification for gradient method, 26
  mean field annealing, 141
  simulated annealing, 27, 69, 140, 143, 241
  statistical mechanics and, 139–140
  theory and objectives, 138–139
Optimized LVQ1, 16–18

P

Papert, S., 21, 28, 65
Parallel distributed systems, 62, See also
      Neural networks
Parker, D. B., 266
Parkinson's disease, 235–237
Path-width standardization, 84
Pattern classifiers, See Classifiers
Pattern recognition, 277, See also
      Handwritten digit recognition
  ALOPEX, See ALOPEX
  biological neural network and, 19
  clustering operation, 138, See also
      Clustering
  dimensionality reduction, 96–100, 136
  dysfunction, face recognition simulation,
      317–321
  feature extraction, See Feature extraction
      methods
  image preprocessing, See Preprocessing
  optimization, See Optimization
  speaker identification, 302–314
  system design, 144–153
  theory and applications, 135–137

Pavlidis, T., 165
Peitgen, H-O., 123
Peleg, S., 124
Pentland, A. P., 124
Perceptron, multilayer, See Multilayer
      perceptrons
Perceptron, single layer, 21–22, 28, 62, 65
Perceptron convergence procedure, 21–22
Performance assessment, 50–53, See also
      Prediction error evaluation
Photoreceptors, 295, 296
π-method, 47
Pinter, R. B., 294–295
POLY, 49
Polynomial interpolation, 214
POLYREG, 48
Potassium and sodium currents, 285–287
Potential energy, 140
PPREG, 48, 49
Prediction error evaluation, 50–53,
      100–105
  bootstrapping estimation, 52, 54–55,
      100–105
  hold-out method, 51
  jackknife estimation, 53, 73
  K-fold cross-validation, 51–52
Preprocessing, 79–87, 136, 166–170
  Bartlett statistic, 85–87
  blurring, 170
  center of mass adjustment, 168
  contrast enhancement, 81
  dilation, 84–85
  fixing to standard size, 168
  line thinning (skeletonization),
      168
  mammogram images, 200
  neurological model, 293
  noise thresholding, 168
  reducing resolution, 169–170
  rotation, 168–169
  segmentation, 81
  size normalization, 85, 87
  skeletonization, 82–84
  smoothing by median filter, 81
  speech recognition system, 307–308,
      314
  translation normalization ("centering"),
      81–82
  unsupervised handwritten digit
      recognition system, 166–170

visual evoked potentials system,
    186
Principal component analysis (PCA),
    96–98, 234
Probabilistic neural networks, 74
Projection Pursuit Regression (PPR), 47
Prosopagnosia, 317
Pseudo steepest descent method, 25
Pyramidal processing, 348–351

Q

Quadratic discriminant classifier, 5
Quadratic discriminant score, 11
Quadratic spline wavelets, 114–116
Quadrature mirror filters (QMFs),
    303–305, 309, 314, 318
Quickprop, 31, 68
Quinlan, J. R., 6

R

Radial Basis Function (RBF) networks,
    5–6
Ramp, 62
Receiver operating characteristic (ROC)
    analysis, 73, 74
Receptive fields (RFs), 70, 229, 231–235,
    241, 244, 347–353
Refraction, 279
Refractory period, 279, 290
Regan, D., 189
Regression methods, 34–40
    comparison of experimental results,
        48–49
    optimal canonical correlation analysis,
        37, 40–41
    optimal scoring and, 36, 37–38
    flexible discriminant analysis, 37–40
    linear discriminant analysis, 37,
        41–46
    recently developed nonparametric
        methods, 47
    software, 48
Response function, 141–142
Response integrators (RIs), 350–351
Retina simulation ("silicon retina"),
    295–297
Retinal damage classification, 221–227

feature extraction methods, 223
image processing, 223
modular neural network, 223–227
Reverse correlation method, 347
Ridge regression, 33
Ripley, B., 48
Roques-Carmes, 124–125
Rosenblatt, F., 21, 62, 64
Rumelhart, D. E., 21, 22, 31, 266

S

Sanger, T. D., 152
Sarkar, N., 124–125
Scree plot, 97
Script recognition, 164
Segmentation, 81
Self-learning layered neural network,
    323–342, See also Neocognitron
Self-organization, 64, 203, See also Unsu-
    pervised learning
Self-organizing map, 6, 14, 15, 20
Sensory image preprocessing, 293
Servo system, 293–294
SGLD matrices, 126–130
Shift invariant neural network, 73
Short-time Fourier transform (STFT), 109,
    112, 302, 303
Sigmoid function, 62, 66, 211, 268,
    281–282
Signature verification, 163
Silicon retina, 295–297
Simulated annealing, 27, 69, 140, 143,
    241
Single Instruction Multiple Data (SIMD)
    architecture, 72, 271
Single layer perceptron, 21–22, 28, 62, 65
Singular value decomposition (SVD), 3,
    42–43, 45
Size normalization, 85, 87
Skeletonization, 82–84, 168
Smoothing by median filter, 81
Sodium and potassium currents, 285–287
Space partition methods, 6
Speaker recognition system, 301–314
    cocktail party effect, 302, 310–311, 314
    data preprocessing, 307–308, 314
    multiresolution analysis through wavelet
        decomposition, 303–305

coefficients for template matching,
308–310
comparison of neural networks with
ALOPEX template matching,
312–313
quadrature mirror filters, 303–305,
309
results, 310–314
Speech recognition, 301–314, See also
Speaker recognition system
Spline wavelet transform, 112–114
Split-and-merge algorithm, 165
Squashing functions, 28
Standard error approximation, 101–102
Standards, 62
Steepest descent gradient step optimization,
15, 22, 25, 27, 29–30, See also
Gradient descent methods
Stinchcombe, M., 28, 29
Stochastic optimization, 27
Suen, C. Y., 165
Sum of squared residuals, 51
Superposition theory, 28
Supervised learning, 19–20, 64, 203,
265–266
feed-forward networks, 19, 20–22
learning vector quantizer, 14
Synapse, 61, 62, 279
Synaptic weight, 61, 279

T

Teague, M. R., 93
Temporal learning, 23
Tesauro, G., 215
Texture analysis, fractal dimensions and,
123, 124
Texture quantification, 74
Three-dimensional neural network
architecture, 229–238
receptive fields, 229, 231
simulations, 231–232
Parkinson's disease, 235–237
receptive fields, 232–235
Threshold logic element, 62
Threshold potential, 279
Tibshirani, R., 3, 37, 48
Top-down neuronal analysis, 279

Total Probability of Misclassification
(TPM), 4, 9
equivalence of LPM and minimum TPM
classifier, 14
minimum TPM decision rule, 11
Tou, J. T., 165, 323
Training, 27–28, 64–72, 147–153,
265–266, See also ALOPEX;
Backpropagation; Clustering;
Optimization
causes of failure in learning, 29
enhancing convergence rate and
generalization, 30–34
Kullback-Leibler distance, 32
Quickprop, 31–32
suggested heuristics, 30–31
weight decay, 33–34
Hebbian, See Hebbian training
inter-network feedback, 149–150
supervised, See Supervised learning
unsupervised, See Unsupervised learning
Training set, 19
Transconductance amplifier, 273, 274, 280,
282, 285, 290, 293
Translation normalization ("centering"),
81–82
Traveling salesman problem, 27
Two-group regression, linear discriminant
function and, 34–36
Tzanakou, E., See Micheli-Tzanakou,
Evangelia

U

Unsupervised learning, 20, 64, 203, 266, See
also Clustering
Adaptive Resonance Theory (ART)
networks, 20
fuzzy C-means routine, 153–159
handwritten digit recognition, 163–182,
See also Handwritten digit
recognition
layered neural network, 323–342, See
also Neocognitron
mammogram classification, 197–218,
See Mammography
pattern labeling, 157–158
visual evoked potentials, 185–193, See

also Visual evoked potentials
U.S. Postal Service, 165

V

Vapnik-Chervonenkis dimension, 215
Vector Quantization (VQ), 5
Very large scale integration (VLSI)
  ALOPEX implementation, 72, 270–275
  analog vs. digital design, 72,
    270–273
    backpropagation comparison, 274
    parallel implementation, 272
  auditory system modeling, 297–298
  lateral inhibition, 293, 294–295
  neurological process modeling,
    292–298
  neuromorphic models, 280–292
    activation function, 280–282, 285
    dynamic model, 282, 283
    frequency-based models, 284–285
    hybrid neural processing element,
      292
    multicompartmental silicon dendrite,
      287
    sodium and potassium currents,
      285–287
    Wolpert-Micheli-Tzanakou model,
      289–290, 295
  operational transconductance amplifier
    (OTA), 280, 282, 285, 290, 293
  servo controller, 293–294
  silicon retina, 295–297
  simulated annealing applications, 27
Vision models, 347–358
  application to template matching,
    350–351
  black box approach, 353–355
  brain to computer link, 351–353, 358
  cost function, 348
  distributed representation, 347–348
  modified ALOPEX, 348–350
  receptive fields, 70, 229, 231–235, 241,
    244, 347–353
  retinal damage classification, 221–227
  silicon retina, 295–297
  visual evoked potentials, 185–193,
    353–358
Visual agnosia, 317–321, 353–358

Visual evoked potentials, 185–193,
  353–358
  ALOPEX trained feature extraction,
    187–188
    results, 188–193
  black box approach, 353–355
  clustering analysis and, 188–191
  data collection and preprocessing, 186
  generalization capability, 193
  results, 188–191
  usefulness of unsupervised pattern recog-
    nition routine, 191–192
Visual Ophthalmologist (VO), 221–227
VLSI, See Very large scale integration
Voltage-controlled oscillator (VCO), 285,
  292
Voss, R. F., 123

W

Warington, E. K., 317
Wavelet analysis, 109–119, 303
  ALOPEX trained speech recognition
    system, 302–314
    coefficients for template matching,
      308–310
    face recognition system, 318, 321
    quadrature mirror filters, 303–305,
      309
    speaker recognition system, 301–314
  discrete B-spline wavelet transform, 114
  discrete wavelet transform, 112
    fast algorithm, 117–119
  mammogram feature extraction,
    200–201
  multiresolution decomposition,
    303–305
    quadratic spline wavelets design,
      114–116
  retinal damage classification system, 223
  spline wavelet transform, 112–114
  use in data compression and reconstruc-
    tion, 314
Wavelet coefficients, 111
Wavelet filters, 111
Weight decay, 33–34
Werbos, P. J., 22, 69, 266
While, H., 28, 29
White, J., 28

Widrow, B., 25, 64, 85
Wilks' lambda, 85
Williams, A., 21, 22, 31
Wolpert-Micheli-Tzanakou model,
  289–292, 295

X

XOR problem, 21, 28, 65, 215

Z

Zadeh's Principle of Incompatibility, 153

Zernike moment invariants (ZMIs), 92–96
Zernike moments, 48, 86, 87–96,
  105
  advantages for pattern recognition, 92
  defined, 87–88
  features from, 92–96
  geometric moments, 90
  modulus value of, 94–96
  order of, 90–92
  reconstruction by, 90–92
  rotational invariance, 92–96
Zernike polynomials, 88–89, 105
Zhang, W., 73
Zheng, B., 73